When Prayer Takes Place

When Prayer Takes Place
Forays into a Biblical World

J. Gerald Janzen

Edited by
Brent A. Strawn *and* Patrick D. Miller

CASCADE *Books* • Eugene, Oregon

WHEN PRAYER TAKES PLACE
Forays into a Biblical World

Copyright © 2012 J. Gerald Janzen. All rights reserved. Except for brief quotations in critical publications or reviews, no part of this book may be reproduced in any manner without prior written permission from the publisher. Write: Permissions, Wipf and Stock Publishers, 199 W. 8th Ave., Suite 3, Eugene, OR 97401.

Cascade Books
An Imprint of Wipf and Stock Publishers
199 W. 8th Ave., Suite 3
Eugene, OR 97401

www.wipfandstock.com

ISBN 13: 978-1-60899-367-3

Cataloguing-in-Publication data:

Janzen, J. Gerald, 1932–

    When prayer takes place : forays into a biblical world / J. Gerald Janzen; edited by Brent A. Strawn and Patrick D. Miller.

    xx + 430 pp. ; 23 cm. Includes bibliographical references and index.

    ISBN 13: 978-1-60899-367-3

    1. Bible—Prayers. 2. Prayer—Biblical teaching. I. Strawn, Brent A. II. Miller, Patrick D. III. Title.

BS680 P64 J36 2012

Manufactured in the USA

In memory of
Frank and Betty Anne

The prayer psalm can, when said, evoke the setting in which prayer occurs. The utterance calls forth what the Bible calls a maqom, *a special place where God is present for the human being. God's name is called, the name that is his presence and bears all the qualities of his person. To "call on the name of God" is to place oneself in his presence. The prayer describes the self, presents the self to God in all its weakness and need. It lets the self be in the real Presence. Through the psalm the self speaks to God. All that is essential to the place of prayer is brought about with the saying of the psalm.*

—JAMES LUTHER MAYS, *THE LORD REIGNS*

When I pray in my chamber, I build a Temple there, that houre; And, that minute, when I cast out a prayer, in the street, I build a Temple there; And when my soule prayes without any voyce, my very body is then a Temple. And God, who knowes what I am doing in these actions, erecting these Temples, he comes to them, and prospers, and blesses my devotions.

—JOHN DONNE, SERMON ON LUKE 2:29–30

The name of the LORD is a strong tower;
the righteous runs into it and is safe.

—PROVERBS 18:10 (RSV*)

For if you keep silence at such a time as this, relief and deliverance will rise for the Jews from another place.

—ESTHER 4:14 (NRSV*)

Christ is building you into a place where God lives through the Spirit.

—EPHESIANS 2:22 (CEB)

Contents

Foreword, by Brent A. Strawn and Patrick D. Miller / xi
Preface / xv
Acknowledgments / xvii
Abbreviations / xix

Introduction: From Plane to Plane / 1

Part I: Orienting Ourselves in the Biblical World

1. "... and the Bush Was Not Consumed" / 17
   Addenda / 23

2. What's in a Name? "Yahweh" in Exodus 3 and the Wider Biblical Context / 24
   Names and Their Meanings / 24
   The Name "Yahweh" and Its Meaning for Israel / 27
   The Name and Its Meaning for Existence Today / 35

3. What Does the Priestly Blessing *Do*? / 38
   The Priestly Blessing and God's Blessing in Creation / 38
   The Vocabulary of P and the Priestly Blessing / 41
   Cosmos, Tabernacle, and the Priestly Blessing / 43

4. Praying in the Space God Creates for the World / 50
   Making Space / 50
   The *Memra* / 52
   Space for God / 53
   Praying in the Space / 55

Part II: Forays into a Biblical World

5. Prayer as Self-Address: The Case of Hannah / 61
   Comfort: God's and the Self's / 62
   Self-Encouragement in the Worship of God / 65
   śiaḥ: The Means of Self-Encouragement / 70
   Hannah's Meditation and Self-Address / 72

Contents

6   The Root *škl* and the Soul Bereaved in Psalm 35 / 77
       škl as Maternal Bereavement / 78
       Compassion Requited and Unrequited / 79
       Suffering the Loss of Matrixal Connections:
           A Psychological Perspective / 86
       Rage and the Bitterness of Unrequited Compassion / 88

7   As God Is My Witness: Another Look at Psalm 12:6 / 91
       What, Precisely, Does God Promise the Psalmist? / 91
       Another Look at the Language of the Promise / 92
       God's Promise and Job's Hope-Against-Hope in Job 16:19 / 96

8   "And Not We Ourselves": Psalm 100:3 and the Eschatological Reign of God / 99
       Is "Not We Ourselves" Grammatical Hebrew? / 101
       The Range of Variation in a Stock Expression / 102
       But Why the Need to Disavow Self-Creation? / 103
       On Divine and Human "Making" / 104
       The Verb *ga'ah* / 109
       Practical Atheism in Psalm(s) 9–10 / 110
       More on the Self-Confident Claim, "I Won't Slip" / 111
       Recurring to Craigie's Remarks on Psalm 9–10 / 112
       "And Not We Ourselves": Resolving the Conflict Drama
           in Psalms 93–99 / 116
       Themes in Psalms 93–100 Bearing on Psalm 100:3 / 117
       The Conflict Drama Resolved in Psalm 100:3 / 126
       Is "Not We Ourselves" Palatable in Today's World? / 129

9   Standing on the Promises of God: On the Thematic Resonance of "No Foothold" in Psalm 69 / 134
       Thematic Ligatures in the Psalms / 134
       On the Social Significance of "Standing" / 136
       One's Standing in Others' Eyes / 137
       Standing in the Face of Reproach in Psalm 69 / 140
       "Standing" as a Ligature throughout the Psalter / 141
       The Case of Jeremiah / 144
       Standing before God: The Case of Daniel / 151
       On Some Hebrew Expressions Involving the Verbs
           ḥazaq and *'ameṣ* / 154
       Standing before God: The Case of Habakkuk / 155

　　　　　Back to Psalm 69 / 156
　　　　　A Last Word, then, on Psalm 69 / 160
　　　　　A Belated Confession / 164

10　　The Verb *ya'ameṣ* in Psalm 27:14: Who Is Strengthening Whom? / 165
　　　　　A Preliminary Review of the Hebrew Text in Psalm 27:14 / 166
　　　　　The Verb *'mṣ* in the Qal Stem / 167
　　　　　The Verb *'mṣ* in the Piel Stem / 169
　　　　　The Verb *'mṣ* in the Hithpael Stem / 172
　　　　　The Verb *'mṣ* in the Hiphil Stem / 174
　　　　　Weighing the Pros and the Cons / 174
　　　　　The Special Case of the Verb *'rk* in the Hiphil Stem / 176
　　　　　Final Assessment / 179
　　　　　A Brief Excursion to Psalm 27:8 / 181
　　　　　Two Modern Afterwords to Psalm 27 / 184

11　　Revisiting "Forever" in Psalm 23:6 / 188
　　　　　Aspects of Experience "in God's House" / 189
　　　　　Experiencing Time, Mundane and Otherwise / 193
　　　　　Connotations of the Phrase, "Length of Days," and Its Cognates / 195
　　　　　On Some Axes of Affirmation Converging in Psalm 23:6b / 199
　　　　　Drawing Matters to a Conclusion / 204
　　　　　Addendum / 207

Part III: The Standpoint of Two Prophets

12　　Solidarity and Solitariness in Ancient Israel: The Case of Jeremiah / 211

13　　Eschatological Symbol and Existence in Habakkuk / 218
　　　　　From Despairing Complaint to Affirmation in Hope / 220
　　　　　On Existing Eschatologically within and for the Present Time / 236

Part IV: An Interlude

14　　Toward a Hermeneutics of Resonance: A Methodological Interlude between the Testaments / 241
　　　　　Richard B. Hays on Intertextual Resonance / 243
　　　　　Patrick D. Miller on Resonance / 250
　　　　　Samuel Taylor Coleridge on Resonance in the Nature of Things / 255

Contents

       Resonance and Alfred North Whitehead's Di-Polar Cosmology / 265
       Rupert Sheldrake on Morphic Resonance / 269
       Hans Loewald on Resonance in Nature and in Human
          Becoming / 270
       Resonance and *The Disenchantment of Secular Discourse* / 282
       Resonance between the Testaments in Proverbs 8
          and Colossians 1 / 289
       On Hymnic Resonance and Community Cohesion / 296

Part V: New Testament Afterword

15   "Hid with Christ in God" / 303
       Praying to the Father Who Is "in Secret" (Matthew 6:6) / 306
       God's "Secret Place" as Temptation and as Reality / 309
       "Your Life Is Hid with Christ in God" / 311
       Prayer in Romans 8 as the Nexus of the Solidarity of Heaven and
          Earth / 312
       Garrisoned in Prayer / 319
       A Postscript to Be Read in Retrospect / 323

16   Faith as a Foothold "within the Veil": Afterwords in the Letter to the Hebrews / 328
       Faith as a Foothold on Things Hoped For / 329
       Reproach and "Standing" in Hebrews and in Psalm 69 / 331
       Jesus as Son on the Throne / High Priest in the Tabernacle / 334
       Jesus as *archegon kai teleioten* of Faith(fulness) / 344

17   Redeeming the Expression "Redeeming the Time" / 348
       Exagorazo in Classical Greek / 350
       Buying Time in Daniel 2 / 353
       The Verb *pa'am* as a "Beating of Times" / 354
       Redeeming the Time in Ephesians 5:16 / 359
       The Prayer of Empowerment in Ephesians 3:14–21 and the
          Empowering Vision in Daniel 10 / 374
       Conclusion / 379

Bibliography / 393

Author Index / 401

Scripture Index / 405

Foreword

Anyone who knows J. Gerald Janzen and his work knows the truth of the aphorism that "everything is connected to everything else." Two vignettes about Janzen may thus serve as enlightening entrées into the present collection of essays—some old, many new, all of which in their present form are fresh and insightful, the latter being qualities that consistently mark the author and his mind. It is noteworthy that both vignettes involve Janzen's beloved professor, Frank Moore Cross.

The first vignette: It was a summer night in 1967. Janzen was back in Cambridge, Massachusetts teaching during a break between his second and third years on faculty at his old seminary, The College of Emmanuel and St. Chad, which was his first academic appointment. He and his wife Eileen were having hors d'oeuvres with the Crosses, and, as the story goes, after chatting about various matters of academic interest, Janzen finally screwed up enough courage to ask his esteemed teacher if he (Cross) thought he (Janzen) might have a future in textual criticism—the subject of Janzen's doctoral work at Harvard under Cross.[1] Cross chuckled and said, "Your mind is much too vivacious." And, with that, they went in for dinner.

The second vignette: Fast forward many years later. Janzen is now an established scholar and giving a paper on Jeremiah at The Colloquium for Biblical Research, an elite group of scholars in Old Testament/Hebrew Bible, begun by Janzen and others from his student days at Harvard. During the paper, a debate broke out among those in attendance with regard to what methodology he was using. Janzen listened to one person say this, and another person that. As is his way, however, Janzen later put that moment together with yet another in which a student once asked Cross what it took to be a great epigrapher and Cross responded with "an *Eiferform*." Janzen, who had not caught the answer aurally, thought Cross was using a German term. He figured it was a compound word, somehow related to

---

1. Subsequently published as Janzen, *Studies in the Text of Jeremiah*.

German *Eifer*, "zeal, enthusiasm," but for the life of him Janzen couldn't recall this no doubt crucial *Gattung* from the Continent. Finally, he leaned over to his colleague (Pat Miller), and asked "What's an *Eiferform*?" At which point Miller corrected him: "He said, '*An eye for form*.'"

Now, in Janzenesque style, for an exegesis of these vignettes that weaves them together into a meaningful unit that is, hopefully, greater than the sum of the parts—"hopefully," because we are not as adept as Janzen in his own brand of exegesis, such that we are quite confident that he himself would do a far better job. (Rare indeed is the interpreter who is so self-conscious, so self-aware, and at every conceivable level!) First, then, there is Janzen the textual critic, who seized upon his dissertation topic when Cross offered it, in no small part because, by his own admission, it was the only one he could begin to fathom methodologically. We suspect this was due not to a lack of imagination, but to an overabundance of the same (recall the "too vivacious" mind). In Janzen's own words, "The concreteness of the issues, the piece-by-piece detail of the data, anchored me—the journey of a thousand miles involving, not the superhuman feat of leaping tall buildings, or mountains, in a single bound, but thousands and thousands of baby steps, minute textual difference after minute textual difference, until finally I was able to finish."[2] Far from a dead end, however, or an endeavor that simply (!) opened up to more of the same kind of work, Janzen did indeed develop, in the process of that dissertation, an *Eye-for-form* in Cross's epigraphic sense, though, for Janzen, it was not about paleography but about text types, recensions, textual families—but above all, about words and words and words and the relationships of, and between, words. That disciplined attention ignited his *Eiferform*, which was, in truth, already present (and thus more fundamental than Janzen's text-critical aspect)—that is, his passion for words and connections between and among them, and not only between biblical words, but between biblical words and all other words, wherever they are found, and however they cast light on the biblical words. This *Eifer* could not be contained or constrained by the standard run-of-the-mill textual criticism; again, Janzen's mind is "much too vivacious" for that. So, while schooled in the *Eye-for*, Janzen's more primal *Eifer* has taken the text-critical *Eye-for* to places, discourses, and disciplines that it rarely goes. Few indeed are the textual critics who cite Robert Frost, Alfred North Whitehead, Samuel Taylor Coleridge, and Hans Loewald . . . *in the same article*. Scratch that: there are no such textual critics save Janzen! And that is not a solitary,

2. Email correspondence, May 4, 2012.

Foreword

virtuoso performance, it is a *matter of course* with Janzen: day in, day out, *Eifer* meets *Eye-for*, and vice versa—not only in his scholarly work, but even in his email correspondence, which is nothing if not an education in scholarly breadth, existential depth, and the art of the English language.³ And let us not neglect to mention the man's memory: he never forgets a thing, and all that is remembered is somehow brought to bear in his *Eye-for-Eifer*.

Janzen first told us the *Eifer* story as a kind of justification for what others might deem a haphazard or altogether random moving among texts (he once described it as "snuffling," as pigs do when they root around for truffles), but we think it is an apology in the best and classic sense of the word: an explanation (*apologia*) of how his vivacious mind (*Eifer*) and insightful eye (*Eye-for*) work. They simply cannot be held in check, and the connections—as organic as embedded hyperlinks in a website (indeed, *more* organic than that)—reach out to the most unlikely places but are, consistently and compellingly, established by Janzen's attention to fine (textual) detail after fine (textual) detail. So it is that the present volume is not restricted to the Old Testament (Parts One–Three), but consistently makes profound recourse to the New Testament (Part Five), and the two parts are united by what amounts to a short monograph on resonance (Part Four) that alone is worth the price of the volume.

Since Janzen has provided a more detailed overview of the book, its parts, and its overall design (see the Introduction), we will not repeat such information here. We content ourselves with saying that the reader of the present collection has in their hands a great gift. Said reader is about to enjoy the benefits of a master exegete, who, in the famous words of Johann Albrecht Bengel (1687–1752)—"applies the text wholly to himself and applies himself wholly to the text." And here, by his own account, is this master exegete's magnum opus: a profound, wide-ranging, passion-filled "snuffling after," ultimately, what it means when people pray and how that

---

3. We cannot resist one example, though it could be repeated *ad infinitum*, from one of Janzen's emails (this one dated February 18, 2010): "But he [speaking of Richard B. Hays's work; see especially chapter 14] nowhere says where it [resonance] lies on the spectrum; and the use of the term in a couple of the authors he draws on, together with a couple of instances in his own use, is teasingly suggestive in ways he may not fully have recognized, and that in any case connect beautifully with what is implicit in Pat's usage [speaking here of Miller's presidential address to the Society of Biblical Literature; see chapter 14]; and, moreover, suddenly taps into a hermeneutical project that had lain, I thought, abandoned within me like a gold-mine shaft in Montana whose mouth had silted up through disuse and become hidden from view by a growth of desert shrubs."

xiii

Foreword

prayer can (and does!) make all the difference in the world: the biblical one and the one here and now.

It has been a great privilege for us to work on these essays with Gerry because working with him is consistently a joy and an education—both an education in joy, and a model of how joyous learning can be. We thank him for his good humor and patience during the editorial process and for teaching us so much along the way. We also thank the publishers who granted permission to reprint chapters 1–7 and 12–13 (see the Acknowledgments for original publication details). Note that even the previously published chapters have been edited to bring them into conformity with the rest of the volume and to update them in certain ways. One such way is the simplified transliteration schema that is adopted throughout the volume. Despite the fact that so much of Janzen's work depends on technical knowledge of the original biblical languages (even the scripts), so as to trace intertextual resonance through similar if not identical words, roots, phrases, and the like, the exegetical points are clear even in simplified transliteration. The simpler form was adopted since we believed that scholars accustomed to the languages would not be bothered by the simplification, whereas other readers who were not accustomed to the languages or to highly technical diacritical markings, might be put off by them, and miss out on the many gifts this collection offers.[4]

Our thanks also go to Deborah Van Der Lande for her help with electronic manipulation of the previously published essays, and, especially to Kevin J. Barbour, who labored tirelessly on the transliterations and in tagging the manuscript for the typesetter. Finally, Henry M. Huberty compiled the indices and Aubrey Buster helped in reading the proofs. The work of Barbour and Huberty was supported by grants from the Candler School of Theology, Emory University, which has our gratitude.

Brent A. Strawn
Atlanta, Georgia

Patrick D. Miller
Louisville, Kentucky

---

4. Note that this practice of simplified transliteration was extended even to citations of other works—once again for the sake of making this material widely accessible. Readers should note, however, that cited works often use the original scripts or technical transliteration in place of the simplified forms found here.

Preface

THE ESSAYS IN THE present volume began life independently of one another, as individual forays into specific locales in the biblical terrain. When Brent Strawn and Patrick Miller encouraged me to publish them in a single volume, it was envisioned simply as a collection of freestanding essays on loosely connected themes occurring largely in the Psalms. In arranging and retouching them I noted that in one foray after another I was crisscrossing trails I had already trodden, and that again and again, explicitly or implicitly, the topic concerned the *topos*, the *place*, of prayer in a world that may or may not make room for such an act. This triggered the question with which the introductory chapter opens: When Jesus prayed, where was he? Following the English maxim, "in for a penny, in for a pound," I then ventured a few high-risk forays into the New Testament, to see how some of the terrain crisscrossed in the Old Testament essays might be mapped within its pages. As I say, high-risk, for the New Testament is nowadays a field well trodden by highly specialized investigators intimately familiar with each other's work. To put the matter in the words of some of Caleb's fellow spies on returning from their foray into the promised land, there are scholarly giants in that field, and one who comes to it from a lifetime spent in the wilderness of the Old Testament does well to feel "like a grasshopper" in their eyes. But I resolved, doggedly, to proceed as under the urging of my own "inner Caleb." The result is three forays into what I have called a "New Testament Afterword" (Part Five) on themes traced in the Old Testament. In these I make no claim to adequate familiarity with the breadth and depth of current New Testament scholarship on the texts and themes they address; in that respect I offer what no doubt are comparatively "naïve" readings. But what I do hear in these New Testament texts is informed by a long formation in the same Scriptures that *in*formed the writers of the New Testament, and this schooling, for that reason alone, may give my readings a peculiar interest.

What is my method? In the Fall of 1978 I began a year-long sabbatical in which I aimed to extricate a method for biblical interpretation from the many-dimensioned writings of Samuel Taylor Coleridge—his theories

Preface

of poetic diction, of imagination as both a mode of perception and the fount of creativity, and of the interplay of subjectivity and objectivity; his philosophical theology; and not least his own prolific practice in reading biblical texts. As a formal effort, the project failed—I discovered that I had neither the appetite nor the competence for developing theories of interpretation. For better and for worse, I find texts themselves more enticing, more engrossing, than any theories put forth about them. This includes Coleridge's own texts, which I have since continued to read in an unsystematic way, and his influence—no doubt refracted and distorted—underlies these essays and surfaces from time to time. Catherine M. Wallace, who wrote a seminal study of Coleridge's *Biographia Literaria*, and who for several years taught courses in the history of literary theory, later in life has written, "what is theory, after all, but explaining what it is you do by instinct, by instinct and through love?" Looking back over the essays in this volume in an attempt to identify my "method," I suppose that I might claim at least to exemplify the meaning implied in that word's Greek root as my *Webster's Collegiate Dictionary* (second edition) puts it: from Greek *meta* ("after") + *hodos* ("way"), "method, investigation following after." In this sense, my method is to follow the trail of words and images as they lead me from one passage to another, sometimes by direct quotation, sometimes by one text echoing another, sometimes simply by the resonance that arises between them. In an attempt to justify this method more explicitly, especially in respect to the notion of "resonance" between texts, I have written a lengthy "interlude" (Part Four), placed between the Old Testament and New Testament essays, in the attempt to clarify for myself what it is I have done "by instinct and through love." For it is through love of biblical texts that, in Coleridge's words, "*find* me at greater depths of my being," that I have been moved to engage in these forays. It is in hope of serving such a love in others that they are offered here.

I am indebted to Brent Strawn and Patrick Miller for their editorial work in fitting these essays for publication, and especially to Pat for the seasoned wisdom and patient tolerance with which he responds to my frequent off-the-cuff email comments on some intriguing nook or cranny in the Psalms. I want to acknowledge also the contagious influence of my colleague, Joe R. Jones, whose penetrating theological judgments, arresting use of language and companionship in faith are a constant stimulus. And, always, there is my wife, Eileen, with whom I am privileged to read a psalm each day at that time when prayer takes place.

J. Gerald Janzen
May 2012

Acknowledgments

Chapter 1: Originally published as "... and the Bush Was Not Consumed," *Encounter* 63 (2002) 119–27 (subsequently republished in *JBQ* 31 [2003] 219–25). Used with permission.

Chapter 2: Originally published as "What's in a Name? 'Yahweh' in Exodus 3 and the Wider Biblical Context," *Int* 33 (1979) 227–39. Used with permission.

Chapter 3: Originally published as "What Does the Priestly Blessing Do?" in *From Babel to Babylon: Essays on Biblical History and Literature in Honour of Brian Peckham*, edited by Joyce Rilett Wood, John E. Harvey, and Mark Leuchter, 26–35. LHBOTS 455. New York: T. & T. Clark, 2006. Used with permission.

Chapter 4: Originally published as "Praying in the Space God Creates for the World," in *"And God Saw that It Was Good": Essays on Creation and God in Honor of Terence E. Fretheim*, edited by Frederick J. Gaiser and Mark A. Throntveit, 112–19. Word & World Supplement Series 5. Saint Paul: Luther Seminary, 2006. Used with permission.

Chapter 5: Originally published as "Prayer and/as Self-Address: The Case of Hannah," in *A God So Near: Essays on Old Testament Theology in Honor of Patrick D. Miller*, edited by Brent A. Strawn and Nancy R. Bowen, 113–27. Winona Lake, IN: Eisenbrauns, 2003. Used with permission.

Chapter 6: Originally published as "The Root *škl* and the Soul Bereaved in Psalm 35," *JSOT* 65 (1995) 55–69. Used with permission.

Chapter 7: Originally published as "Another Look at Psalm XII 6," *VT* 54 (2004) 157–64. Used with permission.

Chapter 8: Previously unpublished.

Acknowledgments

Chapter 9: Previously unpublished.

Chapter 10: Previously unpublished.

Chapter 11: Previously unpublished.

Chapter 12: Originally published as "Jeremiah 20:7-18," *Int* 37 (1983) 178–83. Used with permission.

Chapter 13: Originally published as "Eschatological Symbol and Existence in Habakkuk," *CBQ* 44 (1982) 394–414. Used with permission.

Chapter 14: Previously unpublished.

Chapter 15: Previously unpublished.

Chapter 16: Previously unpublished.

Chapter 17: Previously unpublished.

Abbreviations

AB	Anchor Bible
AIL	Ancient Israel and Its Literature
BASOR	*Bulletin of the American Schools of Oriental Research*
BBB	Bonner biblische Beiträgre
BDB	Francis Brown, S. R. Driver, and Charles A. Briggs. *Hebrew and English Lexicon of the Old Testament.* Oxford: Clarendon, 1907
BHS	*Biblia Hebraica Stuttgartensia.* Edited by Karl Elliger and Wilhelm Rudolph. Stuttgart: Deutsche Bibelanstalt, 1983
CBQ	*Catholic Biblical Quarterly*
CBQMS	Catholic Biblical Quarterly Monograph Series
CEB	Common English Bible
CSR	*Christian Scholar's Review*
EDNT	*Exegetical Dictionary of the New Testament.* 3 vols. Edited by Horst Balz and Gerhard Schneider. Grand Rapids: Eerdmans, 1990–93
FAT	Forschungen zum Alten Testament
GKC	*Gesenius' Hebrew Grammar.* Edited by E. Kautzsch. Translated by A. E. Cowley. 2nd ed. Oxford: Oxford University Press, 1910
HALAT	L. Koehler, W. Baumgartner, and J. J. Stamm. *Hebräisches und aramäisches Lexikon zum Alten Testament.* 5 vols. Leiden: Brill, 1967–95
HALOT	L. Koehler, W. Baumgartner, and J. J. Stamm. *The Hebrew and Aramaic Lexicon of the Old Testament.* Translated and edited by M. E. J. Richardson. 5 vols. Leiden: Brill, 1994–2000
HBT	*Horizons in Biblical Theology*
HSM	Harvard Semitic Monographs
HTR	*Harvard Theological Review*
Int	*Interpretation*
ITC	International Theological Commentary
JB	Jerusalem Bible
JBL	*Journal of Biblical Literature*
JBQ	*Jewish Bible Quarterly*
JCS	*Journal of Cuneiform Studies*
JPS	Jewish Publication Society
JSOT	*Journal for the Study of the Old Testament*
JSOTSup	Journal for the Study of the Old Testament Supplement Series
LHBOTS	Library of Hebrew Bible/Old Testament Studies

Abbreviations

LSJ	Henry George Liddell, Robert Scott, and Henry Stuart Jones. *A Greek-English Lexicon*. 9th ed. Oxford: Clarendon, 1996
LXX	Septuagint (Greek Old Testament)
MT	Masoretic Text
NCB	New Century Bible
NIGTC	New International Greek Testament Commentary
NIV	New International Version
NJPSV	New Jewish Publication Society Version (Tanakh)
NRSV	New Revised Standard Version
NT	New Testament
NTL	The New Testament Library
NTS	*New Testament Studies*
OED	Oxford English Dictionary
OT	Old Testament
OTL	The Old Testament Library
PNTC	Pillar New Testament Commentary
RSV	Revised Standard Version
SBLDS	Society of Biblical Literature Dissertation Series
TDNT	*Theological Dictionary of the New Testament*. 10 vols. Edited by Gerhard Kittel and Gerhard Friedrich. Translated by Geoffrey W. Bromiley. Grand Rapids: Eerdmans, 1964–76
TP	*Theologie und Philosophie*
TWAT	*Theologisches Wörterbuch zum Alten Testament*. 10 vols. Edited by G. Johannes Botteweck and Helmer Ringgren. Stuttgart: Kohlhammer, 1970–2000
VT	*Vetus Testamentum*
WBC	Word Biblical Commentary
ZAW	*Zeitschrift für die alttestamentliche Wissenschaft*

Sigla:

fem.	feminine
masc.	masculine
pl.	plural
sg.	singular
v(v).	verse(s)
/	lines of poetry *or* the MT (Hebrew) reading followed by the LXX (Greek) reading
//	parallel to/with
*	indicates modification to base translation

Introduction

From Plane to Plane

WHEN JESUS PRAYED, WHERE in the world was he? The four Gospels portray him as praying in a variety of places: on a mountain, and in a wilderness or "lonely" place; in the midst of his disciples, and, again alone, in Gethsemane. According to Matthew, when he taught his disciples to pray he counseled them to "go into your inner room and shut the door and pray to your Father who is in secret [*en to krypto*]." Immediately—which is to say, after about fifty-five years of reading this text—I find myself wondering whether this phrase "in secret" may contain more significance than meets the eye. Because, in the Old Testament—the book from which Jesus and his fellow Jews learned their religious ABC's—that phrase, "in secret," or "in a secret place," at times appears to mean more than it says.

Frequently, the Hebrew word, *seter*, "secret," refers to God's sheltering presence in the sanctuary. For example, Ps 91:1 reads, "One who dwells *in the secret place* [*beseter*] of the Most High, who abides in the shadow of the Almighty." So, the psalmist often seeks that place, especially as a refuge from enemies (Pss 27:5; 31:21; 32:7; 61:4; 91:1; 119:114).[1] But once, when Israel calls out in distress, God says, "I answered you *in the secret place* [*beseter*] of thunder." Clearly this "secret place" is not the earthly sanctuary. What, or where, is it? In Ps 18:11, David says that when God came to help him in answer to his prayer, God "made darkness his *covering* [*seter*] around him, / his canopy thick clouds dark with water." Taken together,

---

1. Where the numbering of verses differs in English translations from the numbering in the Hebrew Bible, I shall follow the English text. Readers of the original text will know how to make the adjustment. For the most part I shall quote from the RSV, following the example of the late New Testament scholar, Krister Stendahl. Himself a staunch advocate for the ordination of women to Christian ministry, and holding that for, liturgical readings, inclusive-language translations are preferable, he wrote, nevertheless, that "I still consider [the RSV] to be the most useful translation for study purposes, because it has an adequate closeness to the Greek" (*Final Account*, xi).

these two texts imply that God's presence and activity are at one and the same time revealed and hidden within the various features of a violent thunderstorm.

But what does the verbal form of *seter* mean in Isa 45:15? The passage goes like this: "Thus says the LORD: 'The wealth of Egypt and the merchandise of Ethiopia, and the Sabeans, men of stature, shall come over to you and be yours, they shall follow you; they shall come over in chains and bow down to you. They will make supplication to you, saying: "God is *with you [bak]* only, and there is no other, no god besides him." Truly, thou art a God *who hidest thyself* [*ʾel mistatter*], O God of Israel, the Savior.'"

In this passage, all the human second person pronouns are in the second person feminine singular. God is addressing Zion, currently in ruins, about the fact that the Persian king, Cyrus, will rebuild the city and set the Jewish exiles free to return home (Isa 45:1–13). When other nations and peoples come to bring tribute to God in the restored city, they will confess: "God is *in you [bak]* only." (The Hebrew preposition, *b*-, here echoes the other passages with the expression, *beseter*; and the confessional line following makes that implication explicit: "Truly, thou art a God *who hidest thyself* [*ʾel mistatter*], O God of Israel, the Savior.") If the biblical God is a hidden God, as this passage suggests, God is at one and the same time hidden and manifest within the restored community.

But only the *restored* community? What of the exilic interim? What of those who mourn the destruction of the city and temple, and who would continue to use psalms like Psalm 91? Is God still a *secret place* of prayer and refuge? In the book of Ezekiel, the "glory of the LORD" that departed from the temple in Jerusalem before its destruction (Ezekiel 10) is envisioned as standing, then, "upon the mountain which is on the east side of the city." When the "glory of the LORD" is envisioned as returning to the rebuilt temple, it enters the temple "by the way of the gate whose prospect is toward the east." The implication is that the glory of the LORD during the exile is "hidden" among the exiles in Babylon. And this is made explicit in Ezek 11:16, where God says of the exiles, "Though I removed them far off among the nations, and though I scattered them among the countries, yet I have been a sanctuary [*miqdaš*] to them for a while in the countries where they have gone." Clearly they have no temple in Babylon. So, in the interim between the destruction of the Jerusalem temple whose roots go back to God's instruction to Moses in Exod 28:8 ("let them make me a *miqdaš*, that I may dwell in their midst") and the "house" for God that Cyrus will commission in a Jerusalem to which the exiles are to return—in this interim God will become a stand-in *miqdaš*. When, then, the exiles in Babylon pray to God, *where are they*?

To be sure, in one sense they are praying "in Babylon" (Jer 29:7), a city that occupies a specific meeting-point of GPS coordinates in the four-dimensional world of space-and-time. But are they not also praying *beseter*? Are they not also praying—as Jesus counsels his disciples to pray—"in secret"? What might this mean? Is this *beseter* plottable in the same sense as the city of Babylon? The question I am raising here, and a preliminary clue as to how we may think of the experiential reality it points to, may be indicated by a pair of anecdotes from two personal experiences of momentarily finding myself in a space-time "warp."

It was reunion week at the seminary where I had undertaken my first theological degree. I had walked over to have breakfast at the student union building of the neighboring university, and taken a seat at an otherwise unused, large round table. Shortly, three women from the university's administrative staff took seats opposite me. As I ate unhurriedly, reading the morning newspaper, they engaged in intimate, animated conversation. Although I could hear them quite easily, I did not feel intruded upon by their voices, because the warm intimacy of their voice-tones clearly, somehow, configured a space that included them and excluded me. Suddenly, my concentration on what I was reading was interrupted by a change in the tone of the current speaker, as, pausing in the middle of an anecdote, she shifted from easy intimacy into a neutral, impersonal courtesy, saying, "excuse me—could you pass the sugar?" Clearly her tone signaled that she was speaking to me, a stranger. "Oh yes—of course"; and with that I slid the sugar bowl to the center of the table where she was able to reach it, at which point, with a further polite, "thank you," she lapsed back into the intimate tones of her anecdote while I turned back to my newspaper and coffee. But as I did so, I realized with a start that I had just participated in a space-warp. Simply—and paradoxically—by shifting from a tone of intimacy to one of impersonal courtesy, she had reconfigured her existential space from one that had included her friends and (appropriately) excluded me, to a space that included me, a stranger, and (appropriately) excluded her friends.

Academics are often accused of being absent-minded. ("Has Gerry heard anything I've said?" "No—he's miles away.") My retort, in self-defense, is that I have never in my life been absent-minded; but there have been times when, for what seemed to me like a short while, I was present-minded somewhere else. I think I may have come to this way of putting the matter as the result of a midnight observation in Jerusalem in the fall of 1964—my second anecdote of a space-time warp. My wife Eileen and I were living at the Hebrew Union College Biblical and Archeological School, I as a graduate student, along with a few others, under Frank M. Cross of Harvard as Annual Visiting Director. Eileen and I were returning from a late evening scooter ride, and were now pulling into the school parking lot. I brought the scooter to a stop facing the part of the wall where a half-dozen tall, narrow windows allowed us to see into Professor Cross's study. There he was, seated in the deep study chair, the desk piled with books, some of them open and leaning at an angle for convenient consultation, his legs extended and his heels resting on the edge of his desk, his gaze fixed on a huge tome lying open on his lap and across his legs. Before I turned away for us to return to our lodgings, it struck me that though his body was there in that chair, his mind seemed to be far away (or unfathomably deeper down, in some hidden recess inside him), holding conversation with the ancient, medieval, and modern authors whose written words, together with his receptive concentration on them, brought into being a space-time warp that threw the four-dimensional world of his study and his body into relation with a whole new dimension of interior depth. The temporal texture of that interior world was illustrated the day his wife Betty Anne phoned down at noon to announce that the lambchop lunch she had promised him at breakfast was now ready. He arose from his study chair, bounded up the stairs, sat down at the table and, after one mouthful, exclaimed, "Betty Anne, this chop is cold!" To which she replied, "Well, it was hot when I called you for lunch." He looked at his watch. It was 4:00 p.m.

By these two anecdotes, I mean to suggest that when prayer takes place it always occurs, and can be plotted, at a specific intersection of the coordinates of our ordinary four-dimensional, physical and embodied space-time world. And it also takes place within a human, social world (a world whose relational structures vary from intimacy to impersonal courtesy to unthinking rudeness to outright hostility and even dire conflict). Prayer also takes place in this latter world, whether in the "common prayer" of a synagogue, mosque, or church; solitary prayer in one's inner room or favorite "deserted" place; or internal prayer in the face of antagonistic or hostile individuals or groups. Here too, a sociologist and a psychologist could team up to "plot" such prayer-events on the relational coordinates of a social/psychological analytical grid of their devising. But prayer does not simply occur at either this physical or this social/psychological locus, as though these respective places are given and prayer simply takes up a place that they provide. Rather, at the intersection of these two

From Plane to Plane

sets of coordinates, prayer "takes place," in the sense that the act of prayer shapes a space for itself (the way the bird of Ps 84:3 shapes a nest for itself), and has a standing before God, that does not depend on the space and the standing that those other coordinates provide for it.

To address God in prayer, whether along with a group, alone in a group, or by oneself in a deserted spot (I want to coin a term here and say, whether solidarily or solitarily), is to participate in the emergence of a "place" that comes into existence, or is once again re-entered, through the act of prayer. Think of Psalm 100, the so-called *Jubilate Deo*, especially the words, "Serve the Lord with gladness! / Come into his presence with singing! / .... / Enter his gates with [*b*-] thanksgiving, / and his courts with [*b*-] praise!" In the days when this psalm was sung in relation to the sanctuary standing in Jerusalem, the gates through which one would pass while singing would stand tall and sturdy in their carved beauty, and the courts would open out spaciously to receive them as they continued to sing there. But where there is no such sanctuary, I suggest that we may take the preposition *b*-, not in an instrumental, but in a locative sense, such that it is the very offering of thanksgiving that is the gates through which one enters into the space where God dwells "in secret" (Psalm 91); and it is the very offering of praise that configures a spacious secret place within which one may meet with God.

The nature of this "place" is such that it can be expressed or talked about only by using words that belong and have their meaning first of all in the two other "worlds," the interwoven but distinguishable worlds of four-dimensional physical/embodied space-time and social and psychological inter-personal relationships. When they are used to talk about, and to talk within, this third sort of "space," these words may be said to function figuratively or metaphorically. But in our day it is dangerous to put the matter this way. This is because we have become habituated, by a variety of practices and for a number of reasons, to think that reality and truth can best be expressed literally, and that when words are used figuratively or metaphorically, this is merely a colorful way of juicing up what could more straightforwardly be stated literally. One could, for example, quote a widely used high school textbook on English poetry, where after each poem the students are asked to state what the poem says "in their own words." This is to assume that what the poem is getting at could be got at also in straightforward, "literal" language. Even the great literary scholar, John Livingston Lowes, in writing a monumental study of the sources lying behind Samuel Taylor Coleridge's poem, "Kubla Khan," mistitled his study, *The Road to Xanadu*. The road that Lowe traces leads to the poem; but it is the poem itself that is the road to Xanadu, and if one wishes to enter into the Xanadu that Coleridge is concerned with, one must stay within the poem and trust it to lead one to that enchanted and unfathomable place.

So, if we are asked to think and speak about—and especially to speak within—this third "space" with the aid of metaphors and figures of speech, we must resist the default tendency of our modern minds to think of those words as "merely metaphorical." The way in which this third "secret" space is distinct from and yet connected to our other two worlds calls for words that arise within those two worlds, yet, by the manner of their use, point to a world that is distinguishable from them. The most adequate, the most truth-telling language for this third world is the figurative language of metaphor. An example: When Jesus says, "I have meat to eat that you know not of," he means to say that the reality he is talking about is like, and is connected to, their own physical experience of eating food and their own social experience of eating with others or alone; yet it is distinct in such a way that, if they understand his words only as applying to those two worlds, they will not (yet) know what he is talking bout. The connection between this world and the other two worlds (worlds that we may properly and constructively call "secular") is evidenced in the way that our experiences in this third world may become manifest in the way we come to fresh and new experience, understanding, and action in the "secular" worlds. This, I think, is part of what the Gospel of John is getting at in saying that, in Jesus of Nazareth, "the word became flesh and tented [*eskenosen*] among us"—the Greek word here being the verbal counterpart to the noun, *skene*, that appears in the Greek translation of Exod 25:9 where the Hebrew noun, *miškan*, is translated in the RSV as "tabernacle" in which God after Sinai dwelt among the people *beseter*.

The importance of this third realm is that—like the old farmhand in Robert Frost's poem, "From Plane To Plane," who wouldn't hoe corn rows in both directions, but hoed one way, shouldered his hoe and walked back to the beginning, and only then lowered his hoe for another assault on the weeds, because "a man has got to keep his extrication"—like that farm hand, prayer is one way in which we are enabled to "keep our extrication" from—or within—our physical and social worlds which otherwise could overpower us and either press us into abject conformity or crush us to death.

To turn, now, to the essays collected in this volume: They were for the most part written as separate, self-standing studies. Each essay began when some word, phrase, image or sentence in an otherwise familiar passage suddenly "made strange" and piqued my interest by some fugitive suspicion that there may be more to its meaning than had formerly met my eye. (As a matter of fact, this is what happened with the phrase, "in secret," that for some reason came to mind when I sat down to draft this introduction.) From such a starting-point, I typically moved out (like an internet junkie following up web-links indicated by underlined words in an internet essay or news report) to other places in the Bible where the word was used, the phrase occurred, the image reappeared, or another sentence made a similar or contrasting statement on the same topic. My movements from passage to passage were not initially directed by method or design, but rather followed my nose the way a dog runs ahead of its master, darting into the underbrush first on one side of the path and then on the other, following the zigzag of rabbit trails as they crisscross one another, but always moving forward in the direction of the master's walk. What I discovered was that, just as one can get from Indianapolis to Boston most quickly by the turnpike, but in so doing misses most of the scenery and local color of state highways and local roads, just so, I was not interested in simply rushing to an interpretive conclusion, but rather in following each local trail and examining each local stopping-point with unhurried curiosity and appreciation.

Recently, I have been browsing in a collection of essays by an Assyriologist, Professor William Moran. In one of the essays (the title of which—"The Most Magic Word"—is itself suggestive), this scholar quotes from a letter in which the German poet Schiller writes to Goethe as follows: "The purpose of the epic poet is present at every point of his movement. Hence we do not hasten impatiently to an end, but linger with love at every step."[2] I recall an afternoon when, on sabbatical leave from my seminary duties (sabbatical being itself a form of fruitful "extrication"), I sat in on one of Professor Moran's classes. The student was interpreting a text in ancient Babylonian, and at one point commented on a verb in that text as meaning, "to dig." When he thought he had made his point, Professor Moran patiently asked the student whether he could identify what kind of digging the verb referred to. The student looked blank for a moment, then shook his head. With constructively elegant precision, Professor Moran encouraged the student to appreciate the significance for a full understanding of the passage that lay in the fact that the verb in question meant, specifically, to dig or trench accumulated silt and mud out of an irrigation canal.

In my own biblical studies, I myself have stumbled inadvertently onto the secret that Professor Moran so masterfully practiced himself, and

---

2. Moran, *Most Magic Word*, 59 and reference there.

for which he has now given me words to describe my own mind-and-body feeling as I go about zigzagging through scripture. It is the feeling of "not hasten[ing] impatiently to an end, but linger[ing] with love at every step." This introduction has already given a brief demonstration, and thereby fair warning—*caveat emptor*—that the person who reads this book will often be slowed down to pay particular attention to the nuances and connotations of a particular word or image in the passage at hand and in other passages that it stands ready to connect the reader to. This kind of reading is like picking raspberries—one at a time, while moving carefully so as not to snag one's sleeve or scratch a bare arm on a prickly raspberry cane. The present essays are offered as a collection in hope that their content, as well as their compositional structure, will induce the reader, like Schiller with epic texts, to read and "linger with love at every step." Not with love for what I have to say! But with love for what the quoted and examined biblical texts have to say, as I slow our reading down for the individual words and phrases and their web-link-like connections with other places in the Bible. In other words, this is not a book to be read. It is a book to be studied. Or rather, it is to be used as one "hitchhiker's guide to the galaxy" of scripture. Used in this way, it may assist the reader in the practice of lingering lovingly on text after text along a variety of often-intersecting paths of biblical meaning.

One other thing: Some of the papers are sprinkled liberally with Hebrew and Greek words in English transliteration. This is to enable us to gain a finer appreciation of the precise connotations and overtones of individual words as well as their potential for linkage to related passages elsewhere in the Bible—such linkage not always apparent in English translation. To make the essays user-friendly, or at least user-accessible, by a wider readership than those who read the original biblical languages, these transliterations have been simplified from their more scientific form. Moreover, I shall make every effort to accompany original-language citations with lexical or grammatical comments that clarify the fine points at issue. Those who do read Hebrew and Greek will, I trust, be patient with these comments and forgive the simplified transliterations.

Either explicitly or implicitly, the essays all have to do with prayer. As most of them were written independently of the others, the reader should feel free to read them in any order, as dictated by personal interest. In his book, *The Poetry of Robert Frost: Constellations of Intention*, Reuben Brower quotes Frost as saying that a reader need not read his poems in any particular order, since "[p]rogress is not the aim, but circulation. The thing is to get among the poems where they hold each other apart in their places as the stars do." Whenever Frost set himself to gathering a number of separately composed poems into a publishable volume, "[t]he interest, the pastime, was to learn if there had been any divinity shaping my ends and I had been building better than I knew. In other words could anything of larger design, even the roughest, any broken or dotted continuity, or any fragment of a figure be discovered among the apparently random lesser designs of the several poems?"[3] It is in that sense, I hope, that the essays, as collected here, may display an after-the-fact "constellation of intentions."

A word about the title. I hope that the comments to this point convey to the reader an initial sense of what I intend to mean by "when prayer takes place." As for the subtitle, the *Oxford English Dictionary* (*OED*) says that, in one meaning, a "foray" is an advance troop on an incursion for supplies and to reconnoiter a territory to be taken over by a following military force. (Think of the Israelite spies, in Numbers 13, on a foray into south Canaan, one of the leaders of which is Caleb, a name that—fittingly enough in the context of this introduction—means, "Dog!") More peaceably, a foray may be a scouting party sent out ahead of a wagon train westward. The year before I was born, my parents turned an old army tent and a flatbed wagon into a covered wagon and, with my three older brothers, joined such a train on a 350-mile trek into northern Saskatchewan to enter a land claim under the provisions of the homestead act. (My earliest memory dates to that homestead. All my life I have carried in the back of my mind the scene of our dad dragging a dead horse out of a log barn with another horse, around the corner to the right and into the bushes. When I shared this memory with our mother, at a family reunion forty years later, she gaped in disbelief: "You can't remember that! You were born in August; and before the snow was off the ground the following spring, I stood in the doorway of the log cabin, holding you on my hip, and watching that scene, sick with worry as to how dad was going to be able to put the crops in with only one horse." I suspect that my infant sensitivity to her emotional intensity burned that scene into my brain.) It turned out that our dad, whose health had already been poor before the trek, became even more unwell, and he was forced to abandon his claim on the homestead, which then reverted to the Crown. So all that I have from that attempt of our family to "take place" in the form of a homestead is a single snapshot memory. Now, I take the Psalms as various forays into a space which they may aspire to inhabit "forever" (as in Ps 84:4; and in Pss 23:6 and 27:4, on which see more fully chapters 10–11 below), but which, as most scholars hold, they perforce must relinquish in death if not sooner. So I offer

---

3. Brower, *Robert Frost*, 1–2.

especially the essays on the Psalms and the Prayer of Hannah as Israelite forays into a biblical world. In some of these psalms, as in others, all that sustains the psalmist is a snapshot memory of a brief time spent in that "third world," *beseter*. My own reflections track my own attempts, like that Israelite dog, Caleb, to make brief forays into that world. It is common among biblical scholars to refer to the activity of biblical interpretation as consisting in "exegesis," which, from the Greek verb, *exegeomai*, means, at base, to "lead out" the meaning to where the hearers and readers are. My method, in these essays, may more properly be called "eisegesis"— not in the modern sense of reading one's own meanings into the text and then pretending to have found them there, but in one original sense of the Greek verb, *eisegeomai*, which means, "to lead in, to introduce, to induct by means of religious rites." I seek not so much to deliver the meanings of the text to the reader, as to help the reader enter into the text as a foray into the biblical world and its secret presence.

It remains to comment briefly on each of the essays that follow. In Part One, "Orienting Ourselves in the Biblical World," I begin with ". . . and the Bush Was Not Consumed," a passage foundational for all biblical narrative insofar as it portrays Moses as drawn by a strange sight out of his ordinary space-time world and warped into a world and onto a standpoint that is holy. That holy standpoint is intimately related to the social and political world of Moses' people in Egypt, a world and a regime under which they are being crushed to death. In "What's in a Name? 'Yahweh' in Exodus 3 and the Wider Biblical Context," I explore the implications of God's "exegesis," or drawing out, of the meaning of the divine name that is represented in most English translations by the phrase, "the LORD," and which is understood by many scholars as originally pronounced (when it was still uttered), "Yahweh." This name is the most important word in the Old Testament. In "What Does the Priestly Blessing *Do*?" I explore what three-dimensional "word-event" transpires—what third world takes shape once again—when, in accordance with the instruction in Num 6:22–27, the worship leader blesses the people by "putting" God's name on them, the people receive the blessing, and in and through those acts of blessing and reception it is in fact God who blesses. To this point the essays provide a general context for the specific practice of prayer, while already introducing some aspects of that practice. In "Praying in the Space God Creates for the World," I reflect on how God's creation of the world may be thought of as arising out of how God—who, prior to all worldly beginnings, exists as the ultimate mystery of a life that is self-constituting and whose infinite and eternal "here-ness-and-now-ness" knows no bounds—how such a God acts in such a way as to make a space that is not God within God's boundlessness, and then calls into existence, into that space as it were from out of nowhere, the various components and living beings of creation. The space that God makes for creation subsequently becomes the divine basis for "the place that prayer takes" in the midst of this space-time world.

In Part Two, "Forays into a Biblical World," I offer forays into one woman's prayer and into six psalms, in which, as I have said, I typically begin with a word or phrase or sentence, and from there move back and forth between it and "hyperlinked" other passages, and between those passages themselves, so that gradually a rough-and-ready sense may begin to arise of some of the contours and coordinates that are distinctive to the places that prayer takes, and some of the biblical constellations begin to appear by which we may take our bearings, especially when we find ourselves alone and in various kinds of darkness. The essay, "The Verb *ya'ameṣ* in Psalm 27:14: Who Is Strengthening Whom?" may put the greatest strain on the reader's patience. But it may be in this essay that the tracks running through all the previous essays converge and give their combined payoff. I confess to having written it both for its own sake and yet with all the others resonating quietly in the back of my head. (In fact it arose out of an initial attempt to bring Ps 27:14 into the picture toward the end of the preceding essay.) To the degree that the reader can stay with the slow-moving zigzag in this essay, to that degree, I trust, the biblical text itself, apart from my own comments on it, will reward the trouble expended to do so. The final essay in this group reconsiders the question of whether, in Psalm 23, the psalmist's foray into God's (secret) presence ends at death, or whether the psalm intimates a permanent tenure there.

In Part Three, "The Standpoint of Two Prophets," the two essays are of a more general character, although each in its own way engages a biblical text concerning the prayer-forays of a biblical prophet. Jeremiah has come into the picture already, in the essay on Psalm 69; and the short essay, "Solidarity and Solitariness in Ancient Israel: The Case of Jeremiah," explores this figure as a classic biblical case of a solitary prayer presence before God that is at the same time rooted at the deepest level in the prophet's solidarity with the people. I conclude Part Three with "Eschatological Symbol and Existence in Habakkuk," centering on another prophet who is also introduced in the essay on Psalm 69. This essay recapitulates, yet under a fresh perspective, the way this prophet's prayer not only takes place in a troubled space-time world, but also "takes time" in such a way as to "keep its extrication" in a world whose imperial powers threaten to subject Israel's claim on the future to sudden foreclosure.

This prophet opens his prayer with questions familiar from the Psalter—"Lord, how long?" and "why?" In this connection, it is worth noting that, both in the *OED* and in *Webster's New Collegiate Dictionary* (1981), the first meaning for the word, "space," is given as "denoting time or duration"—as the *OED* puts it, "a lapse or extent of time between two definite points." So it is that, in the KJV, a worship scene in the book of Revelation is described at one point in the words, "there was silence in heaven for the space of half an hour" (Rev 8:1). This capacity to experience time itself as constricting or as expansively hospitable helps us to appreciate how Habakkuk undergoes a "time-warp," in that, having begun to pray by crying out, "How long, O Lord?" he ends with the confession and avowal, "Though the fig tree does not blossom, / nor fruit be on the vines, / the produce of the olive fail / and the fields yield no food, / the flock be cut off from the fold / and there be no herd in the stalls, / yet I will rejoice in the Lord, / I will joy in the God of my salvation. / God, the Lord, is my strength; / he makes my feet like hinds' feet, / he makes me tread upon my high places" (Hab 3:17–19).

At a time of economic crisis, such as has been afflicting much of the contemporary world, one could do worse than seek to join Habakkuk in this his foray, and then, renewed by that period of extrication, put one's hoe to the ground and once again attack the weeds that have sprung up in our midst and threaten to choke off all productive and sustaining growth.

Following these essays, I undertake to reflect at length on an aspect of the relation between biblical texts that I, along with others, have referred to as "resonance," but that to my knowledge has not received the sort of attention that has been given to the phenomenon of inter-textual "echoes." As an interlude (Part Four) before the final three essays, I explore in a preliminary, heuristic way how inter-textual resonance might relate to the phenomenon—and significance—of resonance in human existence and in the natural world, suggesting in this way also the theological implications of this phenomenon.

In Part Five, "New Testament Afterword," I then extend this series of forays into the biblical world by ranging forward to trace briefly how themes central to this study as focused on the Old Testament are taken up or echoed in the New Testament. Perhaps the most central theme concerns the unfathomable mystery of the divine self-naming in Exod 3:14–15 as echoed, in my judgment, in the Alpha-Omega and related formulations in the book of Revelation and, perhaps, in the characterization in Heb 12:2

of Jesus as "pioneer" and "perfecter" of faith. By the exploratory attention I give to this theme, the present volume of forays into the biblical world returns to its starting point in the first two essays. One implication of this return is that, to echo Frost's words quoted earlier, the aim is not so much progress as circulation, a process of reading—in the present instance, the process of reading the various parts of the biblical canon in such a way that each re-reading of a particular passage takes on fresh meaning in light of the other passages read in the meantime, and at the same time, by means of that fresh meaning, sheds still more light on the other passages read thereafter. The final foray, "Redeeming the Expression 'Redeeming the Time,'" broaches the possibility of applying the spatial dimension of the preceding essays also to the dimension of time. When prayer "takes place," it also "takes time." As I propose—playing off of Anthony Powell's epic novel title, "A Dance to the Music of Time," when prayer truly takes place, one finds oneself beginning to dance in time to the music of eternity.

What I hope to show, then, in Part Five, is that the words and images, motifs and themes, exhortations, proclamations, and encouragements in the New Testament take on a depth of richness and a range of application in direct ratio to the reader's prior immersion in and formation by the semantic grammars and existential fields of energy generated and presented in the Old Testament. For, when the author of Hebrews writes, "the word of God is living and active, sharper than any two-edged sword, piercing to the division of soul and spirit, of joints and marrow, and discerning the thoughts and intentions of the heart; and before him no creature is hidden, but all are open and laid bare to the eyes of him with whom we have to do" (Heb 4:12–13), it is of what we have come to call the Old Testament that the writer speaks. It is when we read the New Testament as both an energized and a re-energizing Afterword to the Old Testament, that the above words, originally referring back to the original canonical testimonies, come in time to apply to the New Testament also as partaking of that same living, active power in the lives of their readers.

part one

Orienting Ourselves
in the Biblical World

1

"... and the Bush Was Not Consumed"

THE IDEA FOR THIS essay struck me[1] during a visit to the Jewish Theological Seminary (JTS) in New York. I had been browsing the library at Union Theological Seminary nearby, and had come across a reference to an article in a journal that only JTS carried. As I approached the building, I noticed the "logo" over the entrance—a flaming bush and the words, *wehasseneh 'enennu 'ukkal*: "and the bush was not consumed."

The text was, of course, familiar to me. I had lectured on it and led exegetical discussions of it over a period of thirty-five years, and I had ventured an interpretation of it in my little commentary on the Book of Exodus. But when I saw it over the entrance to JTS, something stirred within the depths of my hermeneutical imagination: a sense that in this setting the bush and its text might offer a new meaning. In a hurry, however, I did not pause to reflect on what that meaning might be, but found my way to the library and lost myself blissfully amid its treasures. On my way out, I did pause to examine the displays and memorials in one hallway, especially those concerning the Holocaust. That is when the bush blazed forth with new meaning. Of course! How could I have been so blind all these years?

I had assumed, with commentators generally, that the fire in the bush is a manifestation of the biblical God who, as a later writer put it, "is a consuming fire" (Heb 12:29, following a reference to the scene in Exodus 19). I had been preoccupied with the fact that, if this fire manifests the presence of the holy and living God, then it is remarkable that so flammable a thing as a bush should burn and yet not be consumed. For the scene stands in contrast to the fearful warnings in Exodus 19, the people's fearful response in Exod 20:19, and the similar dire warning in Exod 33:3. The only thing I could compare with the bush that burned but was not

---

1. I use the verb intentionally.

PART ONE: Orienting Ourselves in the Biblical World

consumed was the scene in Exod 24:9–11. There, following the covenant-making ceremony in 24:3–8, "Moses and Aaron, Nadab, and Abihu, and seventy of the elders of Israel went up, and they saw the God of Israel; and there was under his feet as it were a pavement of sapphire stone, like the very heaven for clearness. And he did not lay his hand on the chief men of the people of Israel; they beheld God, and ate and drank" (Exod 24:9–11). The dire warning, three days earlier, that anyone gazing on God who was to descend on the mountain in fire would run the risk of perishing (19:21), here gave way to an intimacy in which these representatives of the people could, amazingly, see God and live. Was the bush that burned and was not consumed a veiled promise of divine-human intimacy, now unveiled under the sacred canopy of the covenant sealed in blood and solemn words?

Such a reading continues to hold a great attraction for me, and I am loath to give it up. But in the present essay I want to propose a reading provoked by my visit to JTS.[2] In my view, this reading does greater justice to a question that my earlier understanding had left unsatisfied: Is the appearance and voice of God to Moses out of a bush that burns but is not consumed integral to the narrative context? Or is it simply an arbitrary sign whose significance is exhausted in getting Moses' attention?

I begin with the observation that, while fire is a frequent symbol of the Holy One, fire also carries many other connotations. One that bears on the reading I am proposing is exemplified in the following passages:

> (1) When you pass through the waters I will be with you;
> and through the rivers, they shall not overwhelm you;
> when you walk through the fire you shall not be burned,
> and the flame shall not consume you. (Isa 43:2)

Coming from the prophet of the Babylonian Exile, the images of fire and water connote the afflictions that the Judahites suffered in the fall of Jerusalem and its aftermath.

> (2) Behold, I have refined you, but not like silver;
> I have tried you in the furnace of affliction. (Isa 48:10)

Here again, fire (implicit in the image of the furnace) refers to the afflictions suffered at the hands of the Babylonians. In several passages,

---

2. It occurs to me now, in editing this essay for the present volume, that my earlier and later readings may be taken as two dimensions of the same issue. Israel, as a bush that burns and is not consumed under the oppressions of other peoples, is also a bush that—despite the threat of Exod 32:10—is not consumed by God's judgments but sustained by God's mercy and grace, steadfast love and faithfulness (Exod 34:6).

"... and the Bush Was Not Consumed"

however, the image of the furnace with its implied fire is applied to the last stage of the ancestral sojourn in Egypt.

> (3) But the LORD has taken you, and brought you forth out of the iron furnace, out of Egypt, to be a people of his own possession, as at this day. (Deut 4:20)

> (4) For they are thy people, and thy heritage, which thou didst bring out of Egypt, from the midst of the iron furnace. (1 Kgs 8:51)

> (5) Cursed be the man who does not heed the words of this covenant which I commanded your fathers when I brought them out of the land of Egypt, from the iron furnace. (Jer 11:3–4)

> (6) For thou, O God, hast tested us;
> thou hast tried us as silver is tried.
> thou didst bring us into the net;
> thou didst lay affliction on our loins;
> thou didst let men ride over our heads;
> we went through fire and through water;
> yet thou has brought us forth to a spacious place. (Ps 66:10–12)

These last verses follow a passage celebrating Israel's deliverance from Egypt (Ps 66:5–7).

In all these passages, while the affliction occurs within the horizon of the sovereignty of Israel's God, the fire refers most immediately to the human agency of Egypt or Babylon, and characterizes Israel's affliction under that rule. With this in mind, let us turn our attention to the possible connotations of the bush in Exodus, chapter 3.

That a vine, a bush, or a tree can symbolize a human individual or group is evident from numerous passages. In the Book of Genesis it connotes the fecundity of Israel's ancestors. With the birth of Isaac, Abraham plants a tamarisk tree (Gen 21:33); and when Jacob blesses his twelve sons he blesses Joseph in the name of Shadday, giver of the "blessings of the breasts and of the womb," in virtue of which "Joseph is a fruitful bough, a fruitful bough by a spring" (Gen 49:22–26). The blessedness of those who fear the LORD and walk in his ways include the promise that "your wife will be like a fruitful vine within your house; your children will be like olive shoots around your table (Ps 128:3). If a eunuch says, "I am a dry tree" (Isa 56:3), the psalmist reassures Israel that the one who keeps *torah* is like a fruitful tree (Psalm 1; cf. Pss 52:8; 92:12; and Jer 17:3–8 with its contrast between a tree planted by water and a shrub in the desert).[3]

---

3. The reader may want to recall this reference to the tree in later discussions of Psalm 1 in this volume.

PART ONE: Orienting Ourselves in the Biblical World

Ezekiel's extended parable in which vine and tree symbolize human rulers concludes with the generalizing comment, "and all the trees of the field shall know that I the LORD bring low the high tree, and make high the low tree, dry up the green tree, and make the dry tree flourish" (Ezek 17:24). Davidic kings may be spoken of in similar imagery. Isaiah 11:1 announces that "there shall come forth a shoot from the stump of Jesse, and a branch shall grow out of his roots," and Jer 23:5 announces, "Behold, the days are coming, says the LORD, when I will raise up for David a righteous branch." Psalm 80 remembers before God the time when "Thou didst bring a vine out of Egypt; thou didst drive out the nations and plant it" (v. 8). But then, turning to the present, it prays, "Turn again, O God of hosts! Look down from heaven, and see; Have regard for this vine, the stock which thy right hand planted. They have burned it with fire, they have cut it down" (vv. 14-16). In Jer 11:16, the image is again applied to the people of Judah: "The LORD once called you, 'A green olive tree, fair with goodly fruit'; but with the roar of a great tempest he will set fire to it, and its branches will be consumed." Then, to the people in exile, the tree becomes an image of hope: "Like the days of a tree shall the days of my people be, and my chosen shall long enjoy the work of their hands" (Isa 65:22). Finally, in Isa 53:2, the image is used to introduce the Servant of the LORD: "He grew up before him like a young plant, and like a root out of dry ground." If the Servant is the people Israel (perhaps as epitomized in an ideal exemplar and representative; cf. Isa 49:1-6), the image of the young plant as a root out of dry ground (cf. Prov 30:15-16) aptly describes the birth of Isaac to Abraham and Sarah. With this we may turn now to the scene in Exod 3:1-6.

The context is set in Exod 2:23-25: under the oppressive rule of Egypt, the people of Israel cry out for help, and hearing their cry, God remembers the covenant with the ancestors, sees and knows their plight, and comes down to act. How does God do so? Here we may recall the thesis central to Abraham Heschel's magisterial study, *The Prophets*.[4] In his view, central to the prophetic vocation is an experience of, indeed a participation in, the divine pathos. By "divine pathos" Heschel means God's knowing of the vicissitudes of God's people—a knowing that is not simply, so to speak, an intellectual recognition, but a total felt awareness of the people's condition together with a commitment to the divine action (in mercy or in judgment) that condition calls for. The prophet is one who is drawn into God's

---

4. Heschel, *Prophets*, xiii-xix.

pathos, sympathetically feels God's pathos for the people, and speaks and acts out of that sympathetic feeling.

Heschel's analysis aptly applies to the call of Moses at the burning bush. For what Moses comes to know through that bush is God's pathos over the people's affliction in Egypt, a pathos that he eventually comes to share. It is noteworthy that the scene at the bush is sandwiched between Exod 2:23–25 and its repetition in 3:7–8. I suggest that these two bracketing passages are the textual exposition of the symbolism of the bush that burns and is not consumed. In seeing the bush, Moses hears, speaking out of its midst, the God who says, "I am the God of your father, the God of Abraham, the God of Isaac, and the God of Jacob." Put plainly, the bush is emblematic of the descendants of these ancestors, suffering under the fiery trials imposed on them by the new Egyptian regime. It is a bush, however, that is not consumed by those trials, for it is a bush over which God watches to prolong its life and sustain it in fruitfulness. Indeed, it is a bush out of the midst of which God is heard in a voice that, while distinguishable from the cries of the people in 2:23–25, is not separable from them. At the heart of Israel's groaning under bondage is the cry of a divine pathos that moves to come down and redeem.[5] That is the voice and the pathos that lies at the heart of Moses' calling.

Is the burning bush, then, merely an attention-getting device, or is it thematically integral to the narrative context? Clearly, in my view, it is the latter. It is commonly recognized that, form-critically, the call of Moses is to be compared to other passages portraying a prophetic call. In that case, one may note that in the inaugural visions that come to Jeremiah in connection with his prophetic call, the almond rod and the boiling pot that he sees (Jeremiah 1), are integral to his prophetic message and to the political setting in which he is called to speak. The same may be said for what Amos sees in the five visions reported in Amos 7:1–3, 4–6, 7–9; 8:1–3; 9:1, visions that bracket the report, in 8:10–17, of his prophetic call. To

---

5. Readers of Paul's Letter to the Romans may be struck by the intertextual resonance when he writes (8:14), "For all who are led by the Spirit of God are sons of God. For you did not receive the spirit of slavery to fall back into fear, but you have received the spirit of sonship. When we cry, 'Abba! Father!' it is the Spirit himself bearing witness with our spirit that we are children of God, and if children, then heirs, heirs of God and fellow heirs with Christ, provided we suffer with him in order that we may also be glorified with him." For, as Paul goes on to write, the groaning of the children of God, in solidarity with the groaning of all creation, is undergirded by the intercessory groaning—the pathos—of the Spirit of God. One may note that the issue in the Book of Exodus is precisely the status of Israel: Are they slaves of Pharaoh? Or is Israel God's first-born son (Exod 4:22–23)?

be sure, in all these instances the symbolic import of the visions is made explicit through an interpretive divine word, while in Exodus 3, it may be objected, the significance that I am proposing for the burning bush is at most implicit. But is it only implicit? The verbal resonance between God *seeing* the suffering people in Exod 2:23, Moses *seeing* the flaming bush in 3:2, 3, 4, and God's statement in 3:7 that "I have *seen* the affliction of my people who are in Egypt," makes of 3:7–10, with its call to Moses to share God's concern, the explication of the meaning of the bush that burns and is not consumed. Just as Moses is drawn toward the bush by its unconsumed flaming, so he is drawn into God's pathos for the descendants of the ancestors.

The bush, then, is the fruitful but now threatened growth in Egypt of the community springing from Abraham and Sarah, the "young plant" and "root out of dry ground" of which the exilic prophet will later speak in Isaiah 53 (cf. also Isa 51:1–3). It is a bush that burns under the oppression of Egypt; but, by virtue of the God who is with it (cf. Gen 21:22; 24:40; 26:3, 24, 28; 28:15; 35:2; 39:2, 3; 46:2–4; 48:15–16; and Exod 3:12), it is not and will not be consumed.

Such a reading draws the scene in Exodus 3 into luminous thematic accord with later portrayals of the people's sufferings at the hands of their oppressors. Most noteworthy are the portrayals of the three youths in Nebuchadnezzar's fiery furnace (Daniel 3), and those of the aged priest and the widow woman and her seven sons who are put to death in fiery trials under the minions of Antiochus IV (2 Maccabees 6–7; 4 Maccabees). The Greek version of Daniel includes a lengthy addition after 3:23 that appears in the apocrypha under the heading, "The Prayer of Azariah." In the midst of the fire, Azariah (the Hebrew name of the youth whom Babylonians call Abednego) prays to God for all the people, offering himself and his two companions in the fire as a sacrificial offering on their behalf. Tellingly, at the very heart of this prayer for the deliverance of the people, Azariah calls upon God: "For thy name's sake do not give us up utterly, and do not break thy covenant, and do not withdraw thy mercy from us, for the sake of Abraham thy beloved and for the sake of Isaac thy servant and Israel thy holy one, to whom thou didst promise to make their descendants as many as the stars of heaven and as the sand on the shore of the sea" (Prayer of Azariah 11–13).

These portrayals concerning the Jewish martyrs carry forward, I suggest, the picture of the descendants of Abraham and Sarah, of Isaac and Rebekah, and of Jacob and Leah and Rachel, as a bush that though burning is not consumed. The emblem over the door of JTS carries that testimony forward into a post-Holocaust world.

Addenda

1. When I had completed the above study, it belatedly occurred to me to consult some Jewish commentary on the bush. In *Midrash Rabbah* on Exodus 3 I found the following comment: "Why did God show Moses such a symbol? Because he (Moses) had thought to himself that the Egyptians might consume Israel; hence did God show him a fire which burnt but did not consume, saying to him, 'Just as the thorn-bush is burning and is not consumed, so the Egyptians will not be able to destroy Israel.'" In the same vein, Nahum M. Sarna writes, "The bush that remains intact in the face of the flames may be symbolic of the people of Israel surviving Egyptian oppression."[6] I am happy to think that the reading offered in this essay was not only inspired by my visit to the Jewish Theological Seminary but also accords with this ancient and contemporary strand of Jewish interpretation.

2. In view of Abraham Heschel's thesis concerning the pathos of God, and the prophet as one who is drawn to participate sympathetically in that pathos, we may take the fire in the bush to connote not only the oppression of Egypt and the people's affliction under it, but also the fervency of the divine pathos over that situation. Those wishing to reflect further on this theme in relation to the crucifixion of Jesus might take as a starting point the poem, "Final Verdict," by Dilys Laing.[7]

---

6. Sarna, *Exodus*, 14.
7. Laing, *Collected Poems*, 405.

2

What's in a Name?

"Yahweh" in Exodus 3 and the Wider Biblical Context

Names and Their Meanings

WHAT'S IN A NAME? As a title, the question sounds more suitable for a whimsical column in the *Reader's Digest* than for an essay in a scholarly volume. Discovery at an early age that my given name Gerald derives from a Germanic root meaning "spear-wielder" has not provided any deep clue to who I am or what I am to become. In this etymological sense, an appropriate response to the opening question would be "nowadays, not a heck of a lot."

Even nowadays names are not merely labels. What's in the papal name John Paul? In the time of Pius XII, only an unlikely collocation of two distinct papal names. But by October 1978 what was "in" John Paul is both a legacy and a vision. In this sense one might answer the opening question this way: What's in a name is its actual history and the future as foreshadowed or claimed by that history. If, then, in all seriousness one were to ask "What's in the name of Israel's God, Yahweh?" the answer might be made: What's in that name is its actual history and the future as claimed by that history. To say what is in that name, then, would call for telling the history of Israel; or, so far as Israel itself was concerned, it would call for telling the history of all things and the future of all things as claimed by that history. Looked at in this way, the biblical narrative taken as a whole could be read as an explication of what is in the name Yahweh.

What's in a Name?

Yet nowadays (for all our official distrust of etymological naming) we still betray the desire to have the name in its root sense correspond to or express and embody the person in her or his intrinsic identity, in a way that goes deeper than history, in a way that touches the springs out of which flow the energies and powers which help to shape history, be that history cosmic, national, local or only familial and individual. Precisely because our formal naming so seldom serves us here, we resort to nicknames. In such examples as "Tricky Dick" or "The Angel of Dien Bien Phu" it is not so much that a standing name subsequently acquires a meaning given it by the shape of events; but rather, that an intrinsic quality of personal energy, an inherent power of being and style of action, is named as that which helps to give shape to events. To be sure, this intrinsic quality and energy were first intuited from or in connection with the events. And yet, in bestowing such a nickname we mean (consciously or unconsciously) to identify something behind the events, something manifest in them. One may say that in the first type of name-meaning the history explains the meaning of the name, while in the second type the meaning of the name accounts for the history. Is it possible, now, for us to approach the name "Yahweh" in this second sense? Is it possible to say something about that name, not through the telling of the history in which that name is embedded, but through an entry into what the name means in itself, a meaning which is not derived from the history but from which the history itself is seen to be derived?

This form of the question seems to direct us back to an investigation of etymology and into the labyrinthine ways taken by scholarly research and debate concerning the original meaning (perhaps in pre-Israelite settings; among the Kenites or Canaanites or Amorites, or elsewhere) of "Yahweh." Such a history-of-religions and comparative philological approach has led to a variety of solutions. It may be that we are eventually convinced (as I happen to be) of the basic soundness of the view of Frank Moore Cross who understands the divine name as originating in an old sentence-name for the Canaanite High God El, so that biblical Yahweh Sebaoth is to be taken as meaning, originally, "(El who) creates the hosts."[1] But whichever etymological analysis is adopted, it is fair to say that that etymological and pre-Israelite sense is present or reflected only in part of the total biblical tradition, within a wider range of meaning which it has come to have for Israel. Indeed, one cannot rule out the possibility that through its use in Israel this divine name may have come to bear a

---

1. Cross, *Canaanite Myth*, 60–75.

PART ONE: Orienting Ourselves in the Biblical World

distinctive meaning quite removed from its earlier pre-Israelite, and even early-Israelite, meaning. We seem then to be cast back upon the notion that the name Yahweh, insofar as it is more than a mere label (which the translators of the RSV would as soon have done away with), insofar as it has a meaning, has that meaning as a mosaic built up in the course of the history associated with it. But this means that it is the history which sheds light on the name, whereas the name in no way identifies the divine reality manifest in that history. Unlike the names of other ancient Near Eastern deities, then, the divine name Yahweh in itself is no datum for theology but a convenient peg around which to focus the theology derived from the narrative and other biblical traditions. Nor is the name in itself a dynamic point of departure for personal and social existence but a convenient symbol representing the tradition as a whole taken as such an impetus.

But for all its value in the context of a history-of-religions investigation, this kind of etymological approach is not the only one possible; indeed, it may draw our attention away from that etymological approach which is presented before our eyes in the biblical tradition and which, I maintain, is the proper datum for investigation into what the name Yahweh meant for mature Israelite religion. I refer of course to the narrative in Exodus 3. Here we are told, explicitly, what the name of Yahweh intrinsically means, in such a way that we are to understand, not the name from the history, but the history from the name. It is the divine reality identified not just *by*, but *in* the name Yahweh which shapes the story, not only in Exodus but throughout the Bible. But in order to appreciate this we must recognize just what is going on at the crucial point in Exodus 3. What we have here is what has been called "popular etymology." The practice, a form of punning which often turns on sound-similarity, is well known in the Bible; it occurs throughout the Old Testament and even in the New Testament (e.g., "Peter" in Matt 16:18). All too often in modern times, specific instances of such popular etymology have received only the amount of attention needed to point out their inaccuracy from a historical and linguistic point of view. Yet so far as a proper understanding of the biblical narrative is concerned, it is as irrelevant as it is correct to observe that "Babel" in Genesis 11 does not come from a root meaning "to confound"; or to observe that the name "Moses" in Exodus 2 is not formed from a root meaning "to draw out." But I have overstated my point. It *is* relevant to make this observation. But the relevance lies precisely in freeing us—freeing our literary and our religious imagination—from our preoccupation with historical etymology so that we may then attend to the literary etymology, that is, the one required for an understanding of the narrative in its own terms in the text before us. What I am suggesting is that the primary context for our understanding of the meaning of the divine name Yahweh in the Bible is not the history of the religion of Israel and of the ancient Near East generally, but the practice within the Bible of popular etymology. By analogy, the proper context for the understanding of Rachel's naming of Dan, Naphtali, Gad, Asher and Joseph (Genesis 30) is not the general Semitic onomasticon of the second millenium BCE, but rather the literary etymologies as given within the context of plot and structure in the Rachel narrative in Genesis 29–35.

Before I proceed to explore Exodus 3 from this point of view, one other methodological statement is in order. This is that, in order to convey what I consider to be the requisite frame of reference for our understanding of the divine name, I propose to work with the biblical tradition in its mature—that is to say its final—form. This approach obviously is liable to the charge of treating the materials one-dimensionally or anachronistically, through disregard of the complex and lengthy process of formation of the text. I acknowledge this. Yet in taking this approach, I am seeking to do justice to the fact that the faith of Israel, in its drive toward verbal and literary expression, displayed a restlessness with every preliminary form (as recovered by historical-critical investigation) until it came to rest in the form of the text as we have it. I do not deny the value of historical-critical analysis of the text as a means to answer certain specific historical questions. Indeed, part of this value lies in its enabling us to distinguish between the history of Israel and the history of the growth of the text, and the present text as a final mythic construct. But until we address ourselves to that construct as such, we have not really arrived at the mature and proper datum for theological—and even for history-of-religions—reflection.

The Name "Yahweh" and Its Meaning for Israel

The giving or etymologizing of the name Yahweh in Exodus 3 is to be interpreted not just in the context of the scene at the burning bush but in the larger context of the Israelites' plight in Egypt. Yet even to start here is to begin *in medias res*. As the narrative now opens, the plight of the Israelites is to be read against the backdrop, not just of the patriarchal narratives, but of the first creation story. In Egypt, the children of Israel find themselves flourishing in such a way as to exemplify the divine mandate at creation (compare Exod 1:7 with Gen 1:28). This is surprising, in view of the negative connotations which Egypt carries in most of the Old

PART ONE: Orienting Ourselves in the Biblical World

Testament. Yet we are to suppose that the people flourish, not in spite of Egypt, but because of Egypt's hospitality. (This is clear from the preceding narrative and from the reversal announced in Exod 1:8.) A brief review of the Genesis traditions from a theological point of view will help us to place this initial Egyptian situation, and therefore the giving of the name in Exodus 3, in a sufficiently wide context.

One way of giving expression to our sense of the vigorous this-worldly realism of the Old Testament traditions is to say that the world created by Israel's God is a world of efficacious creatures, each enjoying its own existence "after its own kind," and enjoying the power to participate in transmitting and conveying such existence to others of its own kind. In the biblical traditions these powers are spoken of in the concrete imagery of fruitfulness, blessing, multiplication and so on. But for our purposes I think we may generalize this imagery and speak of fruitfulness or blessing as the power of a creature, in its actuality, to exist and to communicate existence in the form peculiar to its kind.

Further, the "host" or multitude of creatures exists in the form of what we may call "structures of actuality." By actuality here I mean, simply, creatures in their concrete actual existence as created and as enjoying existential power. ("Actuality" here may be contrasted with such "potentiality" as may be said to be the case between the divine utterance "let there be X" and the narrative observation "and there was X.") By structures here I mean, simply, the discrimination or separation of these efficacious creatures into coherent orders of actuality. ("Structure" here may be contrasted with such formlessness or *tohu* as was the case prior to creative activity in Gen 1:2.) These orders or structures of actuality are reciprocally efficacious and supportive and are pronounced to be good. Generalizing one step further, we may express the vigorous this-worldly realism of Old Testament religion in these terms: Within the limits of the respective orders or structures of existence, the actual is the ground of the possible.

As considered up to this point, the created order may be said to be a comprehensive world of structures of actuality comprising individual creatures who receive their existence from the creator and from others of their kind, and who with the help of God (e.g., Gen 4:1) impart that existence to others of their kind. But of course the created world does not display this dynamism unambiguously. In actual experience, the structures of actuality take the form of a tangled skein of good and evil, of blessing and curse, of fruitfulness and sterility, tending toward life and toward death. The calamity of creation is evident in the fact that creaturely existential powers are efficacious for life and for death. The story of creation therefore careens perilously along a critical path whose twists and turns are chartered representatively, and with ever-increasing fatefulness, in the portrayals of the so-called primeval history on through Genesis 11.

The patriarchal narratives, then, portray the rise and the first stages of formation of a new structure of actuality in the emergent community identified by the names of Abraham, Isaac, and Jacob. This new structure is envisaged as arising in response to the impasse portrayed in the primeval history, and as an actual agency for the restoration of the primal intent and form of creation. This envisagement is attested in the preoccupation throughout the patriarchal narratives with blessing and the habitation of a land. Though on the one hand the repeated reversal of barrenness by the restoration of fertility is brought about through divine initiative and power (e.g., Gen 15:2; 30:2), yet on the other hand human participation is also emphasized: in the form of the faithful obedience which is a *sine qua non* of the developing structure, and in the form of the efficacious part played by the various patriarchal blessings, culminating in the epochal blessings uttered by Jacob in Genesis 48 and 49. If, now, we wish to reflect on the significance of all this for our question concerning the name of Israel's God, we may begin by observing that the dynamic growth of this new structure of actuality, identified by the names Abraham, Isaac, and Jacob, is matched by a concurrent dynamic development in one of the epithets of the God under whose divine aegis this structure emerges and begins to take shape: He is first "the God of Abraham" (Gen 26:23); then "the God of Abraham . . . and the God of Isaac" (Gen 28:13); and later "the God of Abraham and the God of Isaac and the God of Jacob" (Exod 3:6). The God so named is the divine power disclosed through this new structure of actuality as it emerges in history. Insofar as the history is the clue to the character of the God, the history is the clue to the meaning of the name.

Now, this new actual structure must make its way, must pursue its critical path, amidst the ambiguities of the wider structures which make up the world. In the last long section of the patriarchal narratives, this new community finds itself threatened by the negative efficacy of the famine in the land; and in the face of that threat, it finds itself greeted with the positive efficacy of the hospitality extended by Egypt. Thus, taken as a whole, the actual structure of the world is ambiguous; but taken in its several substructures, the negativity of one part is counter-balanced by the positivity of another. Israel's escape from famine into Egypt and ensuing prosperity (Exod 1:7) attests the continuing positive character of Egypt as a local structure of actuality. In the terms of our generalization above, Egypt is the benign ground of the possibility of Israel's continued existence. Hence

PART ONE: Orienting Ourselves in the Biblical World

the existential powers which the Israelites and Egypt enjoy under God are fruitful powers, and the life thereby made possible seeks its own intrinsic worth without the need for narrative elaboration.

But then there arises a new king over Egypt who does not know Joseph; and the brief narrative resolution dissolves into dissonance. What has happened? The Egyptian power has changed its character. Once life-supporting, that power has become life-threatening. Israel faces a life-and-death crisis, which I would define in the form of a generalized question: When the actual situation becomes deathly oppressive, is the actual the limit of the possible? Does the character and do the resources of the actual present strictly define and determine what the future shall be?

Of course, there is more to the actual situation than Egypt. There is the Israelite community with its own existential powers. And there is Moses who occupies a dialectical position in the situation, being in-formed to some extent by both structures, the Egyptian and the Israelite. The first few episodes in the exodus story portray the mounting struggle between Israel and Egypt, and Moses' preliminary attempts to inject his own power of action into the struggle. But in spite of Israel's initial display of existential vigor (significantly enough displayed in the form of success in childbirth), and in spite of Moses' intervention, the conclusion of these first episodes finds Moses in flight to the wilderness and Israel reduced to that mode of action and power depicted in Exod 2:23–24—crying out in lament. It is not too much to say that, under the pressure of Egypt, Israel's "structure of actuality" has been given the shape or has taken the shape of a structure of lamentation as a mode of being. But the implication and tendency of that structure and mode of being is that, if the actual is the sole ground and limit of the possible, then in the face of what Egypt has become, Israel's future holds only death.

What, now, is the significance of Moses' flight to the wilderness? Considered merely in its material topographical aspect, the wilderness is that "outlaw" region beyond the reach of Egypt's organized power where Moses may find fugitive asylum. In this aspect it is as actual a locus as is Egypt, though of a different character. But as an image and a motif, the wilderness in the Old Testament is much more than this— or should we perhaps say, much less. In Jer 4:23–26, for instance, it stands as the opposite of created actuality. (See also Deut 32:10 with its *tohu*, and Deut 32:13–14.) Relevant to our discussion, the wilderness is a *tohu*—a formless waste. It stands over against city and sown land in some sense as Gen 1:2 stands over against the whole created actuality. The wilderness, we might say, is the realm of the non-actual or the realm of the suspension of the actual. Is then the actual the limit of the possible? Yes, so long as Egypt, the Israelites, and Moses remain locked in struggle in Egypt. But Moses' flight from Egypt into the wilderness—a flight from the rigid determinations of that oppressive actuality—sets the stage for a different answer to that question. It is of the utmost significance, in my view, that that answer, in the form of the divine name and its explication, comes to Moses in the wilderness.

We are ready, now at last, to turn to the question of the meaning of the name Yahweh as disclosed in Exodus 3. The passage opens with the theophany at the burning bush, in which God preliminarily identifies himself to Moses: "I am the God of your father, the God of Abraham, the God of Isaac, and the God of Jacob" (Exod 3:6). Having identified himself in terms of the history in and by which he has become known, he now announces his intention of acting in such a way as to "live up" to that name, to be faithful to that name, in delivering and bringing up his people whose cry has come up to him (3:7–8). This is followed by the announcement that he will effect this deliverance in and through Moses, who is to go and bring forth this people (3:9–10).

But Moses has already tried to intervene on behalf of the Israelites. He knows the actual situation: He knows himself, he knows the Israelites, and he knows the power of Egypt (you can't fight city hall). So he counters with a doubting question: "Who am I . . . ?" Now this question may be considered as a dodge, or as a rhetorical question needing no answer except the obvious one: Moses does not have what it takes to carry out such a mission. But it may also be construed as a genuine doubt, a pinpoint opening in the settledness of his self-knowledge; it may betoken the first beginnings of an existential question.[2] More likely his response contains elements both of dodging resistance and of doubt-filled opening out toward what is being proposed to him, as settled actuality and budding possibility vie within him. Into the opening made by Moses' question comes God's answer: "I will be with you" (*'ehyeh 'immak*). It appears that Moses' question is answered by way of God's implicit re-definition of who Moses is. Who he is can no longer be defined merely in terms of who he had hitherto taken himself to be, or in terms merely of the actual situation from which he is in flight. By virtue of God's answer, who Moses is can henceforth be measured adequately only by including a reference to this God henceforth present with him.

---

2. For an exploration into the difference between *rhetorical* and *existential* questions in the sense here intended, see Janzen, "Metaphor and Reality," esp. 417–22.

But in what character is God with him? As the God of Abraham and of Isaac and of Jacob? From the preceding divine self-designation (3:6) Moses could conclude this—and indeed in a sense he does (3:13). But this dialogue which has thrown his own settled identity into question seems also to leave the old designation of God no longer entirely satisfactory. To be sure, Moses' response in verse 13 in part reflects a continuing desire to evade the call and mission. Yet the *form* of this response—a hypothetical agreement to go, followed by a hypothetical question—betrays also a continuing pin-point of openness to the mission at least in the mode of hypothetical imagination, and openness, as I have suggested, which has accepted the questionability of Moses' identity and which, almost as a reflex of that very dynamic, now presses the question of the identity of God: Is it, perhaps, no longer sufficient to speak merely of the God of Abraham and of Isaac and of Jacob? There is a sense in which this second question is simply an extension of the first: If who Moses is is to be defined henceforth with reference to the presence of God, then who Moses is depends to that degree on who this God is. Consonant with the character of Moses' second question as an extension of his first, God's answer to it begins in the same manner as his answer to the first: "I will be . . ." (*'ehyeh* . . .). What might we expect to follow? A clearly definitive answer in terms of some specific mode of divine power (e.g., "I will be with you as a dread warrior," in the manner of Jer 20:11; cf. Exod 15:3)? Or, perhaps, a lapidary reiteration of the first answer, "I will be with you," as a transcendent rebuff of the thrust of Moses' question? The answer comes: "I will be *who I will be*." The content of this answer will occupy us in a moment, but first another matter requires our attention: my earlier assertion that in this passage we have an instance of popular etymologizing (though either the term "popular" is inappropriate here, or this traditional practice here receives its apotheosis). The rhetorical development of Yahweh's response, while keeping strictly within the dialogical flow of the passage as a whole, takes the form of what we may call a painstaking three-step semantic equation. The three steps are signaled by the three-fold repetition of the narrative rubric (God said to Moses; and he said; God said also to Moses; Exod 3:14–15). This three-fold responsive movement achieves the semantic equation *'ehyeh 'ašer 'ehyeh* = *'ehyeh* = *yahweh*. Whatever the name Yahweh may earlier have denoted or connoted in Israel or for other peoples, from now on it is to be understood in terms of this intrinsic self-definition. Henceforth, "Yahweh" names God, not with reference to this or that specific instance or structure or order of actuality (as the divine power manifest in it or responsible for it), nor even with reference to the whole of actuality (as summed up in the epithet "maker of heaven and earth," Gen 14:22). From now on, "Yahweh" is that name which identifies Israel's God purely in intrinsic terms, as that divine power of existing which is defined or qualified or limited by no principle except the divine existential intention itself. What is so named is the primal reality whose power and efficacy constitutes, by its own intention, the living fount and origin and range of all that is or may be. It is as this living fount of all possibility that God may be envisioned as the creator of all finite and creatural actuality. There is the most intimate connection between the divine mystery expressed in this "I will be . . . ," and the world-creating efficacy of the utterance "let there be. . . ." The finite actualities in creation are grounded in the infinite potentiality and potency of the creator. If these finite actualities enjoy a derived but real existential power which they may transmit to others—so that actuality may beget itself—yet it is only by an idolatry that these actualities may be taken as posing the limit of the possible. The issue of idolatry in the Book of Exodus comes to explicit formulation in the Decalogue, and comes to explicit trial in the incident of the golden calf. But, especially in the latter instance, these subsequent texts only make explicit what is implicit as the fundamental issue in Israel's plight in Egypt: Is the actual the limit of the possible? The idol-polemic of the Old Testament is entailed in the name of Israel's God: Yahweh.

The first person to face this issue is Moses. He is challenged to move beyond who he knows himself to be and to re-define himself as one who is promised the presence of One who "will be who he will be." In principle, Moses' self-understanding now must remain open to possibilities which will be disclosed only in the successive situations in which he will find himself. By virtue of the self-transcendence which Yahweh's presence makes possible to Moses, he may now return from the wilderness (the setting for encounter with the power of the possible) to the actual situation in Egypt—himself no longer merely an actual component in that structure or tangled skein of actuality, but now a bearer—in his new identity as it were an embodiment—of new possibility. That new possibility is offered repeatedly to Egypt, as a possibility in terms of which Egypt may reinterpret itself as a power-structure and so re-define itself. But Egypt's response stands in the sharpest contrast to that of Moses. The first address to Pharaoh in Yahweh's name evokes the question, not as to his *own* identity, that he should let them go, but rather, "Who is *Yahweh*, that I should heed his voice and let Israel go?" (5:2). Moses' initial resistance of Yahweh was gradually overcome by the signs Yahweh granted him (4:1–17); Pharaoh's initial and reiterated resistance is momentarily overcome at several points, as he momentarily moves beyond his settled decision, but finally his resistance is stiffened by all the signs granted in the plagues. Egypt learns to its grief that where the actual seeks to continue as an unchanging definition of the possible, that equation spells death, not for the oppressed, but for the oppressor; and this is not solely by an arbitrary or extrinsic judgment, but by the very rigidity of its own actuality which becomes a form of *rigor mortis*. As for Israel, the people in their response to Yahweh find themselves somewhere between Moses and Egypt, responding, but timidly, and then repeatedly falling back. Only Moses' steadfast commitment to the people and to Yahweh at crucial points turns the situation in the direction of an open future. For, as steadfastly open to Yahweh, Moses himself embodies in the world, and in a finite way, the power of the possible to re-define the actual. And, as steadfastly anchored within the people as one of them (cf., e.g., 32:9–14), Moses himself constitutes the recalcitrant community's pin-point openness to Yahweh and the realm of the possible. He, by embodying both in himself, is the mediator between the actual situation and the power of Yahweh.

But the divine self-definition in Exodus 3 holds yet one more item for comment. If the God of the Exodus is not to be identified merely with that emerging structure of actuality represented by the patriarchs, nevertheless as 3:15 shows, that identity continues to *include* the most intimate association with Abraham, Isaac, and Jacob, forever and throughout all generations. Thus, we are not to drive a transcendental wedge between what might appear to be merely a religion of culture and Yahwism as a transcendental critic of culture and its religious dimensions. From now on, the issue of the faithfulness of God is posed both in terms of his faithfulness to the actual situation and its historical claims upon him, and in terms of his faithfulness to the intrinsic mystery of the divine life as pure unbounded intention. Complementarily, from now on the issue of the faithfulness of Israel is posed in terms of its loyalty to the name Yahweh: in the implications of that name for who Yahweh is and for who Israel is. Like Yahweh, God of Abraham and of Isaac and of Jacob, Israel is called to be faithful to its past. Like Moses, Israel may never again allow itself merely to come to terms with the actual situation—for this is idolatry and death. The secret, the burden, the vocation of Israel lies in the divine name entrusted to it in the Book of Exodus.

The Name and Its Meaning for Existence Today

It remains, now, to indicate some aspects of the bearing of the central thesis of this paper upon our existence today. The preceding remarks have been offered in the mode of exegetical exploration and theological generalization. Now I would like to speak confessionally. I will try to say what difference all this has made to me and leave it to others to make of this essay what they wish in the context of their own actual situations.

1. It so happened (for reasons that are beyond my understanding) that much of my early life was freighted with a sense of my own unreality. Life was something that happened to other people, or that other people enacted; I was there as an onlooker, or I was elsewhere as a daydreamer. In fourth grade we all were to copy out a poem and submit it in a local handwriting competition. On the day of the exhibit, we stood around examining the entries and noting the awards. "Where was my entry?" asked my older brother. I pulled it out of my pocket where I had stuffed it and showed it to him. It was no prize-winner; but it was no worse than some there. Yet it had seemed pointless to submit it—that was only what *other* kids did. This incident embedded itself in my memory as representing how I felt about myself. Now, appropriation of the biblical portrayals of existence—especially with the help of certain contemporary philosophical modes of thought—has worked a change in my sense of myself-in-the-world. Increasingly, as with a sense of "participatory dynamic realism," I am aware of myself as enjoying and exercising specific existential powers in a community of other beings—human and non-human—who display such powers also, in their own way and after their own kind. I become aware that *I* make a difference in a world of *others* who make a difference. And the differences that are made are utterly worthwhile, as measured in terms of the increasing appreciation of all kinds of actual objects and events and persons and structured situations and as measured in terms of my willingness to enter into public exchange through my own concrete action. For example, after writing this paper, I did not stuff the manuscript into my pocket and leave it there. From this personal Egypt I have begun to experience deliverance toward a restoration of my primal creatureliness as a finite efficacy among other finite efficacies.

2. But the difference I make by virtue of my existential powers is discovered to be dismayingly ambiguous. The sense of the hurt and the injury that I bring to others—injecting negative energies into their lives, damaging their sense of self and of the world, converting their situation by that much into negative structures of efficacy—increases and becomes

more problematic with the growing sense of myself as being capable of action. Such dynamics produce situations in which defensive hardenings of guilt and stiffening reactions of offended innocence lock one another in an impasse which becomes mutually imprisoning and destructive. Here that other "exegesis" of the name Yahweh in Exod 33:19 becomes relevant. At that point in Israel's life, the divine mystery earlier named as *'ehyeh 'ašer 'ehyeh* specifies itself, vis-à-vis the situation, as *weriḥamti 'et-'ašer 'araḥem*—"I will have mercy upon whom I will have mercy." Forgiveness here becomes experienced as a mode of the enactment of Yahweh's infinite power of possibility in the face of the actual impasses arising out of the rigidities of guilt and offended innocence. Forgiveness enters into the impasse, on both sides, as a melting of the situation and as an opening—however pin-point and however brief—within which words can be said and deeds done to restore to the relationship and to the situation a possibility of forward movement.

3. Since my early teens, I have been haunted by the sense of the emptiness of worldly values and the futility of worldly achievements in the face of their inevitable annihilation in death and, eventually, the death of the solar system. The passing years have placed more and more of what significance life held for me *behind* me. Nostalgia and resistance to change were sea-anchors intended to secure me against the wind-drift which carries everything toward the edge of the world. But Easter has begun to mean the presence of Yahweh in the face of that actuality to end all actualities. The resurrection has come to represent the treasuring up of the concrete achievements and actual values to which history has given birth, negotiating at the cost of death itself the impasse thrown up by the concrete failures and actual evils to which history has given birth. Under the sign of the name of Yahweh, Easter has led me no longer to resist time and not to a flight from this world but to a positive valuation of and commitment to this-worldly actions in the knowledge that they are "not in vain" in Yahweh.

4. I begin, then, to experience what it may mean to live eschatologically, in myself, toward others, and toward the various structures of contemporary actuality. This does not mean playing off eschatology against history, or the future against the past. For Yahweh is forever the God of Abraham and of Isaac and of Jacob (Exod 3:15). For me, to live eschatologically is to live out of a past that I have incorporated into myself, toward an open future. With Moses I must be willing genuinely to ask "who am I?" And I must again and again overcome the idolatry which desires to

settle on who I know myself so far to be. Much more importantly, I must overcome that idolatry in my attitudes toward others. I take warning from such attitudes shown toward Jesus by his own family, noting that such attitudes possessed sufficient negative efficacy to inhibit even his existential powers (Mark 6:1–6). And I take encouragement from Simon's (initially brief) freedom from such attitudes in his response to Jesus' question "Who do you say that I am?" In the act of envisioning Jesus as Messiah (Matt 16:13–20), he in turn found himself envisaged as Peter—a name wildly at variance with his actual character up to that point, a name later needing reinstatement through forgiveness (John 22:1–23), yet a name whose accuracy was eventually borne out.

5. The dominant focus of the preceding remarks falls on individual existence and inter-personal relations. I have deliberately chosen such a focus as a counterbalance to the social and political character of the situation portrayed in the Book of Exodus. As that book should make sufficiently clear, the central thesis of this paper should be capable of application to the larger social structures in which we find ourselves today. Meanwhile, as an ordained minister, I ponder this question in its bearing upon all dimensions of our existence: when we bless and are blessed in these words,

> Yahweh bless you and keep you;
> Yahweh make his face shine upon you and be gracious to you;
> Yahweh lift up his countenance upon you and give you peace,

what does this mean, and what is happening? Is such an act of blessing a piece of shadow-play? Or is it a dynamic event, after its own kind? And to what extent is its efficaciousness bound up with an understanding of what it means to have Yahweh's name "put upon" us (Num 6:27)? This chapter is offered as an attempt to contribute to such an understanding.

3

What Does the Priestly Blessing *Do*?

THE PRIESTLY BLESSING IN Num 6:24–26 exemplifies what J. L. Austin calls "performative speech," in which words do not simply refer to something, but actually do what they say.[1] "I love you" not only reports the speaker's affections, but enacts them verbally; and their efficacy is felt by one to whom they are agreeably addressed. Likewise, "God bless you" is a verbal act whose efficacy is conveyed in those words. The archetypal instance of performative speech in the Bible occurs in Genesis 1, where God speaks the world into existence: "And God said, 'Let there be X,' . . . and there was X." In my view, the primal speech-act that humans are capable of, "in the image of God," is the Priestly Blessing in Num 6:24–26.

As performative speech, what is it, precisely, that the Priestly Blessing *does*? We may approach this question by examining earlier acts of blessing in the Priestly strand (hereafter P) of the biblical narrative. For the sake of convenience, I will summarize the analysis of Walter Brueggemann.

The Priestly Blessing and God's Blessing in Creation

Brueggemann builds on H. W. Wolff's exploration of the "kerygmatic focus" in, respectively, the Yahwist, the Elohist, and the Deuteronomic Historian. He notes that in each case Wolff has identified "a frequently used formula as the clue to [that] tradition . . . Moreover, in each case he has been able to relate that formula to the historical situation and theological issue being addressed by the tradition." Positing the situation of P as most likely exilic, or perhaps early post-exilic, Brueggemann characterizes P as "a circle of traditions which regarded cultic matters as primary for the reconstruction of the community of faith now understood as Judaism."[2] And

---

1. Austin, *How to Do Things with Words*.
2. Brueggemann, "Kerygma of the Priestly Writers," 397–98.

he finds P's "central kerygmatic assertion" in the blessing-formula running through Genesis. That formula, not surprisingly, is grounded in creation itself, where "God blessed them and God said to them, 'Be fruitful and multiply; and fill the earth and subdue it and have dominion'" (Gen 1:28).

Brueggemann reads this blessing, and the general scenario of "creation . . . out of *chaos*" in terms of Israel's "*exilic situation* of poverty, misery and despair which he now transforms into a situation of joy and *šalom*."[3] Of the first repetition of this formula (Gen 8:17; 9:1, 7), he observes that it comes in connection with the re-statement in Gen 9:6 of the *imago* doctrine of 1:26, and he comments, "the announcement makes a turn in P, for it is new creation after the flood. *It is a new creation again being wrought out of chaos*."[4] The formula recurs in Gen 17:20; 28:1–4; 35:11–12; 47:27; and 48:3–4; but he notes that "[t]he P tradition does not carry the theme into the Mosaic era," saying of its echo in Exod 1:7 that "the theme stands . . . as a programmatic statement for all the rest of the history of the people."[5]

Brueggemann's general analysis is fruitful. But his comment on Exod 1:7 is lame. The report in Exod 1:7 is hardly a programmatic statement. It is important to note the different ways in which P carries forward the kerygmatic theme of Gen 1:28. In 9:1, 7, God *blesses* Noah. In 17:20, God *promises* Abraham to bestow such a blessing on Ishmael. In 28:1–4, Isaac *blesses* Jacob in the name of El Shaddai (the imperatives of 1:22, 28 and 9:1, 7 becoming jussives, or third-person imperatives). In 35:11–12, El Shaddai *blesses* Jacob. In 47:27 (as in Exod 1:7), the narrator *reports* that Israel, in Egypt, enjoys the efficacy of the programmatic blessings. In 48:3–4, Jacob *recollects* for Joseph El Shaddai's blessing at Luz. Programmatically speaking, God's direct blessing punctuates the narrative at three points: (1) on humankind following the creation; (2) on Noah and posterity following the post-flood renewal of creation; and (3) on Jacob and posterity on his return from exile in Paddan-aram. Frank Moore Cross has shown how P "framed the major sections of Genesis" by use of the *toledot*-formula "these are the generations of . . ."[6] This formula appears first in connection with creation story (Gen 2:4a), and finally in Gen 37:2 where it launches the story of the family of Jacob. If Genesis 36 is a "side-bar" to the main narrative, God's blessing on Jacob in 35:11–12 thus forms the programmatic impetus for the story launched in 37:2.

3. Ibid., 401 (italics original).
4. Ibid., 403 (italics added).
5. Ibid., 406.
6. Cross, *Canaanite Myth*, 302.

PART ONE: Orienting Ourselves in the Biblical World

But if God's blessing forms the programmatic core of P's kerygma, it is difficult to suppose that P simply drops it after Genesis. Other sacral institutions given at Sinai are grounded in creation, such as the Sabbath day; and the sanctuary itself is a microcosm of the cosmos (see further below). Surely we should discern some relation between the blessing theme in Genesis 1 and priestly blessings as reported in Lev 9:22–23 and instituted in Num 6:22–27.

But apart from the verb "bless" and the use of jussives, the Priestly Blessing is strikingly unlike Isaac's blessing on Jacob in Gen 28:3. E. W. Davies claims that Num 6:24–26 "is replete with words and idioms that are completely alien to the vocabulary of P."[7] In that case, how are the elements in this blessing related to the elements in that earlier blessing? And how, if at all, is it rooted, as the earlier one was, in the creation account of Genesis 1?

In presenting God's self-revelation (Exod 6:2–8), P draws a clear theological distinction between the universal and ancestral eras and the epoch inaugurated under Moses and Aaron. For P, the era preceding Israel's ancestors transpired under the aegis of ’elohim and the ancestral era under the aegis of ’el šadday, while the Mosaic epoch originated with God's self-identification as YHWH. During those earlier eras, the central concern, thematically, was with the fruitfulness of the earth and its human inhabitants—in the case of the ancestors, with their recurrent problems of barrenness, and of space to dwell amid recurrent famine and peregrination. For P, as for the older epic, the central concern from Exodus 1 forward was with the external threat of political oppression and the internal threat of covenant disloyalty and sin. In the older epic, the external threat was resolved through Yahweh becoming a warrior-deliverer and covenant Lord (Exod 3:14–22; 19–24), while the internal threat, arising through the sin of the golden calf, was resolved through the self-presentation of Yahweh at Exod 33:19, reformulated at 34:6 as "A God merciful and gracious [raḥum weḥannun], slow to anger, and abounding in steadfast love and faithfulness." In "somber, sin-obsessed"[8] P, the internal threat is dealt with through the elaborate cultic apparatus of sacrifice. If Cross is right in arguing that an account of primordial human rebellion is absent in the P strata of Genesis 1–11 because P assumes the older epic account of the Yahwist and the Elohist,[9] we may likewise suppose that P's provision for

---

7. Davies, *Numbers*, 66.
8. Cross, *Canaanite Myth*, 307.
9. Ibid., 306. It is evident from my language that I continue, despite a good deal of

dealing with Israel's sin stands in some inter-textual relation to the epic version in Exodus 32–34.

All of this bears on the relation between the blessing theme in Genesis and the thematics of the Priestly Blessing in Num 6:24–26. Quite simply, the critical issues have shifted. Now what threatens God's creation program is no longer merely barrenness and famine, but threats external and internal to a divine-human relation now conceived on the model of political and sacral relations governed by detailed laws sustained by sanctions for their violation. This shift in the critical issues accounts for the shift in the content of the blessing. Nevertheless, as I will show, the Priestly Blessing continues to be grounded in Genesis 1. Meanwhile, it will be necessary to explore whether this blessing is, as Davies claims, "replete with words and idioms that are completely alien to the vocabulary of P."

The Vocabulary of P and the Priestly Blessing

To identify "the vocabulary of P"—that is, the vocabulary at home in the Jerusalem cult and in Priestly circles—only with what appears in the P stratum of the Pentateuch is too restrictive. We may note, for example, that the terms of Num 6:24–26 are rife in the Psalms. The prayer, "let thy face shine" in Pss 80:3, 7, 19 (similarly, Ps 31:16) is offered to God "enthroned upon the cherubim." How would the priest respond liturgically to that prayer? Psalm 4:1 opens with the appeal, "be gracious [ḥanan] to me," continues with the petition, "let the light of your face shine on us" (v. 6), and ends on the note of šalom. Psalm 67 opens with "God be gracious [ḥanan] to us and bless us and shine his face upon us" (v. 1), that God's saving power may be known (v. 2), and concludes with the affirmation that "the earth has yielded its increase; God, our God, has blessed us." The earth's increase signifies the blessing of šalom, as explicitly in Ps 147:14. This prayer speech cannot be disconnected from the formal liturgical language in which cultic personnel gave assurance of God's presence and help.

In discussing Num 6:24–26, Baruch Levine makes the following methodological observation. "It is reasonable . . . to associate greetings and good wishes with the dicta of official liturgy in the realization that worshipers addressed God in ways similar to the way they addressed one another, and similar to the way they addressed kings and persons in authority, both

current scholarly opinion, to take the non-P epic tradition(s) as earlier than P in their formulation as a narrative.

orally and in writing . . . The formulae and literary processes we are discussing were not, however, unidirectional . . . Normal forms of communication were surely influenced and stimulated by the formulas of liturgy,"[10] and vice versa. Levine's point about the two-way "process" between cultic and non-cultic usages applies also to the relation between cultic scenarios and narrative portrayals. For example, the story of the reconciliation between Esau and Jacob (Gen 32:13–21; 33:4–11) uses terms at home in the Priestly cult. In Gen 32:20 Jacob thinks, "I may appease [*kipper*] him with the present [*minḥa*] that goes before me, and afterwards I shall *see his face*; perhaps he will accept me." And in the climactic scene, Jacob says to Esau, "If I have found favor [*ḥen*] in your sight, then accept my present [*minḥa*] from my hand; for truly to see your face is like *seeing the face of God*, with such favor [*ḥen*] have you received me" (33:10). Again, we may compare P's account of Jacob's name change to Israel, and his blessing by El Shadday (Gen 35:9–11) with the older epic account in Gen 32:22–30 where Jacob sees God face to face and receives a new name and a blessing. And what is the relation between the formula for God as *raḥum weḥannun*, "merciful and gracious" (or, in reverse, *ḥannum weraḥum*) frequent in the Psalms and elsewhere, and its appearance in Exodus 33–34? Canonically speaking, these other occurrences appear as instantiations of the programmatic proclamation in Exod 34:6 and its anticipation in 33:19. Developmentally speaking, it may well be that Exodus 33–34 offers an etiology of the cultic expression, tracing the adjectival terms of 34:6 back to the verbal forms in 33:19. By the way the sentences in 33:19 echo 3:14 (itself a theological exegesis of the divine name, see Chapter 2), the divine mercy and grace are presented as enactments of a divine freedom whose self-constitution goes hand in hand with its performative utterances.

This is to argue that we must cast our net beyond the confines of the literary stratum of P if we are to gain a full appreciation of the intention implicit in the Priestly Blessing of Numbers 6. Given P's periodization of the sacred history at Exod 6:3, we may analyze Num 6:24–26 not simply in terms of its differences from the Priestly blessings in Genesis, but also in terms of its theological dialogue with them. And if the P stratum is a systematic Priestly glossing of the older epic tradition, we may analyze Num 6:24–26 not only in its connections with other elements in that stratum, but also in its dialogue with relevant aspects of the epic tradition such as Gen 32:22–30 and Exodus 33–34.

---

10. Levine, *Numbers 1–20*, 238.

Cosmos, Tabernacle, and the Priestly Blessing

However we are to understand the relation between Aaron's act of blessing in Lev 9:22–23 and the programmatic Priestly Blessing in Num 6:22–27, this much is clear: Both instances are associated with the tabernacle (see Num 7:1–2). This in itself implies a connection between the Priestly act of blessing and the creation story in Genesis 1. For, as Jon Levenson has argued, the tabernacle functions as a microcosm of the cosmos. From his analysis I will draw only three observations: First, as Levenson observes, P's instructions about the Tabernacle in Exodus 25–31 come in the form of seven divine speeches (Exodus 25–31). Second, the narration about building the sanctuary (Exodus 35–39) is followed by their assemblage and the erection of the tabernacle on "the first day of the first month," itself a date with cosmogonic significance. Third, the subsequent narrative is punctuated by seven occurrences of the expression, "just as the LORD had commanded Moses," a rhetorical effect "similar to the sevenfold appearance of the words, 'And God saw that it was good' . . . in Genesis 1." Levenson concludes, "The temple and the world both result from the perfect realization of divine commandments, and nothing that God has commanded falls short of his expectations."[11]

There is an important distinction between the two accounts. Whereas in Genesis 1 the efficacy of God's performative speech is immediate ("God said . . . and it was so"), the relation between God's commands in Exodus 25–31 and the emergence of the tabernacle in Exodus 35–40 is mediated through the actions of Moses and all those whose heart or spirit makes them willing [*nedib*] (Exod 35:5, 21, 29; cf. 25:2). If the tabernacle is to be constructed out of the materials of creation (25:3–7) according to a design given by God (25:9), its foundations rest in human freedom. If in Genesis 1 humans find themselves inhabiting a material cosmos hospitable to human habitation (compare Isa 45:18), God now seeks to inhabit the sanctuary as a microcosmos founded in human freedom. In seeking to dwell within a hospitality freely offered, God seeks, one may say, to dwell within human freedom itself.

The implications of this theology of "mutual indwelling" are explored in 1 Chronicles 29, a passage that stands squarely in the Priestly tradition. This chapter narrates David's provision for Solomon's erection of the temple in Jerusalem. Offering state financing and personal funds, David, in the spirit of Exodus 25 and 35, calls on others to contribute as their hearts

---

11. Levenson, *Creation and the Persistence of Evil*, 83–84.

"make them willing" (*mitnaddeb*; 29:5, 6, 9). Then he offers a prayer at the heart of which (1 Chr 29:14) comes a cry of amazement: "But who am I, and what is my people, that we should be able [*naʿṣor koaḥ*] thus to offer willingly [*lehitnaddeb*]? For all things come from thee, and of thy own have we given thee." How is it that humans—mere transients and aliens on the earth (29:15)—*possess the power* to give to God *freely* what is God's first and last? The phrase *naʿṣor koaḥ* is pivotal. The verb *ʿaṣar* means, "to restrain, prevent, shut up, withhold, retain." Given that "all things come from you" and that "of your own have we given you," what makes anything truly theirs, something to possess, and so to give or withhold, and the contribution therefore a genuine gift freely given? What arises between the "from thee" and the "to thee" is the mystery in which humans, alien and transient as they are on the earth, have the power to hold as their own, the power to withhold or to give what they have received from God. The mystery is that this very power—to withhold or freely give as their hearts make them willing—is part of the "all things" that come from God.[12] Because God gave David this freedom and power to give or to withhold, his gift to God is a genuine gift; and because his act of free giving rests in God's sustaining gift of freedom and power, he finds himself caught up in an unfathomable mystery that fills him with a wonder that can be known only through participation.

To return to the Priestly Blessing in Numbers 6: It is to be performed in a sanctuary that has the character of a sacral microcosm of the cosmos. In that connection, we may begin to see what it *does* there. In a mediating way ("they shall put my name on the Israelites, and I will bless them"), it does what God does in unmediated fashion in Genesis 1. The opening verb "bless" and the concluding clause "give you *šalom*" gather up the result of God's creative acts in Genesis 1—a universe charged with fruitful, blessed life in all its rich, full wholeness.[13]

The Priestly Blessing, then, is a mediated renewal of God's cosmos-ordering performative speech. Each time it is performed and received, there is, in Brueggemann's words, "a new creation again being wrought

---

12. The Davidic speaker in Ps 51:12, caught in the toils of sin and guilt, invokes this mystery in praying: "Restore to me the joy of your salvation, / and sustain in me a willing [*nediba*] spirit."

13. Psalm 29 celebrates God's blessing in the form of a thunder-and-lightning rainstorm that sweeps across the land from the western sea to the eastern interior. The effect of such a rainstorm is to renew the fruitfulness of a land parched after a dry summer. Fittingly, the psalm ends, "May the LORD give strength to his people! May the LORD *bless* his people with *šalom*."

out of chaos."[14] For the world in which this blessing is uttered is a world marked not simply by the "goodness" and "fruitfulness" and "blessing" and "holiness" that mark the creation story, but also by "badness" and "barrenness" and "cursing" and "defilement." The whole Priestly cultic apparatus, laid out so painstakingly in Exodus–Numbers, can be taken as an elaboration of the theology packed into the Priestly Blessing.

But I believe we can draw this perspective on the Priestly Blessing into sharper focus, by attending to the language at its center. For what intervenes between "bless you" and "give you *šalom*" comes as the sacral antidote to the evils that would intervene in the world to frustrate the programmatic blessing of Genesis from issuing in *šalom*. "The LORD *keep you*"—preserve, protect you—from all that would intervene to destroy the blessing and end in loss. But *how* does God "keep," in such a manner that "bless" can end in *šalom*? This is set out in the following three clauses.

Remarkably, in a blessing so compactly formulated, one motif, concerning God's face, is so important that it must be reiterated, suggesting that this motif—along with the interposed "and grace [*ḥanan*] you"—identifies the critical issue for the Priestly tradition in exilic and post-exilic times. The importance of the face is that it can beam and smile in approbation or darken and scowl in disapproval. In Exod 32:11, for example, God's response to idolatry is such that Moses attempts, not simply to "beseech" (RSV) or "implore" (NRSV) God, but, as the Hebrew idiom has it, "sweeten God's face." In Mal 1:9, corrupt priests are exhorted to, as the Hebrew idiom has it, "sweeten the face of God, that he may be gracious [*ḥanan*] to us."

But if these three clauses concerning God's shining, graciously uplifted face are central to the blessing, how are we to connect them to the creation story? In Genesis, Isaac's blessing of Jacob, delivered in jussive verbs, mediates God's programmatic, unmediated blessing, delivered in imperatives and appearing in Genesis 1, 9, and 35. If we seek an unmediated divine utterance as the programmatic basis for the mediated priestly language of Num 6:25–26a, should we not look for it in the older epic narrative of Exod 33:19 and 34:6? Does the imagery of God's gracious, smiling face originate, then, in the saving history as P interprets it?

---

14. Brueggemann, "Kerygma of the Priestly Writers," 403. Writing of P's correlation between creation and Sinai, for instance in regard to the Sabbath, Frank Moore Cross states (*Canaanite Myth*, 295): "This creation-redemption typology is reminiscent of myth, and adumbrates as well the proto-apocalyptic periodization of world history." The Paul who wrote 2 Cor 4:6 would agree.

PART ONE: Orienting Ourselves in the Biblical World

I cannot imagine that P, who etiologized the Sabbath in creation, would be content to associate the institution of something so important as the official Priestly Blessing only with what happened at Sinai. For one thing, the P redaction arises in the aftermath of Jeremiah's terrifying oracle of a divine judgment that will plunge creation back into *tohu wabohu*, "formless and void" (Jer 4:23–26). Only a blessing as radical as that radical judgment—only a blessing rooted in Gen 1:3, and not simply in 1:28—will suffice. How then may we associate the Priestly Blessing with creation? By comparing these two clauses: *ya'er yhwh* (*panaw 'eleka*), "Yahweh shine (his face on you);" and *yehi 'or*, "Let there be light."

The verb *ya'er yhwh* in the Priestly Blessing, a form of the verb *'or*, "to shine," is cognate with the noun *'or*, "light," in Gen 1:3. Hiphil in form, the verb refers to something or someone making something or someone else to shine. In Eccl 8:1 it is said that "the wisdom of a man *makes his face shine* [*ta'ir panaw*]." What in Num 6:25 makes God's face shine? As indicated in the following clause (*wihunneka*, "grace you"), it is the fundamental fact that God is freely and programmatically gracious (*hannun*, Exod 34:6).

If the verb "shine" (*ya'er*) in the Priestly Blessing echoes the noun *'or* ("light") in Gen 1:3, what of the relation between the name *yhwh* and the verb "let there be" (*yehi*) in the creation story? Is it possible that, in hearing *ya'er yhwh* against the background of *yehi 'or*, we are to hear the blessing clause as a mediation of the unmediated primal word of creation, through a grammatical inversion in which the verb in the first instance becomes a noun in the second while the noun in first instance becomes a verb in the second?

The Priestly redactor cannot have been ignorant of the theological etymology in Exod 3:14, where the divine name is interpreted in terms of the verb *haya*, "to come to pass, to be," and is explicated as "I will be who I will be."[15] For the Priestly writer, I suggest, the God who there is thus both revealed and hidden—revealed as, and hidden within, the mystery of divine freedom for self-constitution unbound by any creaturely structure or circumstance—is the same God who in Genesis 1 creates what is not God by the simple expedient of turning the primordial self-performing first person "I will be" into a third-person indirect imperative "let there be." By this inversion, "I will be" creates the space—the "thereness," so to speak—for non-divine otherness to emerge, and then into that non-divine space calls forth light as the creaturely analogue to the light of God's face (Pss 44:3; 89:15; 90:8).

---

15. See chapter 2 above.

Such a connection between Num 6:24–26 and Gen 1:3 through Exod 3:14 finds support, from a hermeneutical point of view, in the interpretive tradition in Targum Neophiti. In a study of the divine name in the Aramaic targums, Robert Hayward takes up the question of the origin and meaning of *memra*, literally "utterance" (from Aramaic *'amar*, "to say, speak"). Noting that "there is still no satisfactory account of *Memra*'s origins," and that the term "is found only in Targumic literature,"[16] Hayward summarizes recent studies suggesting that in fact *memra* "is neither an hypostasis, nor a simple replacement for the Name YHWH [as generally held], but an exegetical term representing a theology of the name 'HYH."[17] In this analysis, *memra* is introduced first at Exod 3:12, where "I will be [*'ehyeh*] with you" is taken to anticipate the sentence in 3:14 (*'ehyeh 'ašer 'ehyeh*) and then the sentence immediately following ("Thus you shall say to the Israelites, '*'ehyeh* has sent me to you'"). In Targum Neophiti, Exod 3:12 is rendered, "I, namely my *Memra*, will be with you." "We conclude that *Memra* stands for the divine *self-designation* 'HYH, which is viewed by the Targum as a divine name."[18] Hayward goes on to note that in Neophiti the 'HYH in 3:14–15 is glossed this way: "he who said [*mn d'mr*] and the world was there [*whwh*] from the beginning, and who is to say [*lm'myr*] to it 'Be there' [*hwwy*], and it will be there [*wyhwwy*], he has sent me to you."[19] Clearly the Targumists at this point associated *'ehyeh* in Exod 3:12–15 with *yehi* in Gen 1:3, not only philologically but theologically.

What, then, does the Priestly Blessing *do*? It mediates God's performative word of world-creation *and* world-blessing uttered unmediatedly in Genesis 1. In this connection, I may note two recent comments on the Priestly Blessing. Of the *wa'ani*, "and I (will bless them)" in 6:27, Jacob Milgrom writes, "Rather, 'and it is I' or 'I Myself.' Hebrew [*wa'ani*] is emphatic, thrusting home the point that not the priests but the Lord is the sole author of the blessing." He goes on to say, "The claim that the blessing . . . has inherent powers of fulfillment is not to be found in Scripture except in Gen 27:34–38. There a vestige of an earlier, pre-Israelite view is still present . . . Otherwise, all utterances, whether blessing or curse, are in effect prayers: Their efficacy depends upon the acquiescence of God."[20]

---

16. Hayward, *Divine Name and Presence*, 7.
17. Ibid., preface.
18. Ibid., 17.
19. Ibid., 18.
20. Milgrom, *Numbers*, 52 and 360.

PART ONE: Orienting Ourselves in the Biblical World

In similar vein, Patrick D. Miller writes, "One must not assume that what we have here is some sort of primitive magic . . . despite the clear signs that blessing in the Old Testament has an effecting power." Noting the three-fold repetition of the divine name, and the emphatic personal pronoun in v. 27, he comments, "The role of the priest or minister figure is there, but it is functional and representative. *He or she enacts nothing.* The provider of these dimensions of life is the God who is worshiped in the service."[21] These comments raise the question as to whether there is any middle ground between automatic magic on the one hand and utterances that, for all their *form* as performative speech, are in *function* barren–whether, for all that the Priestly tradition celebrates humankind as in God's image, that image is in spirit a eunuch.

May we not understand the relation between the priests' actions and God's, in this blessing, on analogy with the relation between God's unmediated creation of the world in Genesis 1 and God's mediated construction of the tabernacle in Exodus 25–31, 35–40? (Exod 25:8, "Let them make me a sanctuary," and Ps 78:69, "he built his sanctuary," identify human and divine dimensions of the same process.) The key may lie in the mystery that evokes David's cry of amazement (1 Chr 29:14), the mystery of human freedom embedded in the freedom of God, a divine freedom that creates by communicating itself to its creatures and entrusting into their hands the capacity to communicate it to other creatures and back to God. Whether we are speaking of tabernacle building or the Priestly Blessing— or for that matter, anything that we understand ourselves to be doing by way of response to divine promise or command[22]—the mystery of efficacious human freedom and action, as received from God and enacted in the sustaining grace of God, is unfathomable to human analysis and known only in participation.

Perhaps one way of expressing the inexpressible is to stammer about the human mediation of God's unmediated grace. To speak of human mediation is to give nature and history the gravity they deserve. To speak of God's unmediated grace is to relieve nature and human history of the burden of an ultimacy they should not be made to bear, a burden which, when we impose it on them, turns them into idols.

Once more, then: what does the priestly Blessing *do*? For those who receive it as a word definitive for their existence, it radically reconstitutes

---

21. Miller, "Blessing of God," 249 (italics original).

22. The issue here is explored further elsewhere in this volume, especially chapters 8 and 10.

the world within which they live, grounding it in the shining face of God vis-à-vis all that threatens to fill their world with *tohu wabohu* (Jer 4:23–26). And it reconstitutes the problematic world within their hearts, a world of clashing passions, motivations and machinations whose force threatens to turn freedom into a servant of destruction—Hebrew *'abaddon*, apocalyptic *Abaddon*. It is no wonder that the psalmist's radical plea, "Create in me a clean heart, O God, / and renew a right spirit within me" (Ps 51:12), is initiated by words that echo the Priestly Blessing: "Grace [*ḥanan*] me, O God, according to thy steadfast love; according to thy abundant mercy [*raḥamim*] blot out my transgressions" (51:1).

4

Praying in the Space God Creates for the World

Several years ago, Terence E. Fretheim sent me an offprint of his essay "Prayer in the Old Testament: Creating Space in the World for God."[1] I read it with much profit, appreciating the way in which he had drawn on the process thought of Alfred North Whitehead. In one of his writings, Whitehead remarked that the history of Western philosophy may be thought of as a series of footnotes to Plato. I offer the present essay in celebration of Terry's many contributions to biblical understanding, and in particular as a footnote to his essay on prayer. He opened his essay with a reflection on experiences of space in ordinary relationships, and I will open this one in the same way, before moving on to matters more formally exegetical and hermeneutical.

Making Space

It was a summer evening, and we were sitting quietly on the screen porch, reading. Suddenly my wife Eileen asked, "Gerry, what word?" "What do you mean, what word?" "You know—'In the beginning was the Word.' *What* Word?" I began to talk about the meaning of *logos*, in Greek thought and Philo, and about the meaning of the Hebrew word *dabar*, when she interrupted with, "No, no! A word is something spoken, something said. What word was it that was spoken in the beginning, that was with God, was God, and through whom all things were made?" Muttering to myself about the limitations of untutored Bible knowledge, I returned to the magazine I was reading.

---

1. Fretheim, "Prayer in the Old Testament," 51–62.

Praying in the Space God Creates for the World

Some months later, I found myself wondering: What if she were right? What if we should be looking for something uttered, that could be understood as God, as being with God, and through which all things were made? Where should we look for such an utterance? In the Old Testament, of course. But where? Asked this way, the question answered itself: God's word to Moses at the burning bush—ʾehyeh ašer ʾehyeh, "I will be who I will be" (Exod 3:14). This utterance can refer, in the first instance, to the internal life of God without any necessary reference to anything outside of God. Formed from the verb hyh, "to come to pass, to become, to be," this utterance would name the unfathomable mystery of the Living God as a fountain of life arising out of itself and falling back into itself, self-constituting and self-constituted. But this is also the word through which all things were made. For in Gen 1:3 God said, yehi ʾor, "Let there be light." Simply by inverting the first-person verb into a third-person verb, the word "I will be" (as Exod 3:14 goes on to abbreviate the utterance) becomes "let there be."[2]

From then on, I would ask students in my introductory Old Testament class this question: If God creates the world by the power of the divine word, of which "Let there be light" is the primordial instance, which of these four words is the explosively powerful word, the word that initiates the creative process? Students would vote for the first, or the third, or the fourth, but no one ever voted for the second, "there." And that, I would then assert, is the initiating word. How so?

According to Jewish and Christian understandings, God is infinite and eternal. One Jewish term for God is ʾEn Sof, "the Boundless One." Put in terms of subjective consciousness, every finite creature distinguishes between "here," as where it is, and "there," as where other creatures are, and between "now," as when it presently is, and "then," as when it once was or later will be. As ʾEn Sof, God is bounded by neither time nor space. Not only "the Eternal Now" (Tillich), God is also "the Infinite Here"—"Here" being the infinitude of God's unbounded presence to God's own self. In terms of Exod 3:14, one could say that God as ʾEn Sof is "I AM THAT I AM," or simply "I AM," as in KJV (following the Latin translation, ego sum qui sum). But that translation of ʾehyeh tends to undergird static conceptions of God as unchanging and immovable (as in Aristotle's "unmoved mover"). The living God of the Bible is more adequately indicated through the dynamic "I will be." But what happens, then, when the ʾEn Sof, the "Boundless Here-Now," by inversion of the dynamic first-person

2. See also chapters 2 and 3.

self-utterance, says, "Let *there* be"? The very *thinking* of third-personness opens up a space in God's first-person consciousness for "otherness," a space in God that is not God, a space into which God can then call what is not God into existence.

Having offered this thought to my students, I would then illustrate it through another domestic experience. As a seminary student I enjoyed a room all to myself. For the first time in my life I was able to arrange the contents of my room as I pleased, and to consider its living space as my own. When I married Eileen, I rediscovered (but now on a parity basis) what it meant to share a living space. Allowing for our respective, respected personal spaces, the two of us inhabited the whole apartment. When our first child was born, that child was welcomed into our apartment. But as she grew and came to have a room of her own in the house we eventually acquired, that room became her space. When her door was closed, we knocked and waited to be invited in. And increasingly, the contents and arrangement of the room took on her character. The house might be in our name, and we might be paying the mortgage and the taxes, but that one room in the house was not simply part of "our" house, it was also "her" room. Because our parental relation with her included our intention and desire to respect her and care for her, it was her room precisely because it was part of our house. It was in coming to this awareness of our evolving domestic space that I came to appreciate the idea of God's creating the world by first making a space for it. Then a deeper analogy occurred to me. All of us begin our lives in an embodied space that, belonging to another person, comes to be given over to us, to shelter and nourish us. So I began to think of God creating the world by the self-limiting act in which the *'En Sof*, the divine *'ehyeh*, said, and goes on saying, "Let *there* be . . ." We arise and flourish within a space freely offered to us—a space that encompasses us within the bosom of the divine un-boundedness. (More recently, I have begun to connect this with Paul's word, "predestined," in Rom 8:29, 30, where the Greek verb *proorisen* means, literally, "to mark off [a space] as with a horizon.")

The *Memra*

When I shared this line of thought with Jay Southwick, a former student, she drew my attention to an essay in which Robert Hayward understands the *logos*, of John 1:1–2 in terms of the Aramaic word *memra*, "word," as used in the Targumic paraphrases of the Old Testament.³ Such an understanding had earlier been proposed by other scholars, but more recently had fallen out of favor. Now, with the discovery of Targum Neophiti, Hayward rehabilitates that interpretation. His argument, in this essay and in a subsequent monograph,⁴ may be summarized as follows:

(1) Actions of God attributed to YHWH ("Yahweh," "the Lord") in the Hebrew Bible are in the Aramaic Targums attributed to the *memra* of YHWH. (2) Close study of Targum Neophiti indicates that this practice is rooted in the periphrastic exegesis of Exod 3:14, where *memra*, "word," serves as a replacement for *'ehyeh*. (3) In Gen 1:3, where the Hebrew Bible reads, "God said, 'Let there be light,'" the Targums attribute this utterance to the *memra*. (4) This hermeneutical move seems to be guided by the occurrence of the verb *yehi*, "let there be," a form of the divine name *'ehyeh*, "I will be," in Exod 3:14. As Hayward says, "the *Memra* is interpreted of past and future creation, and is none other than the first-person name of God."⁵

Space for God

What does all this have to do with prayer as, in Fretheim's words, "creating openings (relational space) for God in the world"?⁶ Consider the tabernacle that God instructs Moses to make in Exodus 25–31. In that place, the high priest bears the names of the twelve tribes of Israel upon his shoulders and his heart in prayer (Exodus 28). But first, it is to be the place for God to dwell (*šakan*, "tent") in Israel's midst (Exod 25:8). This sanctuary, this "space for God," is to be constructed out of the materials of creation—animal, vegetable, mineral. What do humans contribute? "Let everyone whose heart makes him or her willing" (Exod 25:2) contribute these materials. They are not to be collected as a levy, but given as a freewill offering. If the tabernacle is built of the materials of creation, it is founded in human freedom, a freedom that makes space within itself for God's indwelling presence. The physical sanctuary is an "outward and visible sign of an inward and spiritual grace" (*Book of Common Prayer*), a sign of God dwelling within the freedom of the people.

---

3. Hayward, "Holy Name," 16–32.
4. Hayward, *Divine Name and Presence*.
5. Ibid., 19.
6. Fretheim, "Prayer in the Old Testament," 52.

PART ONE: Orienting Ourselves in the Biblical World

Jon Levenson has discussed the way in which the Exodus sanctuary is a microcosm of the cosmos.[7] If that is the case, then to build a sanctuary is to shape a physical space for God within the physical world, a world that God has called forth within the space God has made for it. These physical spaces, cosmos and microcosm, are expressions of two freedoms, divine and human, which now may be understood—and experienced—in terms of mutual indwelling. That mutual indwelling comes to its innermost reality when the people hear God's word proclaimed in the sanctuary (Exod 25:22) and the high priest prays for the people there.

In 1 Kings 8, however, when Solomon prays to God during the dedication of the temple, he says (v. 27), "But will God indeed dwell on the earth? Behold, heaven and the highest heaven cannot contain thee; how much less this house which I have built!" According to this prayer, if God is indeed present in this sanctuary, it is as dwelling (*liškon*, "tenting") within the "thick darkness" of the holy of holies (v. 12). More specifically, it is God's name that is to be "there" (v. 29); and it is because God's name is there that the people (and even foreigners; vv. 41–43) may offer prayer in, or oriented toward, that place. By the third century CE, a Midrashic tradition has it that God's *šekinah*—God's tenting presence—dwells in the holy of holies by concentrating and contracting God's power to a single point, in a divine action called *ṭimṭum*. Fast forward to the sixteenth century and the figure of Isaac Luria. Before his time the common Jewish and Christian doctrine of *creatio ex nihilo* was conceived to occur by a going forth of God's creative word. But where could it go forth to, if God is *'En Sof*—boundlessly infinite and eternal? So Luria inverted *ṭimṭum*, in what Gershom Scholem has called "one of the most amazing and far-reaching conceptions ever put forward in the whole history of Kabbalism."[8] Instead of God contracting *to* a point in the inner sanctum, Luria envisioned God as contracting *from* a point in the divine boundlessness, in a free act of "self-limiting the limitless," so as to make a space that is not God within which to call into existence what is not God. For Luria, this twofold action of withdrawal and going forth is the dialectic by which creation originally arose and by which God continually sustains it.

The fruitfulness of this conception for Christian theological reflection is suggested by Jürgen Moltmann's use. In *The Trinity and the Kingdom*, Moltmann writes, "In order to create something 'outside' himself, the infinite God must have made room for this finitude beforehand, 'in

---

7. Levenson, *Creation and the Persistence of Evil*, 78–99.
8. Scholem, *Major Trends*, 260.

himself'. . . Has not God therefore created the world 'in himself,' giving it time *in* his eternity, finitude *in* his infinity, space *in* his omnipresence and freedom *in* his selfless love? . . . God creates the world by letting his world become and be in *himself*. Let it be! . . . God has released a certain sector of his being, from which he has withdrawn—'a kind of primal, mystical space.'" Again Moltmann writes, as a Christian, "The Trinitarian relationship of the Father, the Son and the Holy Spirit is so wide that the whole creation can find space, time and freedom in it."[9] Elsewhere, he writes, "God makes room for his creation by withdrawing his presence. What comes into being is a *nihil* which does not contain the negation of creaturely being (since creation is not yet existent), but which represents the partial negation of the divine being, inasmuch as God is not yet Creator. The space which comes into being and is set free by God's self-limitation is literally a God-forsaken place."[10]

Praying in the Space

How, then, does this bear on our praying? I turn first to the high-priestly prayer of Jesus in John 17. This prayer has its proximate foundation in John 1:14, where the incarnate Word's dwelling (*eskenosen*, literally "tented") among us in glory, grace, and truth is to be understood against the background of the sanctuary theology of Exodus 25–31 and 35–40. But it has its ultimate foundation in John 1:1–2, understood against the background of the Targums' *memra*-theology connecting Exod 1:14 and Gen 1:3. For Jesus prays, "Father, glorify thou me in thine own presence with the glory which I had with thee before the world was made. I have manifested thy name" (John 17:5–6). What is the unfathomable mystery of that glory and that name? As he continues to pray, Jesus intimates that that glory and that name consist in the unity of the Father and the Son as a unity of mutual indwelling; and the burden of his prayer is that his followers, and all who are drawn by the testimony of his followers, may enter into the mystery of that unity, that mutual indwelling (17:11b, 20–23). It is of the utmost significance for both theological reflection and the practice of prayer that this mystery of unity as mutual indwelling is embodied in an act of prayer, the prayer of Jesus as high priest bearing on his shoulders and his heart the names of his followers and, ultimately, of his whole creation. To pray as a Christian, then, is to enter with Jesus into that space, as the space God

9. Moltmann, *Trinity and the Kingdom*, 109–10 (italics original).
10. Moltmann, *God in Creation*, 87.

has freely opened up for the world to be, a space within which it is safe to invite God, and the company of God, into the space of one's own internal freedom.

I turn finally to Romans 8. According to this chapter, Christian prayer stands in solidarity with the whole creation as it "groans" (*sustenazei*) together in travail and pain, often with a sense of futility (Rom 8:20–22). But Christian groaning in prayer (8:23) is not *despite* the gift of the Spirit, for it is also a participation in the unutterable groaning (*stenagmos*) of the Spirit who "intercedes for us" (8:26). And the intercession for us by the Spirit who dwells *within* us has its counterpart *within* God, as Christ Jesus, "who died, yes, who was raised from the dead," intercedes for us at the right hand of God (8:34). What is the character of our participation in this prayer, the conjoint prayer of the Spirit within us and of the risen Christ in God?

A brief reflection on the groaning of creation in futility and hope: A large portion of that groaning, and that futility, arises through the clash of conflicting desires, aims, understandings, and agendas, a clash that all too often issues in physical conflicts in which each side seeks victory over the other. In such a context, prayer can become just one more weapon in the conflict. The more certain one is of the Rightness of one's Cause, the more fervent one's prayer on behalf of that cause, and the more zealous in pursuit of that cause, the more one's prayer serves simply to strengthen one's certainty and intensify one's zeal. In this way prayer becomes another divisive expression of the spirit of fearful defensiveness that in Gen 11:1–9 builds walls that divide the original unity of humankind into mutually antagonistic groups incapable of understanding one another's speech. In this context, how shall we understand the Spirit's intercession as "helping us in our weakness" (Rom 8:26)? When "we do not know how to pray as we ought [*katho dei*]," are we to hope that the Spirit will support and confirm our knowledge of what specific outcome to pray for "according to the will of God [*kata theon*]" (8:27)?

I suggest another way to understand what is at issue here. Consider Rom 8:15–16, especially the words, "When we cry [*krazo*], 'Abba! Father!' it is the Spirit bearing personal witness with our spirit." In both testaments of the Greek Bible the verb *krazo* again and again refers to a cry of distress or a plea for help in a critical situation. (In 8:15–16 it is "our" part in the *groaning* of creation and of the Spirit.) But Paul's phrase, "Abba! Father!" echoes what we overhear in Gethsemane, where, in the shadow of his enemies, Jesus prays (Mark 14:36), "Abba, Father, all things are possible to thee; remove this cup from me; yet not what I will, but what thou wilt."

Praying in the Space God Creates for the World

Earlier, he had groaned (*estenazen*) in a wordless prayer of solidarity with a deaf-mute (Mark 7:34). Now, praying for himself, Jesus, as we might say, *does not know how to pray as he ought*. What does this mean? Not that he is a neophyte deficient in the practice of prayer! The *katho dei* of Rom 8:26 means literally, "in accordance with what is needed," or, paraphrased, "taking into account what the situation calls for." Jesus has no doubt that he himself would as soon "let this cup pass from me." But he submits this intense desire to the wider purposes of the One whom he calls Father, purposes that have to take into account the ultimate welfare of the whole creation. So Jesus lodges his own specific request with God, places it alongside God's creation-wide concerns, and rests it there, along with the pledge to accept and undergo what God deems best for the creation-wide situation.

It is *this* Jesus who now at the right hand of God makes intercession "for us"—for a world that in its mutual antagonisms continues not to find room for him or for one another. The heavenly intercession of this Jesus continues, I suggest, *in the same spirit* that we see in Mark 14:36. And, I suggest, the intercession of the Spirit of Jesus within us on our behalf (Rom 8:2, 9b, 15–16, 26–27) arises *in that same spirit*. Our "weakness" (Rom 8:26) does not consist in our not knowing *what* to pray for. Our "knowing what to pray for" is all too often the problem! Knowing *how* to pray is a matter of knowing how to place our intensely felt concerns and our specifically focused hopes in the spacious bosom and horizon of God's providential love, along with those concerns that are so deep in us as to stir within us only in the form of inarticulate groanings of our spirit—what the psychotherapist Christopher Bollas calls "the unthought unknown"—and to entrust them to God's providential wisdom.[11]

The world of time and space, with its physical processes and human agendas, is often a place of mutually opposed, grinding necessities, issuing in what the New Testament calls *thlipsis* ("distress, tribulation"), a kind of gridlock that severely constricts any freedom to maneuver, let alone imagine fresh possibilities and find new solutions. To pray in the spirit of Romans 8 is to enter into the space that God has predestined—"prehorizoned"—in creating the world and calling us to be "conformed to the image of his son" (v. 29). To enter that space is to enter into a challenging and yet freeing state of "unknowing." That space, that unknowing, that *nihil*, at times can feel like the God-forsakenness of Mark 15:34, a cruciform space (like traditional church architecture). But precisely as such it

---

11. Bollas, *Shadow of the Object*.

is the arena within which the divine *creatio ex nihilo* works continually to "make all things new." This may mean that we find new possibilities for a way forward out of the *thlipsis*. It may mean (as with Jesus) that we find strength to undergo the *thlipsis,* in trust that "in everything God works for good with those who love him" (Rom 8:28). For neither *thlipsis,* says Paul, nor anything else in all creation, can separate us from the love of Christ, which is the love of God in Christ (vv. 35–39).

To pray in this space is to pray in the Spirit and so to participate in the life of Christ on behalf of the world (v. 17). It is to pray in the space that God creates for the world.

part two

Forays into a Biblical World

5

Prayer as Self-Address

The Case of Hannah

A MONG BIBLICAL WOMEN, HANNAH is most often celebrated for the song she sings after the birth of Samuel (1 Sam 2:1–10). But her earlier prayers in distress (1 Sam 1:3–18), though indicated only indirectly, are the human foundation of her subsequent song. Her movement from prayer in distress to song of praise parallels the movement in psalms of individuals (e.g., Psalms 22; 116), and the movement of her ancestors from the cries and groanings of the Israelites in Egypt (Exod 2:23–25) to the hymn of praise after the deliverance at the sea (Exod 15:1–21). This same movement marks the book of Psalms as a whole. Although in this last example the movement oscillates between complaint and praise, nevertheless, after the opening two psalms, the psalter continues with the next dozen or so psalms voicing various kinds of individual or communal cries for help, and it closes with a virtually unbroken series of psalms of thanksgiving and praise.

Hannah's life, then, as it comes to expression in her prayers, exhibits in miniature the pattern that marks the life of Israel as a whole. In this paper, I wish to examine one aspect of her prayer, namely, her speaking to herself in 1 Sam 1:13, and to bring that into sharper focus through close attention to the Hebrew idiom, *dibber ʿal leb* (RSV: "speaking in her heart"; NRSV: "silently"). As I will show, this idiom, though distinctive in its application to Hannah, fittingly connects this aspect of her prayer to the frequent appearance of "self-address" in the Psalms and in other biblical prayers.

PART TWO: Forays into a Biblical World

Comfort: God's and the Self's

The notion of self-address in prayer appears within the context of one of the recurrent images for the divine-human relation, God's parental care for people. In Ps 27:10, the psalmist cries out, "If my father and mother forsake me, / the LORD will take me up."[1] In exile, Zion cries out, "The LORD has forsaken me, my Lord has forgotten me," and God answers,

> Can a woman forget her nursing child,
>   or show no compassion for the child of her womb?
> Even these may forget,
>   yet I will not forget you. (Isa 49:14–15)

The image of parental divine compassion and care, implicit in Psalm 27 and clear in Isaiah 49, is elaborated in Isa 66:10–13:

> Rejoice with Jerusalem, and be glad for her,
>   all you who love her;
> rejoice with her in joy,
>   all you who mourn over her—
> that you may nurse and be satisfied
>   from her consoling breast;
> that you may drink deeply with delight
>   from her glorious bosom.
> For thus says the LORD:
> I will extend prosperity to her like a river,
>   and the wealth of the nations like an overflowing stream;
> and you shall nurse and be carried on her arm,
>   and dandled on her knees.
> As a mother comforts her child,
>   so I will comfort you;
> you shall be comforted in Jerusalem.

This parental image of the relation between God and Israel accords with the relation between God as 'El Šadday and Israel's ancestors in Genesis. There, 'El Šadday is the giver of the blessings of heaven above and the earth beneath, blessings of breast and womb (Gen 49:25), and is invoked (Gen 43:14; answered in 43:29–30) as the divine source of compassion. These themes of nurture and compassion converge in Isa 66:10–13. The act of nursing in response to the infant's cry of hunger is itself an act of compassion. So the divine compassion extended to Israel in exile is

---

1. In this essay, I use the NRSV; modifications are identified by an asterisk.

movingly imaged as a mother soothing a distraught child at the breast and playing with it on her knees.

Psalm 131 presents this image of divine parental care, as well as the issue of "self-address" in prayer, that lies at the center of my argument. The psalmist finds herself in a state of perplexity that threatens to overwhelm her. Frequently in the psalms, such perplexity arises because the psalmist, though protesting innocence, is accused by others of suffering affliction on account of some wrongdoing. If that is the case with this woman, she may be compared with Job.[2]

For his part, Job counters the accusations of his friends with protestations of innocence and accusations against God—against *'El Šadday*, the name by which God is so frequently designated in this book.[3] When God finally answers him out of the whirlwind (in part as *'El Šadday*; see Job 40:1), Job realizes that the issues he has been wrestling with cannot be humanly fathomed. So he confesses—in words like those in Psalm 131—"I have uttered what I did not understand, / things too wonderful for me, which I did not know" (42:3). In saying this, he may be said to rediscover the wisdom hidden in the proverb, "Trust in the LORD with all your heart, / and do not rely on your own insight" (Prov 3:5). The wisdom tradition repeatedly calls on people to forsake naïveté and vigorously to seek wisdom. But, as this proverb warns, the danger is that humans may come to rely on the wisdom they have acquired, and forget that human understanding must sooner or later give way to unreserved trust in God's wisdom and goodness.

How long has it taken the woman of Psalm 131 to arrive at the resolve, "I do not occupy myself with things too great and too marvelous for me" (v. 1)? And how does she arrive there? Perhaps by hearing God's direct spoken response to her repeated cries, the way Job finally did? I think not. Psalm 131 traces a different path than the straightforward one of human

---

2. Patrick D. Miller discusses connections between Psalm 131 and Job 42:1–6 in his *They Cried to the Lord*, 239–43. This follows immediately on his discussion of Hannah's prayer and hymn in 1 Samuel 1–2 (ibid., 237–39). I am indebted to Miller especially for his translation of Psalm 131, which is adopted in this paper. For further recent scholarly work on this psalm, see his references (ibid., 414–15).

3. Of its forty-eight occurrences in the Old Testament, *Šadday* (or *'El Šadday*) occurs seven times in Priestly contexts relating to Israel's ancestors, and thirty-one times in the book of Job. The other ten occurrences are scattered in other contexts. The two occurrences in Ruth 1:20–21 relate to existential issues similar to those in the ancestral narratives, 1 Samuel 1, and the book of Job. For a brief exploration of Job and Naomi as existential counterparts, and for further discussion of the significance of *Šadday* themes and their bearing on Job's dilemma, see Janzen, "Lust for Life and the Bitterness of Job," 152–62.

distress, direct divine response, and human consolation derived from that response. That path is laid out in v. 2: "I have calmed and quieted my soul, / like a weaned child with [ʿal] its mother; / my soul is like the weaned child that is with [ʿal] me." The psalmist does not speak of God calming and quieting her soul. She speaks of calming and *quieting her own soul*.

She accomplishes this calming through the activity of self-address. The image of the *weaned* child is critically important to understanding how such self-address can provide comfort for a soul in a situation of distress. In Isaiah 66, the comfort that comes to Zion is likened to the comfort extended to a nursling (*yoneq*). A weaned child (*gamul*) does not cease to seek comfort from its mother, but it receives that comfort in a different way. It may come to its mother and still be taken up into her bosom, but something has begun to replace the consolation of the nursing breast. At one stage, a "pacifier" or "soother" serves as a transitional means of comfort. The mother (or father or older sibling) may extend the soother to the child, but eventually the child will place the soother in its own mouth. Later still, the child may resort to some personally selected equivalent, such as a blanket.

What is happening here? Psychologists speak of the process by which the child internalizes the inter-personal dynamic of soothing, so that what originally was the mother's soothing of the child becomes the child's *self*-soothing. As the mother had soothed the child, so now the child sooths itself; yet the self-soothing remains in some sense a drawing upon the soothing extended to it by the mother.[4] A dramatic example of this occurred in the midst of a near-tragedy in Midland, Texas in 1987 when 18-month-old Jessica McClure fell down an old abandoned well. Because of the danger of a cave-in, the rescue dragged on for hours. In the meantime, rescuers lowered a two-way microphone into the well to offer reassurance to the presumably terrified child. To their amazement, instead of a terrified child, they found a reasonably calm little girl who they heard singing to herself songs that her mother had often sung to her.

So it is, I suggest, in Psalm 131. The woman says to God that she cannot fathom the issues that trouble her; but she has calmed and quieted her soul "like a weaned child with its mother; my soul is like the weaned child that is with me." As Miller notes, the preposition (ʿal) repeated in

---

4. In *Playing and Reality*, D. W. Winnicott discusses weaning as a psycho-social process involving, among other things, the child's movement from being soothed in breast-feeding to its ability to soothe itself by means of what he calls "transitional objects." As the following story shows, the "transitional object" may take the form of songs or words learned from the mother and now recited to oneself.

these two lines has the basic meaning, "upon," so that the child here may not be simply "with" its mother, but *on* her lap or bosom.⁵ That, indeed, is precisely how this preposition works in Isa 66:12: "you shall nurse and be carried *on* [*'al*] her arm, and dandled *on* [*'al*] her knees." The weaned child of Psalm 131 no longer nurses, but it can still be held; and as the mother holds it, she observes it becoming quiet.

Deeply troubled by something she cannot understand, the psalmist notices what any observant parent notices in such a situation: while, on the one hand, the calming and quieting is an interpersonal process, on the other it is also a process internal to the child. As the mother reflects on how the child calms and quiets itself within her embrace, she calms and quiets herself, as it were, on God's bosom, trusting in God with all her heart, and relying not on her own insight. Then she turns to her community, calling on it to "hope in the LORD / from this time on and forevermore" (Ps 131:3). If she thus extends her comfort to others, it is by calling on them to calm and quiet their own soul in times of distress. This is not simply an exhortation to "boot-strap" self-help; nor is it simply passive acquiescence in God's sustaining help. It is a mode of that fathomless mystery where the human spirit and the divine spirit commune together, at a depth that lies beyond human understanding.⁶

Self-Encouragement in the Worship of God

This way of viewing Psalm 131 may illuminate a similar dynamic of self-address in prayer found in Psalm 42–43 (understood here as a single prayer-text). The psalmist begins by voicing a soul-thirst for God and a longing to see God's face. But in place of God's satisfaction of that thirst (as in Isa 66:11), the psalmist can only drink his own tears. Yet even as he does so, he assuages his thirst also with the memory of his participation in processions to God's house, processions filled with so many people's glad shouts and songs of thanksgiving that he was buoyed along joyfully by them. Encouraged by this memory of communal buoyant song, he may be said to calm and quiet his soul like the woman in Psalm 131:

---

5. Miller, *They Cried to the Lord*, 239.

6. The best discussion of this mystery known to me occurs in Samuel Taylor Coleridge, *Aids to Reflection*, 78–79. In this passage, Coleridge explores the implications of Rom 8:26, and arrives at his famous description of the "unknown distance" between the deepest reach of human consciousness and the depth of the human spirit where the "first acts and movements of our own will" interact with the Spirit of God.

PART TWO: Forays into a Biblical World

> Why are you cast down, O my soul,
>   and why are you disquieted within me?
> Hope in God; for I shall again praise him,
>   my help and my God. (Ps 42:5)

The exhortation to "hope in God" which the woman in Ps 131:3 directs toward Israel, the speaker in Psalm 42–43 directs to himself. He is both the encourager and the one being encouraged; and the source of the encouragement is the sustaining recollection of previously being borne along and participating in the worship that ended in God's house where his soul in the past had slaked its thirst and he had beheld the face of God. Having internalized that earlier communal dynamic, he is able to reenact it within himself when he is cut off from his community and feels cut off from God. What is especially noteworthy is the oscillation between distress and self-comfort, as the refrain of 42:5 recurs in 42:11 and 43:5.

This repeated self-address is not simply a soliloquy, but occurs within a psalm that begins as address to God (42:1) and approaches its end in the same way (43:4). The self-address and the third person references to God are embraced within, and thereby are an integral part of, the address to God. Just as the weaned child's self-calming in Psalm 131 draws on the internalized calming of the nursing mother, and occurs on her bosom, so the very ability of this psalmist to call on himself to hope in God is derived from earlier hope-filled communal songs of assurance, and takes place within the bosom of his address to God. These two Psalms, then, 131 and 42–43, should alert us not to interpret third-person awareness of God in the midst of prayer, or self-address in that context, as lapses from the spirit of prayer, but as integral aspects of it. Such prayer dynamics are one aspect of the "weaning" that is part of the spiritual growth that biblical religion calls its adherents to, a weaning in the course of which, as the mother's comfort is internalized, distress is transmuted into trust.[7]

What we may call the self-relation in prayer is present graphically in Psalms 131 and 42–43. This aspect of prayer is present much more pervasively, if less obviously, in the psalms than often meets the eye of readers of modern English translations. We have already seen that Psalm 131 ends with a call to Israel, in the imperative voice, to hope in the Lord; and we have seen that in Psalm 42–43 this call is self-directed. It is typical in the Psalms that parallel to this self-directed call, the speaker may, by the use of the plural imperative, call likewise upon others. So, for example, the exhortations in Psalm 100 are all in the plural imperative. But in another

---

7. Viewed in this context, the book of Job testifies to the stringent conditions under which such weaning is sometimes undergone.

standard form, the speakers include themselves among the persons being exhorted. This form uses verbs in the cohortative mood, or first-person indirect imperative. The plural form of the cohortative mood (e.g., "Let us sing unto the LORD") is used to exhort one's cohorts *and oneself* to engage in some action. Psalm 95 provides a good example. A second-person imperative, "O come" (vv. 1, 6) is followed by reiterated first-person plural cohortatives, "let us" (vv. 2-3, 6). But the cohortative can also occur in the singular, "let me. . . ." In such a usage, "the *cohortative* expresses the direction of the will to an action and thus denotes especially self-encouragement (in the 1st plural an exhortation to others at the same time), a resolution or a wish, as an *optative*."⁸ The singular form of the cohortative occurs frequently in psalms of individual complaint or praise. From over fifty instances, we may note the following examples.

Psalm 146 opens with the plural imperative call to "Praise the LORD" (in earlier parlance, "Praise *ye* the LORD"). But whereas in Psalms 148, 149, and 150 this opening call is followed by repeated verbs in the plural imperative, in Psalm 146 the call is re-directed inward to the speaker, with the singular imperative, "Praise the LORD, O my soul," followed immediately by the singular cohortative, "Let me praise the LORD as long as I live; let me sing praises to my God while I have any being."* Psalm 103 opens with the reiterated singular imperative, "Bless the LORD, O my soul; / and all that is within me, bless his holy name." At the end, after a series of plural imperatives calling various creatures to "bless the LORD," the psalm ends as it began, "Bless the LORD, O my soul." Psalm 104 opens like 103, but unlike Psalm 103, at the end it shifts to a series of singular cohortative verbs:

> Let me sing to the LORD as long as I live;
> Let me sing praise to my God while I have being.
> May my meditation [*śiaḥ*] be pleasing to him,
>   for I rejoice in the LORD.* (Ps 104:33-34)

One common call to praise involves the imperative plural use of the verb "sing" (*šir*). It first appears on the lips of Miriam, who calls to Moses and all Israel, in Exod 15:21, "Sing [pl.] to the LORD, for he has triumphed gloriously." As I have argued elsewhere, it is this call which evokes the song in 15:1-18, a song which begins, "Let me sing unto the LORD, for he has triumphed gloriously."*⁹ In this instance, the imperative call of Miriam

---

8. GKC §48e (their italics).
9. Exod 15:19-21 is introduced in Hebrew (and in earlier English translations) by the conjunction "for," and functions as an *analepsis* which belatedly gives the reason for the singing in 15:1-18. This *analepsis* (working like "for" in Gen 20:18) is obscured

PART TWO: Forays into a Biblical World

and her sister song-leaders evokes an individualized response in Moses and the people of Israel. We have already seen how, in Psalm 146, this call-and-response can occur within the individual psalmist. When we see this verb, "let me sing," sprinkled some ten times throughout the Psalms, and this same cohortative form, "let me," of other verbs likewise so sprinkled, we may conclude that the act of self-encouragement in worship of God, as an internalization of the call from others to such worship, is characteristic of biblical prayer.

Whether directed to others or to oneself, the call to praise most naturally arises in response to some helping action of God. But it can also occur in the midst of trouble or distress, prior to the appearance or the assurance of such help. Psalm 57 opens with a plea for God's help when surrounded by enemies, a plea punctuated by an assertion of confidence in that help (v. 3) and ending, even before help arrives, in a self-directed call to praise (vv. 7–9). The imagery in this call is striking:

> My heart is steadfast, O God,
> > my heart is steadfast.
> Let me sing and make melody.
> > Awake, my soul!
> Awake, O harp and lyre!
> > Let me awake the dawn.
> Let me give thanks to you, O Lord, among the peoples;
> > Let me sing praises to you among the nations.*

The imagery shows the psalmist engaging in an act of praise while still in an embattled situation. The soul will not be wakened to praise by the dawning of God's help. Rather, while the psalmist still lies in existential darkness, he will rouse his soul from its sleep; and his soul, moved to praise, will waken his instrument; and by this means he will awake the dawning of the Lord's favor.

As this psalm shows, the self-directed call to praise is not simply a response to divine action, it can become a prayer initiative in the midst of trouble. We see this again in Psalm 13, which begins with the desperate cry, "How long, O LORD? Will you forget me forever?" Acknowledging that God alone can intervene lest he sleep the sleep of death (v. 3), the psalmist somehow finds the ability to rouse himself to praise (vv. 5–6):

---

in the NRSV which omits the "for." For fuller discussion, see Janzen, "Song of Moses, Song of Miriam."

> But I have trusted in your steadfast love;
> > my heart shall rejoice in your salvation.
> Let me sing to the LORD,
> > because he has dealt bountifully with me.*

Sometimes the self-directed call in the midst of distress is not a call to praise but a call not to lose heart or lose confidence in God. Psalm 25 begins,

> To you, O LORD, I lift up my soul.
> O my God, in you I trust,
> > let me not be put to shame;
> > let not my enemies exult over me.
> Let none who wait for you be put to shame;
> > let them be ashamed who are wantonly treacherous.[10]
> (Ps 25:1–3)

The cry to God for help is accompanied by an assertion of trust in God. And this assertion is followed by a self-directed cry of encouragement not to become ashamed, not to allow the enemy's treacherous and hateful exultations against the psalmist (vv. 2–3,19) to become overwhelming. It is they, instead, who should be ashamed. It is easy to allow the hateful tauntings of one's enemies to prevail over one's own sense of acceptance before God. So the continuing appeal to God for help is punctuated again by the call to oneself not to become ashamed by them, but rather to take refuge from them in God (v. 20).[11]

The same pattern is seen in Jeremiah. Beleaguered by those who resist his prophetic message, Jeremiah calls on God for help, and affirms God as his refuge (Jer 17:14–17). Then he cries out:

> Let my persecutors be put to shame,
> > but let me not be put to shame;
> let them be dismayed,
> > but let me not be dismayed.* (Jer 17:18)[12]

This is his faithful response to God's exhortation to him in the context of his call to the prophetic task:

---

10. The third-person Hebrew verbs translated "Let none/Let them" are grammatically jussive and function as third-person indirect imperatives. The NRSV's "Do not let me" obscures the self-directed address here.

11. Other examples of this pattern include Pss 31:1–2, 11–18; 71:1.

12. Again, the NRSV's "Do not let me be ashamed/dismayed" obscures the self-directed address.

> But you, gird up your loins; stand up and tell them everything that I command you. Do not be dismayed by them, or I will dismay you before them. And I for my part have made you today a fortified city, an iron pillar, and a bronze wall, against the whole land—against the kings of Judah, its princes, its priests, and the people of the land. They will fight against you; but they shall not prevail against you, for I am with you, says the Lord, to deliver you.* (Jer 1:17–19)

On the one hand, Jeremiah is given God's assurance of protection against his enemies. On the other hand, he must appropriate that assurance; and one way in which he is able to do that is to refuse to allow his enemies' opposition and plots against him to throw him into such dismay that he abandons his calling. When, then, in chapter 17 he affirms God as his refuge, he also calls on himself not to be dismayed and not to be put to shame.

śiaḥ: The Means of Self-Encouragement

I have been exploring ways in which the act of prayer can include aspects of self-address that have the aim of encouraging one's self in a stance of faithful trust in God despite circumstances. One Hebrew word that includes this connotation (a term which appears on Hannah's lips) is the verb *śiaḥ* and its cognate nouns. We have already encountered this word in Ps 104:33–34:

> Let me sing to the Lord as long as I live;
> > Let me sing praise to my God while I have being.
> May my meditation [*śiaḥ*] be pleasing to him,
> > for I rejoice in the Lord.*

BDB offers the following meanings for the verb: "1. Complain; 2. Muse, meditate upon, study; 3. Talk about, sing."[13] As these summary definitions indicate, the word can be used to indicate a variety of moods and activities in various situations. Most often, what one muses or meditates on, or sings and talks about, is either God's *torah* (so, repeatedly, in Psalm 119) or something God has made or done in the past that evokes thanks and praise (Judg 5:10; Pss 105:2; 143:5; 145:5; Prov 6:22). But frequently the speaker is in distress, and here the word is typically translated

---

13. BDB 966–67. The definitions and discussions in *HALOT* 3:1319–21 and *TWAT* 8:757–61 are similar, but fuller, with bibliography of more recent analyses of the root.

"complaint" or "meditation" (in which case I take the psalmist's troubles to be the *focus* or the *basis* of the meditation) (see Pss 55:2, 17; 64:1; 77:3, 6, 12; 142:2; Job 7:11, 13; 9:27; 10:1; 21:4; 23:2; Prov 23:29). In these passages, I would suggest *śiaḥ* does not refer simply to the psalmist's complaint or trouble as such, but also to the psalmist's appeal to the character and the prior actions of God as a basis for God's hoped-for action in the present. The occurrences in Psalm 77 are particularly noteworthy. The psalmist, deeply distressed over some personal situation, has been crying out to God for a long time, and refuses either to give up or to accept false comfort. Three times he uses the verb *śiaḥ*; and the contexts suggest its rich range of meaning (cohortative singular verbs are marked with a †):

> I †think of [literally: *remember*] God, and I †moan;
> I †meditate [*śiaḥ*], and my spirit faints. (77:3)

> I consider the days of old,
> and †remember the years of long ago.
> I commune with my heart in the night;
> I †meditate [*śiaḥ*] and search my spirit. (77:5–6)

> I will call to mind [literally: *remember*] the deeds of the LORD;
> I will †remember your wonders of old.
> I will meditate on all your work,
> and †muse [*śiaḥ*] on your mighty deeds. (77:11–12)

Here, the verb *śiaḥ* describes a deep inner wrestling that moves the psalmist at times to cry aloud (v. 1), at times to moan inarticulately (v. 3), and at times to descend into soundless agony (v. 4). At the heart of the agony is the tension between the psalmist's experience and conviction of God's steadfast love and compassion and the absence of any sign of that love and compassion in the present situation. What is it that sustains the psalmist, that moves him to cry out unwearyingly and to refuse any false comfort (cf. Jer 6:14)? It is the recollection of what God has done in the past (Ps 77:13–20). This recollection is not a passive recall—it is an intentional act, carried out in a spirit of self-exhortation and self-encouragement. This is indicated in part by the repeated use of the singular cohortative form of the marked verbs. It is indicated also by the psalmist's self-dialogue, as he communes with his heart, mediates, and searches his spirit. When his heart or spirit questions whether God has spurned and forgotten him forever (vv. 7–9), he comforts and encourages himself with recollections of God's wonders of old (vv. 11–20).

That such activity, however plaintive, is an act of loyalty and trust in God is underscored in the book of Job. This sufferer repeatedly uses terms that we will hear on Hannah's lips, as he says, "I will not restrain my mouth; I will speak in the anguish of my spirit; I will complain [śiaḥ] in the bitterness of my soul" (Job 7:11; similarly 10:1; 23:2; cf. 7:13; 21:4). From the connotations of śiaḥ in Psalm 77, we may suppose that Job's śiaḥ included both references to God's past benevolent actions (e.g., Job 29), and the absence of any sign of God's benevolence in the present (e.g., Job 30). And the pervasive presence of such complaints in the psalms and elsewhere would suggest that they, no less than praise, are a sign of one's loyalty and trust in God.[14] But Job's friends think he has gone too far. As Bildad asserts in 15:4, "You are doing away with the fear of God, / and hindering meditation [śiaḥ] before God." Job is not the only one to have his śiaḥ misinterpreted. As the psalms attest, and as we shall see in the case of Hannah, the struggles of the soul with God and with itself before God can be badly misread by the very persons one might have hoped would read them aright.

Hannah's Meditation and Self-Address

Like Sarah, Rebekah, Rachel, and Tamar before her, Hannah is childless. Her plight is economic, religious, and social. Without children, she is vulnerable to economic hardship if her husband dies, she lacks the most palpable evidence of the favor of the God who is giver of the "blessings of breast and womb," and she is vulnerable to the condescending pity and belittling contempt of other women—especially her co-wife, Peninnah—who have themselves received such blessing in abundance.

The occurrence of several terms in the standard vocabulary of complaint, both in the narrative and on Hannah's lips, identifies her prayer as typical of anyone in distress. She weeps and she fasts (1 Sam 1:7-8); she is sad in heart (v. 8), bitter in soul (v. 10; RSV and NRSV: "deeply distressed"), hard pressed in spirit (v. 15; RSV and NRSV: "sorely troubled"), and afflicted (v. 11). She persists in prayer (v. 12) with great intensity, weeping copiously (v. 10) and pouring out her soul before God (v. 15), until, finally, she makes a vow, promising that if God will give her a child she will dedicate it in lifelong Nazirite service (v. 11). As she says in response to Eli's obtuse challenge, "I have been speaking out of my great anxiety

14. On the psalms of complaint as expressions of loyalty and trust, see Mays, *The Lord Reigns*, esp. 23-39.

[*śiaḥ*] and vexation all this time" (v. 16). This vexation (*ka'as*) may arise over her economic vulnerability and her sense of neglect at God's hands; but the sharpness of its bite comes from Peninnah, her rival (*ṣarah*, v. 6). In its masculine form (*ṣar*) this latter word occurs over two dozen times in psalms of complaint or psalms of thanksgiving for deliverance, where it is a term for the psalmist's adversary (e.g., Pss 3:1; 13:4; 27:2; 78:42). We may suspect that, like many adversaries in the psalms, and like Job's friends, Peninnah falsely accuses Hannah of some personal failing as the reason for her plight.

In all this, Hannah is presented as typical of the faithful Israelite at prayer amid prolonged distress. But two items of description are distinctive to her. First, her adversary's severe provocation[15] causes her, not simply to become irritated (v. 6 RSV, NRSV), but literally, "to thunder."[16] A bold image! In the Bible, the figurative use of the verb, "to thunder" aptly images God's self-revelation to creation or against God's foes (e.g., 1 Sam 2:10; 11:10). But it can also characterize creation's address to God in praise (Pss 96:11; 98:7; RSV and NRSV, "Let the sea roar"). Hannah's thundering is not in praise but in complaint.[17] By this bold image she appears as a soul mate of Job who is portrayed as a roaring lion (Job 3:24, where NRSV "groanings" translates *ša'agah*, a word usually indicating the roaring of a lion), a braying ass, and a bellowing ox (Job 6:5).

Before we examine the second peculiar phrase, we may note that Hannah applies to herself the word *śiaḥ*. RSV and NRSV render this noun, "anxiety" (v. 16). I take the term to refer not simply to her anxiety but to the whole range of her prayer activity, an activity whose range has been canvassed above and is exemplified in Psalm 77. In this activity, the person in distress appeals to God for help, questions God's faithfulness, asserts continuing trust in God, calls to mind God's goodness as manifest in previous acts of blessing or deliverance, and, addressing to oneself

---

15. Or "vexing," v. 6. Note verbal *ka'as*, like nominal *ka'as*, "vexation," in v. 16.

16. I see no reason to take *har'im* (Hiphil of *r'm*) as textually problematic. It is a vivid figure for Hannah's cry of complaint. As such, it resembles the usage of the verb *ša'ag*. The latter verb and its cognate noun refer literally to the roaring of a lion; but figuratively it can refer to the activity of rapacious rulers, foreign invaders, and Israel's God in judgment or redemption. The storm theophany in Job 37:4 portrays God as both roaring (*yiš'ag*) and thundering (*yar'em*). If intense, perhaps even aggressive human cries to God in distress can be portrayed with the verb *ša'ag* (Ps 38:9) and its cognate noun (Pss 22:2; 32:3; Job 3:24), they can surely be portrayed also as "thundering."

17. When George Herbert, in his sonnet on "Prayer," characterizes prayer as "reversed thunder," one may wonder whether he has Hannah in mind.

PART TWO: Forays into a Biblical World

communally-learned words of reassurance and comfort, rouses oneself to continue in prayer and even praise in the midst of affliction. That her use of the word *síaḥ* carries so rich a connotation—including self-soothing—is supported by the second phrase applied to her in a novel way, *dibber 'al leb*.

The narrator tells us that Eli observes Hannah's mouth. The NRSV translates, "She was praying silently," while RSV, closer to the Hebrew idiom, translates, "She was speaking in her heart" (v. 13). Neither is precise. The Hebrew idiom, *dibber 'al leb*, in all its other occurrences conveys a quite specific connotation. A speaker, seeking to assuage the grief, allay the anxiety, reassure the guilty fearfulness or mollify the outraged indignation of another party, literally "speaks upon [*al*] the heart" of that party. At times (as in Gen 50:21; Isa 40:1–2) the accompanying verb *niḥam* describes the speaker as seeking to convey "comfort." In its other ten occurrences in the Old Testament, this idiom indicates the reassuring address of one party to another.[18] Only here, in 1 Sam 1:13, are speaker and addressee *one and the same* person. Conventional translations such as "speaking in her heart," or "praying silently," obscure what Hannah is doing here.[19] In using this idiom to characterize her interior speaking or praying, the narrator identifies its specific point. Like the woman in Psalm 131, and like the speaker in Psalm 77 and repeatedly in Psalm 42–43, *Hannah is comforting and encouraging her complaining self.*

But this is not simply a bootstrap type of "self-help" for it is grounded in and appropriates the community's experience of and testimony to the goodness and active (if sometimes hidden) presence of God in the midst of trouble. In Psalm 77, the psalmist draws on the tradition of God's saving acts through the leadership of Moses and Aaron in the exodus from Egypt.

---

18. Gen 34:3; 50:21; Judg 19:3; Ruth 2:13; 2 Sam 19:8; 2 Chr 30:22; 32:6; Isa 40:2; Hos 2:16.

19. In its full listing of the idiom, *dibber 'al leb*, BDB 181 follows its citation of 1 Sam 1:13 with the comment, "seemingly from context *'al* for *'el, to her heart, to herself*" and references Driver, *Notes on the Hebrew Text of the Books of Samuel*, where Driver writes, "not, of course, as Is. 40, 2 al. in the sense of *consoling*" (14). Driver takes the usage to be "another instance of *'al = 'el* and cites Gen 24:45 (*dibber 'el leb*); and Gen 8:21 and 27:41 (*'amar 'el leb*) in support of the meaning, "to herself." But even this construal provides no ground for the translation, "speaking *in* her heart," which would presuppose the standard idiomatic preposition *be-*. Whether the preposition be *'al* or *'el*, it identifies not *the locus* of the speaking but *the one spoken to*. Driver's "not, of course ... in the sense of *consoling*" begs the question. In view of the self-consoling aspects of prayer canvassed above, there is no reason to rule out such a meaning here; and, in view of Eli's obtuse misconstrual of Hannah's prayer as inebriated self-address (on which see below), such a meaning is actually quite appropriate here.

Prayer as Self-Address

In Psalm 42, the psalmist draws on past personal experiences of participation in acts of worship that ended on a note of glad shouts and thanksgivings in the house of God. In Psalm 74, in the face of the desecration and destruction of the sanctuary, the psalmist draws on traditional imagery for God's activity in creating and ordering the world as a habitable place (vv. 12–17). And the exilic prophet, in obeying the heavenly summons to "comfort, comfort my people" and to "speak tenderly to Jerusalem" (*dibber 'al leb*, Isa 40:1–2), recalls to the exilic community's mind the richly diverse traditions of God's activity in creation, promises to and blessings upon the ancestors, deliverance in the exodus, leading and provision in the wilderness, entry into the land, and covenanting promises concerning David and Zion.

What traditions and stories would Hannah meditate on in her complaint over her barrenness in the face of Peninnah's prolific brood of children? Stories like those told about Naomi and Ruth? About the midwives in the midst of oppression in Egypt (Exodus 1)? About the female ancestors in Genesis—Sarah, Rebekah, Leah and Rachel, and Tamar? Especially, perhaps, the story of Rachel vis-à-vis her sister and co-wife, Leah?[20]

But Eli mistook her for a drunken woman. He may have seen others whose festive drinking during communal celebration brought them to such a state. But how could he mistake her *śiaḥ*, in particular her self-comforting, for inebriation? Perhaps because the two behaviors have a common aim. The Bible celebrates the capacity of wine to "gladden the human heart" (Ps 104:14). More somberly, the mother of King Lemuel in Prov 31:2–9 depicts the way in which those "afflicted" and in "bitter distress" (literally, "the bitter in soul," like Hannah in 1 Sam 1:10), attempt to drown their troubles in strong drink. Who has not witnessed a person "in his cups" speaking to himself in sympathetic aggrievedness over his lot, both lamenting his situation and consoling himself in it? In terms of both motivation and behavior, the actions of self-comfort in prayer, and self-comfort in wine or strong drink, may be confused in the eyes of the imperceptive bystander.[21]

---

20. In this paragraph, I am taking the biblical corpus as a total narrative world as presented by the final redactors, inviting us to reflect on the thematic interplay between its various parts. Historically speaking, during the last years of the Tribal Period a woman in Hannah's plight would have been familiar with the stories of earlier women in the community whose similar plight, brought to God in prayer, had been happily resolved.

21. Interestingly, drunkenness and Spirit-filled singing are contrasted in Eph 5:15–20 as ways of coping with "evil days." Given the equivalency, in Mark 2:6–8, of the

But the eventual outcomes are profoundly different. And one may ponder the question as to the spiritual legacy which Hannah bequeaths to Samuel, through her refusal to join the others in their eating and drinking until God answers her prayer, and through her vow that the son granted to her will become a Nazirite, forswearing wine and strong drink as well as the razor. Both her own song in 1 Sam 2:1–10, and Samuel's epoch-making vocation in Israel, are humanly grounded in her faithful, prayerful travail in 1 Samuel 1. In a retrospect opened up by Paul's words in Romans 8, one may hear her *śiaḥ* as her plaintive yet hopeful participation in the groaning and travail of the whole creation. In its internal thundering, inaudible to the ears of Eli, one may also hear her *śiaḥ* as the echo within her own heart of the Spirit of God who, with unutterable groanings, makes intercession for her and her people. The outcome, of course, in Hannah's case, is the song of thanksgiving in 1 Sam 2:1–10.

---

expressions "in their/your hearts" and "within [or to] themselves" (RSV; cf. NRSV), and especially in view of 1 Cor 14:28, "if there is no one to interpret, let them be silent in church and speak to themselves and to God," perhaps Eph 5:19 may be translated, "singing psalms and hymns and spiritual songs *to* (or *within*) *yourselves,* singing and making melody to the Lord *in your hearts.*" Shades of Hannah!

6

The Root *škl* and the Soul Bereaved in Psalm 35

THE PLAINTIFF IN PSALM 35 appeals to God in the face of certain persons who have requited him "evil for good" (v. 12a), so that, as RSV translates v. 12b, "my soul is *forlorn*." So translated, the form *šekol* names an intensely felt loss of the society of those to whom the psalmist had formerly been very close. In vv. 13–14 he describes what his feelings and actions had been toward them when they were sick: He had donned sackcloth, fasted, and prayed "as though for my friend or my brother." But even these similes are not adequate to describe the intimacy or the intensity of his felt relation to them then, so he goes on to say, as translated in RSV (v. 14), "I went about as one who laments his mother, bowed down and in mourning." Yet now that he is in trouble, his own "stumbling" (v. 15) has evoked from them only glee, slander, and mocking. Thus, the "good" he had shown them was his sympathetic feeling and prayer for them in their troubles; and the "evil" they have shown him in return for this "good" is that they have failed to show him like sympathy and support, but instead have taken his troubles as an occasion to turn on him.[1]

In this chapter, I want to explore the precise character of the plaintiff's former felt relation to those whom he now experiences as his foes. Specifically, I want to explore the connotations of the two self-descriptions, *šekol lenapši* (v. 12b) and *'abel-'em* (v. 14b), which RSV translates, "my soul is forlorn" and "one who laments his mother." I shall argue that these self-descriptions are to be understood in closest relation to one another, each phrase illuminating the other, that their combined meaning conforms to the general use of the root *škl*, and that that general use does not in fact

---

1. The book of Job plays out this scenario in exhaustive detail, beginning in Job 4:3–7 with Eliphaz's initial appreciation of Job's former sympathetic concern for those who had stumbled.

authorize a translation of *'abel-'em* as "one who laments his mother." I shall begin, then, with an examination of the occurrences of the root *škl*.

škl as Maternal Bereavement

In current English parlance the term "bereavement" refers to any painful loss through death, whether of parent, spouse, child, or close friend. But *škl* in the Old Testament refers, with one exception,[2] to the parental loss of progeny through miscarriage or sword. More specifically, in a great number of instances it is explicitly mothers who are bereft in one way or the other.[3] 1 Sam 15:33 is noteworthy for the way it emphasizes not only the specific identity but also the specific social context of the bereaved: "As your sword has made *women* childless (*škl*), so shall *your mother be* childless (*škl*) among women." If a mother's joy in child-bearing is celebrated with a particular sort of sympathetic joy among other women (Ruth 4:14-17), her earlier bereavement has its primary sympathetic response in the same circle (Ruth 1:19-21).

In eleven instances the sword is the means of bereavement. In 1 Sam 15:33, and in Jer 15:7, 18:21, it is explicitly the mothers who are so bereaved. Ezekiel 5:17 comes in a passage where the bereaved community can be addressed in the (inclusive?) masculine plural or, where the city as such is in view, in the feminine singular. The verse itself reads, "I will send *famine* and *wild beasts* against you [masc. pl.; LXX sg.], and they will bereave you [fem. sg.]; *pestilence and blood* shall pass through you [fem. sg.]; and I will bring the *sword* upon you [fem. sg.]." Significantly, the shift from masculine plural to feminine singular addressee comes with the introduction of the theme of bereavement. As in Isa 47:8-9, the city is depicted as a mother bereaved of her children. Ezekiel 14:15 comes in a passage concerning God's action against any "land" (*'ereṣ*), an action, as in 5:17, carried out by means of *famine, wild beast, sword, pestilence and blood*, where the pronominal references to that land are all feminine singular. Insofar as this passage culminates in God's four acts of judgment against the city Jerusalem, the feminine singular pronouns in 14:15 are not merely in grammatical accord with the noun *'ereṣ*, but pertain again to the specific figure of maternal bereavement. The occurrences in Ezekiel

---

2. Mal 3:11; see below.

3. This is the case in the following passages: Gen 27:45; 31:38; Exod 23:26; 1 Sam 15:33 (2x); Job 21:10; Prov 17:12; Song 4:2, 6:6; Isa 47:8, 9; 49:20, 21; Jer 15:7; 18:21; Hos 9:14; 13:8.

36 come at the end of a passage that begins as an address to the mountains (masc. pl.) of Israel, then refers to "my land" (ʾarṣi) in v. 5, then reverts to mountains and a masc. pl. pronominal reference. With the threefold reference to the land's bereavement (vv. 12, 13, 14), the pronouns and verbal forms initially are masc. pl., but then soon shift to fem. sg. and remain that way, again consonant with the gender-specific connection of škl. Jeremiah 50:9 similarly portrays the city Babylon in feminine terms as attacked by bereaving warriors. In Lam 1:20 it is feminine Zion who cries out, "In the street the sword bereaves; / in the house it is like death." In Deut 32:25 there is no prominent clue to the subject of bereavement; yet the similarity to Lam 1:20 is suggestive: "In the open the sword shall bereave, / and in the chambers shall be terror." In both instances, apparently, it is the mothers at home who, on learning of their bereavement, experience it like their own death. Given all these feminine bereavements by the sword, and given the wider context in vv. 11–14, Hos 9:12 seems also to have mothers in view.

In 2 Kgs 2:19, 21, it is unclear as to whether the bad water of the city's spring is thought to cause "the land" as such to be unfruitful, or whether it is thought to cause miscarriages in the women who drink from it. In any case, the usage here conforms generically to the passages surveyed above. In Mal 3:11 God says, "I will rebuke the devourer for you, so that it will not destroy the fruits of your soil; and your vine in the field shall not miscarry." Again (especially in view of such figures as Ps 128:3) the usage here conforms to type. Thus, the occurrences of škl surveyed to this point all explicitly or implicitly refer to mothers bereaved of their children. Only in three instances are the bereaved identified explicitly as male: Jacob in Gen 42:36 and 43:14, and the psalmist in Ps 35:12. In my view, this use of škl involves an extension of its customary reference. The keys to understanding this extension of meaning, in its application to males, lie in the imagery of the psalm and in current psychoanalytic understandings of the developmental basis of human felt relations.

Compassion Requited and Unrequited

The essence of the psalmist's complaint against his enemies is that "they requite me evil for good" (v. 12). While this expression may be used where the social context for the "good" action is not specified (Prov 17:13), or where that context is not one of close kinship (1 Sam 25:21), it occurs in Ps 38:20 vis-à-vis "my friends and companions . . . and my kinsmen" (v. 11, ʾohabay wereʿay . . . qerobay). In Psalm 109, the enemy are not named as kith or kin; but that they are closely related may be implied in vv. 4–5:

> In return for my love they accuse me, even as I make prayer for
> them. So they reward me evil for good, and hatred for my love.

The psalmist's action toward them has been (as in Ps 35:13–14) the action of one who is intimately related to them, a relation within which one in turn ought to be able to expect similar treatment. Similarly, in Jer 18:20 the prophet claims to have "stood before thee to speak good for them, / to turn away thy wrath from them." Those who have requited evil for good are here his fellow citizens; but elsewhere they are his own close kin (Jer 11:21–23). In Gen 44:4 the speaker, Joseph, ostensibly appeals to nothing more than his prior benevolent action, in charging his hearers with returning evil for good. But the reader knows—and by Gen 50:17 Joseph's hearers confess—that they have returned evil to a brother; and this in retrospect gives Joseph's "good" the character of an action performed within the bonds of an intimate kin relation. These passages, then, suggest that the psalmist in Ps 35:12, in saying "they requite me evil for good," is not thinking simply of a cold calculus of *do ut des,* of unit action and unit response, between parties who have no other basis of relation (one might say, a *Gesellschaft* relation), but of an organic kin relation (a *Gemeinschaft* relation) within which one does for the other whatever the other's situation calls for, and likewise one ought to be able to assume unreflectively that the other will do for one what is appropriate to one's own situation. Writing of the character of kinship (and, by analogical extension, of covenant) relations, Frank M. Cross quotes Meyer Fortes who says, "Kinship predicates the axiom of amity, the prescriptive ... altruism exhibited in the ethic of generosity.... Kinsfolk are expected to be loving, just and generous to one another and not to demand strictly equivalent returns of one another." Cross writes that *ḥesed,* "as used in early Israel, a society structured by kinship bonds, covers precisely this semantic field. On the other hand, when extended in use outside the kinship group, behavior required by or appropriate to a kinship relationship becomes 'gracious' or 'altruistic' behavior."[4] As is implicit in what Cross goes on to show, this "gracious" behavior consists in extending to the outsider the sort of organic relation and mutuality characteristic of kinfolk.

This organic relation is portrayed in Ps 35:12–14 in a most suggestive manner. When these other persons were sick, the psalmist did what his relation to them implicitly called for: "I wore sackcloth, / I afflicted myself with fasting." The action here is one of sympathy in the original (not the contemporary jaded and sentimentalized) sense of the word: the

---

4. See Cross, *From Epic to Canon,* 5.

psalmist *suffered with* them, and expressed that sympathetic identification both bodily (sackcloth and fasting) and in spirit (prayer). To quote Cross again: "Kinship was conceived in terms of one blood flowing through the veins of the kinship group.... Kindred were of one flesh, one bone" (Cross then quotes W. Robertson Smith, "The whole kindred conceives of itself as having a single life").[5] Of the requirement to love a covenant partner as oneself, and of related covenant terms, Cross writes, "The language of covenant, kinship-in-law, is taken from the language of kinship, kinship-in-flesh."[6] In Ps 35:13 we see this from within the social consciousness of the speaker. The others' sickness was his own sickness; and in turn his prayer was not only "for" them but "with" them as giving their own cry of suffering and need its social amplitude.

These acts of "good" were undertaken "as for my friend or my brother." But that intimacy, close as it is, does not yet go to the heart of the psalmist's sense of identification. To express that, the psalmist must resort to the language of the mother-child relation, which, as contemporary psychoanalytic studies are making increasingly clear, is the primal human relation and the matrix of all other social relations. But has the psalmist lamented for others in their sickness as one who laments his mother (so, e.g., RSV)? Such an image does not make sense. That we ought to translate, "I went about like a lamenting mother,[7] / bowed down and in mourning," is suggested by v. 12: "They requite me evil for good; / my soul is *bereft* [*šekol*]." That is to say, v. 12b and v. 14b present one and the same mother-child figure at two stages of that relation. The psalmist had cared for them as a mother would care for her ill child—to the point of lamenting the feared prospect of the child's death. Now he feels their desertion with an intensity which only the language of maternal bereavement can adequately name: "My soul is bereft."

Such a reading of the three similes, in increasing intimacy and intensity of relation—friend, brother, mother—may now suggest a way of understanding the last line in v. 13, *utepillati 'al ḥeqi tašub*. BDB (in its article on *šub*) renders this line literally, "my prayer turned upon my bosom," but RSV reflects commentators who propose "My prayer was uttered with bowed head"—a strained reading at best. I suggest that the word "bosom"

---

5. Ibid., 3. The citation from Smith is from *Kinship and Marriage in Early Arabia*, 46.

6. Cross, *From Epic to Canon*, 11.

7. Literally, "like the lamenting of a mother," taking the second noun in *ka'abel-'em* as a subjective rather than an objective genitive.

refers to the locus of the prayer, or rather, the locus of those ill others in the psalmist's prayer for them. That is, to pray for others in need is to bear them on one's bosom. (The classic formal and symbolic biblical embodiment of this experience lies perhaps in the provision for priestly garments in Exodus 28. In particular, the two shoulder-pieces are to include twelve stones inscribed with the names of the twelve tribes [six on each shoulder], "and Aaron shall bear their names before the LORD upon [*'al*] his two shoulders for remembrance" [Exod 28:12]. The breastpiece likewise is to contain twelve stones similarly inscribed. So Aaron shall bear the names of the sons of Israel in the breastpiece of judgment upon [*'al*] his heart, when he goes into the holy place, to bring them to continual remembrance before the LORD" [28:29].) Such an act is of course not peculiar to mothers; but its primal instance is the mother's sympathetic bosom-comfort of the troubled child.[8]

We may compare another plaintiff's recourse to the imagery of this primal relation, in Ps 22:9–10, translated literally:

a. Thou art my bursting[9] from the belly;
   b. thou didst entrust me upon my mother's breasts.
   b'. Upon thee was I cast from the womb,
a'. and from my mother's belly my God art thou.

---

8. This is imaged vividly in Psalm 131, on which see further below.

9. The form *goah* is participial, from the root *giah*, "burst forth," of waters gushing from an opening, or figuratively of childbirth (elsewhere, Job 38:8; 40:23; Ps 71:6; Mic 4:10; Ezek 32:2) or the snorting of nostrils (Job 41:20). The figural associations are suggestive for our understanding of the presence of the spring Gihon (*gihon*) just outside historical Jerusalem, and within the Garden of Eden in Gen 2:13. It is as though that garden is the mythical reflex of the primal human space constituted by the relation between mother and child. In psychoanalytic terms, the garden images the shared "space" of the primal relation between mother and child, prior to differentiation into objectivity and subjectivity, the locus of shared "illusion" out of which, as Winnicott and Pruyser among others argue, human culture develops, and in continuing dynamic relation to which culture renews itself. (See, e.g., Winnicott, *Playing and Reality*; and Pruyser, *The Play of Imagination*.) Even after the presumably unwalled village beside historical Gihon became the walled city of Jerusalem, the city continued to draw on Gihon, by construction of the Siloam tunnel which brought its waters through more than a thousand feet of limestone rock, inside the walls and into the Pool of Siloam. It was when the city no longer relied on these waters, but resorted to other means of survival in its current international context (Isa 8:5–8), that Isaiah announced judgment on it. This sociological correlate of the psychoanalytic themes of the primal mother-infant relation is understandable within the sort of perspective Peter Homans offers in *The Ability to Mourn*.

The Root škl and the Soul Bereaved in Psalm 35

The image here is so dense and pivotal as to call for extended comment. The four lines form a chiasm, as indicated in several ways:

(i)	(a) from	(ii) belly	(iii) thou art
	(b) upon	breasts	entrust me
	(b') upon	womb	I was cast
	(a') from	belly[10]	art thou

Especially noteworthy is the almost obsessive focus on the physiological imagery taken together with the prepositions. The psalmist's sense of critical separation from a formerly intimate nurturing and protecting presence,[11] and corresponding cry for re-connection, is deeply resonant with the embodied memory of the primal crisis of separation from the mother and subsequent re-connection at her nursing breast. Of course, the connection at the breast is not constant in the same way as in the belly; for in the latter location the fetus is nourished continuously through the umbilical cord, and the surrounding walls are ever there, whereas the breasts offer the punctuated constancy of regular feeding after the break of sleep and resting or fussing wakefulness.

The verb in line (b), then, identifies the psycho-physiological context in which trust is first required and exercised. Developmentally, then, trust is initially an implicit reliance upon the reappearance of the breast when needed. The chiastic structure of these four lines highlights the semantic and existential interaction in the middle two lines:

    b. thou didst entrust me upon my mother's breasts.
    b'. Upon thee was I cast from the womb.

The deepest analogy the psalmist can appeal to for his need to trust in God, is the separative crisis of his birth and his reconnection to his mother at the breast: "upon my mother's breasts // Upon thee." In this psalm, then, prayer is the act of entrusting oneself upon God, as God had entrusted one upon one's mother's breasts at birth. The social character of this piety is reflected in vv. 3–5: Yahweh is enthroned on the praises of Israel, whose

---

10. For breasts and womb as a poetic pair, see, e.g., Gen 49:25.
11. See also Ps 22:1, 11, 19. The verb "forsake" in v. 1, 'azab, can describe a variety of situations of abandonment. For its use in the maternal relation, see, e.g., Job 39:14; Ps 27:10; Jer 14:5; and, of God, Isa 49:14–16; 60:15–16; cf. Isa 66:10–11 as complemented by 66:12–13.

PART TWO: Forays into a Biblical World

faith in turn is grounded in the trust which their ancestors placed in God for deliverance.

The same image is taken up at a later stage of the mother-child relation, in Ps 131:2 which we may translate, following Patrick D. Miller[12] and others as follows:

> I have calmed and quieted my soul
> like a weaned child upon its mother.
> Like the weaned child upon me is my soul.

In such a reading, the psalmist, speaking as a mother, quiets her soul in prayer the way a weaned child quiets itself upon its mother's bosom—indeed, like the child who at that very moment is quieted upon her own bosom. That the child in question is not a nursling is indicated by the fact that we have here, not a "sucking child" (*'ul* or *yoneq*), but a *gamul,* one who has been weaned. If birth is the first great separative crisis, weaning is the second.[13] For during breast-feeding, the infant at least still takes its food directly from its mother's body; but thereafter, the maternal body connection is maintained only through holding and the physiological resonances implicit in the mother's soothing and encouraging voice.[14] But if, as in Psalm 22, the first instance of prayer is the infant's cry for the mother who is "far off" to draw near to nurse, Psalm 131 suggests in retrospect that the first instance of intercession was the mother's response to that cry in draw-

---

12. Miller, "Things Too Wonderful," 244–45.

13. On the significance of "weaning," whether literally in the instance of breast-nursing mothers, or in any case as a critical transition in one's psycho-social development, see, e.g., Winnicott, *Playing and Reality,* 6–14.

14. Compare Homans' remark (*The Ability to Mourn,* 48) that "hearing engages the unconscious far more directly than seeing." One is reminded of Whitehead's comment, in *Modes of Thought,* that the eye sees bodily surfaces without (for the great majority of people most of the time) the subject's awareness of the physical stimulation of the eye, whereas the ear resonates with perception of sounds which originated as reverberations within the bodily depth of the speaker (one may compare John 20:11–16, where Mary does not recognize Jesus at sight, but only at his speaking her name). The foundation, then, for the continuing connection between mother and child after weaning may lie in the universal practice of holding and crooning lullabies, especially in response to the infant's cries. The crisis comes when such a response is not heard (cf. Ps 22:3). A remarkable perspective on all this is given in the terrible accident in which a child in Texas fell down an abandoned well shaft, and was alone there in the cramped dark for days until rescuers were able to tunnel to where they could rescue her. In the meantime they lowered a microphone so that her family could speak to her. She later said that, when she was most frightened, she calmed herself by singing songs her mother had taught her. (Cf. chapter 5 above.)

The Root škl and the Soul Bereaved in Psalm 35

ing near and nursing the child. If the infant pre-consciously extrapolated from "upon my mother's breasts" to "upon thee" (Psalm 22), the mother at that time for her part extrapolated from "upon me" to "upon thee." Thus the mother's feeding of her infant became her intercessory prayer for it; and that intercessory prayer became the spiritual matrix and invitation for it to move from its mother to God. Thus the infant's maternal horizon becomes the *limen* or threshold for its move to a divine horizon. Since weaning, a similar interaction continues across a wider separation—especially where comforting and reassuring words contain references to the reliability of God despite immediate conditions and circumstances.[15] It is such primal imagery, I suggest, that also surfaces in Ps 35:13c.[16]

---

15. It must be acknowledged, of course, that fathers may and often do lullaby their children, and participate in spoon-feeding and other aspects of infant care, and that thereafter they too offer important vocal reassurance and encouragement. The imagery in 1 Thess 1:7–12 is very much to the point here. Even in this passage, however, the movement from maternal to paternal figures mirrors both the temporal and the (psycho-)logical relation between maternal and paternal offer of "encouragement" and "exhortation." In passing, it may be noted that the textual question at v. 7 may be illuminated from current psychoanalytic understandings. The text reads, depending on the manuscript witness, "we became (a) gentle/(b) babes among you, like a nurse taking care of her children." Hans Loewald writes of the need for the parents to provide the infant and child with "approximately appropriate psychological conditions" for its development, conditions that must take into account the fact that the child's external world, including its parents, "are of a degree of integration and differentiation utterly incongruous with and superior to the level of integrative functioning of the infant and child." In order to bridge this gap, the parents must engage in what, in a theological context, John Calvin spoke of as the divine accommodation and condescension. "The parental, and in the early stages especially the maternal, supply of satisfaction of needs and of support and channeling of maturation processes constitute *a regressive movement on the part of the parents* that minimizes the objective discrepancy and allows the infant to remain in integrative interaction with the environment" (Loewald, *Papers on Psychoanalysis*, 22, emphasis added). Needless to say, Paul had not read Loewald; but like all adults who try to get close to infants when not observed by other adults, he was surely familiar with parental "baby talk." All of this is implicit in the textual reading "babes," which RSV relegates to a footnote, but which (no doubt primarily on the strength of P65) Nestle-Aland, *Novum Testamentum Graecae*, 26th ed., adopts in the text.

16. Compare the connotations of Hebrew *raḥamim*, "mercy, compassion," as cognate with *reḥem*, "womb." On the relation, in ancient Near Eastern traditions, between maternity, compassion, intercession, and lamentation, compare such scenes as: (a) the appeal of Tiamat to Apsu and against his vizier, on behalf of the young gods who, however noisy and rebellious, are their children and therefore should be attended to kindly—*tahiš*; (b) the repeated refrain, in Sumerian and Akkadian laments spanning almost 2000 years, wherein the goddess of the city weeps over her destroyed city like a cow for her calf and a ewe for her lamb (in the so-called Balag lamentations published

It may be objected that if the masculine singular pronouns in Ps 35:10, 25 are any guide, that psalmist is a man. Does this not call the above construal in question? Or if not, how are we to understand a man's use of this language?

Suffering the Loss of Matrixal Connections: A Psychological Perspective

Here we may be helped further by contemporary studies in human psycho-social development. Whereas Freud concentrated his analysis chiefly on the oedipal crisis, centering on the relation between father and child (especially between father and son), his successors in the psychoanalytic tradition have taken up his preliminary work on pre-oedipal relations, and in fact have come to consider the oedipal situation—for all its undoubted importance—secondary in significance to the pre-oedipal dyadic relation between mother and child.

What is of particular interest for our topic is that the pre-oedipal relation, especially in its earliest phase, is marked by the psycho-social "oneness" of mother and child. This oneness is spoken of, variously, as involving "merger" of identity, or as at this stage "undifferentiated" (especially, but not solely, on the part of the child) into self and other. At such a phase, "sympathy" has the character Whitehead describes, in which one feels the feelings of the other, indeed, in which the bonds of one's relation to the other are the feelings one shares with the other, arising in the other and entering into one to become one's own feelings. To recur to the language of W. Robertson Smith, quoted above, the "one blood" of the mother and child is transmuted into a psycho-social sympathy which interweaves with the continuing physiological interaction in the feeding and holding at the bosom. This bond becomes the matrix for all other social bonds,

---

by M. E. Cohen [*The Canonical Lamentations of Ancient Mesopotamia*], frequently it is the goddess who appeals to Enlil the storm god on behalf of her city); (c) the Canaanite king Keret's word to his son not to weep or lament for him, but to call his sister to weep and lament for him, and this for two reasons—she is a daughter whose something-or-other [*ḥmḥh*] is strong, and she is pitiful or compassionate—*rḥmt*, and (d) similarly, the Canaanite gods' lamentation over Baal's death at Mot's hands, coming to a climax as the divine damsel (that is, *rḥm*) Anat searches for him "like the heart of a heifer for her calf, like the heart of a ewe for her lamb." Such traditions suggest that a social feeling in which all may participate finds its most intense location, its radical matrix, in the compassion of "a cow for its calf, a ewe for its lamb," and, as Tiamat primally exemplifies, a mother for her child.

such that all subsequent instances of care, whether from one's father, older siblings, wider kin or community—or even from a stranger—are experienced and "known" within the hermeneutical matrix established in the primal dyad.[17]

Insofar, now, as the infant in the primal relation learns not only to receive, but also to respond and even to initiate, it does so by a sympathetic reaction that has the character of imitation (again, in the original and not the contemporary faded sense of the word). Long before there is sufficient differentiation, let alone capacity, for reflective calculation of the sort presupposed in calculated *do ut des* or "tit for tat" (unit-for-unit exchange), the interaction has the nascent character of mutuality and reciprocity which in transmuted forms marks adult kin relations as characterized in the quotation above from Fortes. Insofar as male as well as female infants begin their psycho-social formation (or "imprinting") in this dyadic relation, it provides the hermeneutical matrix for male no less than female actions of sympathetic identification with and care of others.

But if the primal relation is one of unity and identification,[18] the first psycho-social crisis subsequent to the birth trauma comes with weaning, in both the physiological and the psycho-social sense. Sooner or later the infant and the mother must accept their own and the other's separate-ness and individuality. Even where this acceptance, this weaning, is achieved optimally, it is marked by deep feelings of mourning. Again, it appears, this transition and its concomitant deep feelings—as of the rupture of primal bonds—are matrixal, hermeneutically speaking, for all subsequent experiences of separation and loss. Thus it is not surprising that, on hearing of the loss of all his property and all his children, Job cries out, "Naked I came from the womb, and naked I shall return there." If these words point even behind the separation at weaning, to the birth-trauma itself, this means that they evidence at the deepest embodied level the way in

---

17. It may well be, of course, that in a given individual instance an infant or child may receive "mothering" from one not its biological mother. It remains the case that the hermeneutical matrix of all such social care-giving is the typical mother-child relation.

18. Recent studies suggest that differentiation between infant and mother, in a stage-appropriate mode, is present already at birth and even, it would seem, prenatally (see Stern, *The Interpersonal World of the Infant*). This does not affect the basic point here, except to indicate that the primal relation is not one of mere undifferentiated unity and identification, but of a unity that arises by virtue of rapport at the level of embodied feeling.

which the mother-child relation is matrixal for subsequent social relations and the pain of their rupture.[19]

Rage and the Bitterness of Unrequited Compassion

Such a contemporary understanding of the etiology of social felt relations, I suggest, helps us to appreciate how the male psalmist in Psalm 35 can resort to — or is driven to — the language of maternal care and bereavement in penetrating to the core of his relations with his erstwhile intimate associates who have now become his enemies. If all separation is marked by mourning, such mourning is peculiarly bitter over a separation marked by the moral and spiritual betrayal of requiting evil for good.[20] The only adequate analogue for such mourning is that of a mother who had borne the pains of her children upon her own bosom, but who now, in her own troubles, finds her children turning against her. Such a perspective on Psalm 35 allows us to understand Jacob's two references to his bereavement in Gen 42:36 and 43:14, not as qualifying or subverting my argument for *škl* as in origin and in primary social location mother-specific, but as illustrating how men too may find themselves drawn so deeply into the felt character of their most intimate social relations as to be driven to maternal language in order to give it adequate voice.

In this last connection the image of a she-bear bereaved (*šakul*) of her cubs is of particular interest. In the first instance acting in her own maternal right, she can also become a figure for male rage. The figure occurs three times in the Bible:

1. The proverb cautions, "Let a man meet a she-bear robbed of her cubs, / rather than a fool [*kesil*] in his folly [*'iwwalet*]" (Prov 17:12). The full force of this specific piece of proverbial wisdom may be gained from another proverb that states the matter generically, "Wisdom builds her house, / but folly [*'iwwalet*] tears it down" (Prov 14:1). According to this proverb, the essence of folly is that it is in no sense constructive or home- and

---

19. These words of Job are often taken to be a conventional form of response to sudden loss and grief, rather than Job's own spontaneous coinage. Insofar as Job is the epitome of typical human suffering, and insofar as human conventions of speech inscribe common experiences in forms of language that go more deeply to the heart of things than most individuals' linguistic abilities can take them, the question of convention versus spontaneous feeling is perhaps moot.

20. Cf. Job 3:20 after 3:12, and the elaboration in 6:4-7. See further Janzen, *At the Scent of Water*, esp. chapter 5: "Lust for Life and the Bitterness of Job."

society-building, but it only deconstructs and leaves matters in disarray. In contrast, what is at the heart of all of wisdom's specific actions is the generic concern to build a house for life.[21] When, then, one is advised to wish to meet a bereaved she-bear rather than a fool, a number of things are implied.[22] First, there is something even worse than a bereaved and enraged she-bear. Secondly, what is worse falls outside the realm of recognizably social feeling and behavior, and goes by the name of folly.[23] Nowadays, we might call such a person a psychopath or sociopath, whose destructive rages have no redeeming positive aspect, no socially constructive end in view. Thirdly, the feelings of the bereaved she-bear, bordering on those of the fool, mark the limit of a rage that does not become insanity. Fourthly, the she-bear's rage, however temporarily punitive, is grounded in, and a situationally specific secondary form of, those primally positive maternal feelings which are the matrix of all social relations, the masterplan by which "wisdom builds her house" (Prov 14:1).

When Absalom conspires against his own father, and succeeds in driving him temporarily from his royal seat in Jerusalem, David and his mighty men are said to be enraged, not like a he-bear robbed of his prey or his territory, let alone of his mate or his cubs, but like a she-bear (2 Sam 17:8). Not simply a bare analogy, the image exegetes David's feelings with profound insight not only into their character but also into their matrix and development. For he who now is king in Jerusalem was once a shepherd who defended his sheep against lions and bears, and who later defended Israel against Goliath (1 Sam 17:33–37). But David first learned such feelings and their concomitant actions by feeling his mother's maternal feelings, directed at others on his behalf, and at times no doubt directed at him!

It is not surprising to find such a proverb invoked in 2 Samuel, a book where wisdom erupts frequently in pithy sayings and parables. So too, the

21. Thus, for example, Prov 9:1–6; cf. Isa 45:18–19; 55:1–5.

22. Do I press the imagery of Proverbs too hard? Given how wisdom sayings convey the most profound organic insight into complex social dynamics and relations through what on their face seem only tired homely saws, it is unlikely that one can *over*-read them, however badly one may *mis*-read them.

23. Sociologically, the distinction between "crime" and "abomination" is the distinction between acts that are wrong and punishable but do not banish the perpetrator from the community, and acts that mark the perpetrator as "among us but not of us," as "beyond the pale" of the social matrix with its common values and identity. The "fool," likewise, is not just a wrongdoer, but rather one who in conduct and understanding is manifestly oblivious to the "wisdom" that gives the community its identity and meaning.

proverb enters fittingly into the book of Hosea which is laced with wisdom language and perspectives. At one point, for instance, the historical example of Jacob's deviousness and change at Jabbok is held up for Ephraim to emulate (Hos 12:2–14). As I argue elsewhere,[24] in Genesis 28 Jacob's nocturnal struggle and change of name echoes his original struggle in the womb with Esau and his first naming, and so at Jabbok he may be said to "enter the womb a second time when he is old and be born" (cf. John 3:4). But according to Hos 13:13, when "the pangs of childbirth come for him," that is, for contemporary Ephraim, unlike his ancestor "he is an *unwise* son; for now he does not present himself at the mouth of the womb." Recourse to proverbial analogies is evident a few verses earlier, when Yahweh says, "I will be to them like a lion, like a leopard I will lurk beside the way. / I will fall upon them like a bereaved bear, / I will tear open their breast" (Hos 13:7–8). Does Yahweh's rage border here on divine insanity?[25]

The repeated testimony in this book to divine redemptive love on the other side of historical wrath suggests that Yahweh's wrath is not folly.[26] It remains a situationally specific secondary and temporary form of positive social (whether covenantal or kinship) feeling.

The bearing of this idiom on my discussion of Psalm 35 is that the image of a she-bear bereaved of her cubs identifies rage in its most extreme form. Such rage is a measure of the mourning that finds its primal exemplar in a mother's loss of her child. The biblical word that comes closest to naming the relational matrix of such rage and such mourning is perhaps "zeal," *qin'ah,* at its base a love that is so strong and fierce that not even death and the grave can take its measure (Song 8:6). It is such love, I suggest, that bound the psalmist to his erstwhile associates, a love that, now betrayed, takes the form of mourning on the one hand and a petition for Yahweh's vengeance on the other.

---

24. See Janzen, *Abraham and All the Families of the Earth.*

25. In the ancient world such a question would not be frivolous—compare the figure of the Mesopotamian god Erra, thematically summarized and historically contextualized in Jacobsen, *The Treasures of Darkness,* 226–28; and Roberts, "Erra—Scorched Earth."

26. This is precisely the point in the Epilogue to Job, where Yahweh's wrath is kindled against the three friends for the nonsense they have spoken to Job in his affliction. According to Job 42:8 (and in spite of most translations), Yahweh says, literally, "I will accept his prayer not to deal with you according to folly." The idiom here is, "to deal with someone according to X", where X is an abstract noun (*hesed, tob, ra',* or the like) that characterizes the subject of the verb and not the prepositional object.

7

As God Is My Witness

Another Look at Psalm 12:6

What, Precisely, Does God Promise the Psalmist?

As HAS LONG BEEN recognized, the line ʾāšît bĕyēšaʿ yāpîaḥ lô in Ps 12:6 is problematic. The construction *šyt* + *bĕ-* has been taken to indicate that the object of the verb (in this instance unexpressed) is to be placed somewhere (in this instance "in safety").[1] On the assumption that *yāpîaḥ* is a Hiphil form of the verb *pûaḥ*, "breathe, blow," the expression *yāpîaḥ lô* has been thought to indicate that the deliverance which the psalmist has asked for (v. 2, *hôšîʿāh*) is something the psalmist has been "panting" or "longing" for. Alternatively, *yāpîaḥ lô* has been connected with those who "speak *šāwʾ*"[1] against the psalmist (v. 3) and taken to mean "breathe out" in rage or with malign intent.[2] Some proposals involve emendation.

In recent years, the form *yāpîaḥ* in other passages has been construed on the basis of Ugaritic *yph*, as meaning "witness," whether as a verb or as a substantive.[3] On this basis, Patrick D. Miller has offered a fresh analysis of the line in Psalm 12, which he translates, "I will place in safety the witness in his behalf."[4] Miller's explication of the meaning of the line may be summarized as follows: The psalmist is being subjected to "lies, deceit, and boasting." In particular the complaint, "they utter lies [*šāwʾ*] to each other," is "reminiscent of the prohibition in Deut v. 20, 'You shall not bear *ʿēd šāwʾ*'

---

1. As in, e.g., Ruth 4:16, a child *in* a woman's bosom; Ps 88:7, the psalmist *in* the Pit; 2 Sam 19:29, a servant *among* diners.

2. A sampling of proposals is given in HALAT 3:917, under *pûaḥ* II.

3. See ibid.; also L. Alonso Schokel, *Diccionario biblico hebreo-espanol*.

4. Miller, "*yāpîaḥ* in Psalm xii 6," 495–501.

against your neighbor,'" and of the concern in Deut 19:18. "Over against any such *'ed šaw,*'" according to Miller, Yahweh promises "to protect the one who is a witness in his or her behalf." In support of the fact that such witnesses would need protection, Miller cites Amos 5:7, 10, with its reference to those who hate and abhor the one who speaks the truth in the gate; and he appeals to what he takes to be implicit in the situation surrounding the writer of the Yabneh Yam letter: potential witnesses on the writer's behalf—such as his fellow workers—would be subject to the wrath of the powerful person at whose hands he is currently suffering. In such a reading of Ps 12:6, Yahweh's assurance to the psalmist is that any who testify in his or her behalf will (to use a contemporary American expression) come under a divine Witness Protection Program.

I am persuaded by Miller's arguments for the construal of *yapiaḥ* as "a witness." But I find less convincing his interpretation of *'ašit beyešaʿ* as giving divine assurance of the safety of the witness. That God would respond to the cry, "Save me," by assuring the psalmist of the safety of the witness on his behalf, strikes me as a peculiar piece of indirection. It is much more likely that *'ašit beyešaʿ* conveys, as previous interpretations have assumed, an assurance of the psalmist's safety and, as such, constitutes a direct response to the petition in v. 2. But if Miller and others are right in taking *yapiaḥ* as "a witness," how is the syntax of the line as a whole to be construed? It is the purpose of this paper to answer that question.

Another Look at the Language of the Promise

It is clear that the construction *šyt* + *be-* can indicate the action of placing something or someone somewhere. But I propose that the syntax of Ps 12:6c is better analyzed in terms of the well-attested construction, (*šyt*) + (direct object) + (*le-* recipient of direct object). In this construction, the verb carries the connotation, "provide, appoint," and the sentence has to do with "providing something for someone." Consider the following examples:

Gen 4:25	*šat li . . . zeraʿ ʾaḥer*
Hos 6:11	*šat qaṣir lak*
Ps 9:20	*šitah . . . morah lahem*
Ps 140:5	*moqešim šatu li*
Job 14:13	*tašit li ḥoq*

Construed in terms of this construction, *'ašit . . . yapiaḥ lo* would mean, "I will provide/appoint a witness for him."[5] How, in such a sentence, may we construe the function of the phrase, *beyeša'*?

I propose that the preposition *be-* functions here as a *beth essentiae*, and that the phrase *beyeša'* functions in the sentence as a secondary, predicate accusative. In my view, such an analysis receives support from the following types of usage:

1) The first is the repeated occurrence of the *beth essentiae* in the phrase *be'ezer* where, in most cases, God is appealed to, or acts for the devotee, in God's role as a helper. The texts are as follows:

Deut 33:26	There is none like God, O Jeshurun, who rides through the heavens *as your helper* [*be'ezreka*]
Exod 18:4	The God of my father was my help [*be'ezri*].
Hos 13:9	I will destroy you, O Israel; who can help you [*be'ezreka*]?[6]
Ps 146:5	Happy is he whose help [*be'ezro*] is the God of Jacob.

Compare also the participial construction:

Ps 118:7	The LORD is on my side *as my helper* [*be'ozray*].

In the following passage, the affirmation without *beth essentiae* is paralleled by a phrase with it, possibly a case of the preposition doing double duty for both lines:

Ps 54:4	But surely, God is my helper ['*ozer li*]; the Lord is *the upholder of my life* [*besomeke napši*].

---

5. In Gen 4:25 and Job 14:13 the sequence, as we would expect, is (verb) + (pronoun indirect object) + (noun direct object). In Hosea 6 and Ps 9:20 the noun direct object follows the verb and precedes the pronoun indirect object. In Ps 140:5 the indirect pronoun object follows immediately after the verb, but it has to because the noun direct object comes before the verb. This is to say that the sentence structure I am proposing for Ps 12:6 is not unusual in having *lo* come *after yapiaḥ* in a sentence meaning, "I will provide a witness for him."

6. The translation follows those versions which presuppose *mi* for MT *bi*; see *BHS in loc*. Following MT here would not affect the relevance of the verse for my argument.

PART TWO: Forays into a Biblical World

To be sure, in all of the above instances the *beth essentiae* phrase occurs in the nominative case. But the close semantic connection and frequent collocation of the roots *'zr* and *yš'*[7] invite the suspicion that the phrase *beyeša'* may function like *be'ezer*.

2) In a number of instances, the verb *šyt* takes a double accusative, in which the secondary accusative (italicized in the following examples) indicates what character, status or function is accorded to the primary accusative through the action indicated in the verb.

i) 2 Sam 22:12 He made darkness around him *a canopy.*

ii) 1 Kgs 11:34 I will make him *ruler.*

iii) Isa 26:1 He provides *as salvation* walls and bulwarks.[8]

iv) Ps 84:6 They make [Baca] *a place of springs.*

v) Ps 88:8 You have made me *a horror to* [*le*-] them.

vi) Ps 110:1 I will make your enemies *a footstool* for [*le*-] your feet.

7. One may note especially the instances where both roots occur in the same verse: Deut 21:29; Josh 10:6; Job 26:2; Pss 37:40; 109:26; Isa 49:8; and Isa 63:5. Two of these passages are especially relevant to the analysis of Psalm 12 offered in this paper. Psalm 109 contains a prayer of one who claims to be unjustly accused by "lying tongues" (vv. 2–3; compare Ps 12:3–4; and note the idiom *dibber 'et*, "speak against," in Pss 12:2 and 109:2). In the face of these accusations, the psalmist appeals to God as one "poor and needy" (Ps 109:22, also in part v. 31; compare Ps 12:5). In this appeal, the psalmist prays, "But you, O LORD my Lord, / act on my behalf for your name's sake; / because your steadfast love is good, deliver me [*haṣṣileni*] . . . Help me [*'ozreni*], O LORD my God! / Save me [*hoši'eni*] according to your steadfast love. / Let them know that this is your hand; / you, O LORD, have done it" (Ps 109:21, 26–27) It is unclear whether the supplicant in this psalm looks for divine help through a fellow Israelite who will arise as a witness for the defense. In Isa 63:5 the picture does seem clear: In the absence of any human agency God will intervene directly: "I looked, but there was no helper [*'ozer*]; / I stared, but there was no one to sustain me;/so my own arm brought me victory [*toša'*] and my wrath sustained me."

8. Isa 26:1 is generally rendered, "he sets up salvation as walls and bulwarks." But in view of the fact that Isa 60:10 envisages the rebuilding of Jerusalem's physical walls, and that in Isa 60:18 those walls are called *yešu'ah* and its gates *tehillah*, it is to me more likely that in Isa 26:1 the primary accusative, indicating what has been provided, is "walls and bulwarks," while the secondary accusative is "salvation," indicating their significance for the people. In both Ps 12:6 and Isa 26:1, then, as I analyze them, God is said to provide (*šyt*) something *as salvation*. Moreover, the *lo* phrase, "for him," in the psalm is paralleled by the earlier *lanu* phrase, "for us," in Isa 26:1. In other words, *yešu'ah* in Isa 26:1 and *beyeša'* in the Psalm are syntactic and semantic equivalents.

As God Is My Witness

We may note that, just as in the previous set of examples where *šyt* is followed by an accusative direct object and an indirect object introduced by *le-*, indicating that something is being provided for someone, so here in the last two passages the action in the verb makes the primary accusative object serve as something for someone. Insofar as a *beth essentiae* functioning as a secondary predicate accusative falls in the same syntactic slot and performs the same semantic function as a secondary accusative, the above examples (especially iii, v and vi) give some indirect plausibility to my rendering of Ps 12:6.

The preceding analysis may help us to uncover a precise, if partial, syntactic parallel to Ps 12:6 in Jer 3:19. In RSV it is translated,

> I thought how I would set you among my sons [*'ašitek babbanim*],
> and give you a pleasant land,
> a heritage most beauteous of all nations.
> And I thought you would call me, My Father,
> and would not turn from following me.

As W. L. Holladay observes, such an understanding of *'ašitek babbanim* poses the question, "who would the 'sons' be with whom the 'wife' is included?" He goes on to say, "given the vocative 'sons/children' in v. 22, it is best to translate 'treat as' (so also NAB). Yahweh wants to bequeath to the people land, just as a father bequeaths land to his sons."[9] Though Holladay does not do so, we may support his interpretation by identifying a *beth essentiae* in the phrase *babbanim*.

With the above examples we may compare Deut 1:13, *wa'ašimem berašekem*, "and I will make them *your heads*" where the verb *śym* (a close synonym of *šyt* in this instance) is followed by a pronominal accusative and then a *beth essentiae* phrase as secondary predicate accusative. If instead of *berašekem* the text were to read the semantic equivalent, *lakem beraš*, the parallel to Ps 12:6 would be complete.

It is possible that 1 Sam 8:11 also reflects such a usage. RSV translates, "These will be the ways of the king who will reign over you: he will take your sons and appoint them to his chariots [*bemerkabto*] and to be his horsemen [*beparašaw*], and to run before his chariots." This translation construes *beparašaw* as involving a *beth essentiae,* in which case we may isolate the syntax as follows: *weśam lo <'et benekem> beparašaw,* where the inserted component is understood elliptically from the preceding clause. Such a sentence would parallel my construal of Ps 12:6 exactly. However, the preceding *bemerkabto* is problematical, as apparently containing the

---

9. Holladay, *Jeremiah 1,* 122.

same preposition with a different connotation. If *merkabah* could be taken in the sense of *rekeb* where the latter term means "chariot crew, war-chariot troop," in conjunction with *paraš* as meaning "cavalry,"[10] then both phrases might be taken to contain *beth essentiae*, "and appoint them as charioteers and horsemen." But such a connotation in *merkabah* would be unique to this passage; so the identification of a *beth essentiae* in *beparašaw* is not certain.

3) To this point, I have been searching for phrasal and syntactical usages that would give direct or indirect support to my construal of the line in Ps 12:6 as having the structure, (verb) + (direct object) + (*le*-beneficiary) + (*beth essentiae* phrase identifying the significance of the direct object for the beneficiary). Such a sentence structure is unambiguously attested in the idiom used to speak of God's gift of land to Israelites as an inheritance. Numbers 36:2 may serve as an example: *latet 'et-ha'areṣ benaḥalah ... libne yiśra'el*.[11] To be sure, this idiomatic expression is used only in reference to the motif of God's gift of the land to Israel as an inheritance. But that its syntax reflects a more general usage is suggested by Ezek 40:16, which shows that the construction can occur in a looser, non-idiomatic fashion, and in a social rather than theological context: "If the prince makes a gift to any of his sons out of his inheritance, it shall belong to his sons, it is their property *benaḥalah*." The past phrase, it seems to me, should be translated, not "by inheritance," but "as an inheritance."

Taking all these usages together, it seems clear that the sentence structure I am proposing for the line in Ps 12:6 was sufficiently common in Israel for it to be used by this psalmist and for it to be recognized without difficulty by contemporary hearers and readers of the psalm.

God's Promise and Job's Hope-Against-Hope in Job 16:19

The fact that a prohibition against bearing false witness should be included among Israel's "top ten" commandments is by itself sufficient indication that the social health of a community, insofar as it rests on the relations between its individual members, turns critically, in part, on the reliability, the truthfulness, of what people say about one another, whether formally

---

10. For these senses, see respectively HALAT 3:978, nos. 2 and 3, and HALAT 3:1233, no. 2.

11. Other instances: Num 18:26; 36:2; Josh 13:6, 7; 21:12; 23:4; Ezek 45:1; 46:16; 47:14, 22; 48:29. The verb may be a form of *natan*, *ḥalaq* (Niphal or Piel), *napal* (Qal or Hiphil), or *hayak*.

in legal proceedings or less formally by way of interpretive commentary on or gossip about one another's actions and character. That false witness and slander were an all-too-common reality is attested in many other ways as well. It is reflected, for example, in the six proverbs in which the word *yapiaḥ/yapeaḥ* appears along with *'ed*.[12] But it comes to most poignant expression in the Book of Job, and there especially in chapter 16 where, searching in vain within his community for one who will take his side against the accusations of his friends and the divine accusation which he takes to be implicit in his physical afflictions (Job 16:8), Job, hoping against hope, cries out in a desperate act of blind imagination, "even now my witness is in heaven,/ and he that vouches for me is on high . . . that he would maintain the right of a mortal with God,/ as one does for a neighbor" (16:19, 21).[13] Such, I suggest, is the plight of the speaker in Ps 12:2: he can find no one who is faithful to him in his circumstances. (Compare the collocation of *ḥesed* and *'emunim* in the sceptical query in Prov 20:6.) But *'emunim* here may carry a more specific nuance. In Prov 15:5 the phrase *'ed 'emunim* characterizes a reliable witness, one who speaks the truth; and in Prov 13:17 the phrase *ṣir 'emunim* characterizes a reliable, that is truthtelling envoy. These passages, together with the phrases *yapiaḥ 'emunah* (Prov 12:17), *'ed 'emet* (Prov 14:25), *'ed 'emet wene'eman* (Jer 42:5) and *'edim ne'emanim* (Isa 8:2), may suggest that the complaint in Ps 12:6 is over the absence of anyone who will counter the lies and slanders with reliable and true testimony.

In Ps 72:12 it is said of the king that "he delivers the needy [*'ebyon*] when he calls, / the poor [*'ani*] and him who has no helper [*'en 'ozer*]." But

---

12. Prov 6:19; 12:17; 14:5, 25; 19:5, 9.

13. In characterizing Job's words in this passage as a desperate act of blind imagination, I mean to suggest that it may be misguided to confine the reach or reference of his words to religious conceptions assumed to be conventional in the time in which the passage was written. Religious aspiration, especially under extreme duress, may outstrip the conventions of belief. Robert Browning has the painter Andrea del Sarto, in the poem by that name, exclaim, "A man's reach should exceed his grasp, / Or what's a heaven for." In a metaphysical vein, Alfred North Whitehead concluded a justly famous paragraph on the nature of consciousness with the following lines: "Consciousness is the feeling of negation: in the perception of 'the stone as grey,' such feeling is in barest germ; in the perception of 'the stone as not grey,' such feeling is in full development. Thus the negative perception is the triumph of consciousness. It finally rises to the peak of free imagination, in which the conceptual novelties search through a universe in which they are not datively exemplified" (*Process and Reality*, 245). If Job 16:19-21 rises to such a peak of imaginative outreach, it is no wonder that efforts to pin down the meaning or reference of this passage precisely in terms of the religious categories and understandings current and familiar in its day are so various and so inconclusive.

PART TWO: Forays into a Biblical World

the poor and needy person in Ps 12:6 has no helper, royal or otherwise. As in Isa 63:5, God apparently is appalled that there is no helper; so God promises to provide a witness as deliverance; and in contrast to the lying words of the psalmist's adversaries, God's promises are affirmed as utterly trustworthy (v. 7).

The identity and whereabouts of this witness are not specified. It may even be that, like the speaker in Job 16, the conviction as to the existence of a supporting and saving witness, a conviction arising in the psalmist's awareness as a very word from God, is voiced in the teeth of every datum in the psalmist's world to the contrary. In such a case, the shift in consciousness (a) from v. 2 (b) through v. 6 (c) to vv. 7 and 8a, is like the shift in Habakkuk (a) from 1:1-4, 14-17 (b) through 2:2-4, with God's proffer of a vision for a later time, a vision that "does not lie" because it is reliable,[14] (c) to the prophet's prayer in chapter 3, which begins with a plea that the intervening interval of wrath be shortened and ends with an affirmation of joy in God's salvation in the teeth of all concrete evidence to the contrary. Indeed, one may wonder if, in the last analysis, it is not the promise that God makes to the psalmist in Ps 12:6, as affirmed in v. 7, that constitutes (in heaven, so to speak) the *yapiaḥ* which the psalmist otherwise seeks in vain. For the psalm ends as it began, with the psalmist still surrounded by the wicked, and the human community (*bene 'adam*) still, as in v. 2, united in vile opposition (v. 8). In such a situation, the psalmist's hope is in God alone, as protector and guard (v. 9), and even as witness.

---

14. For an analysis of Hab 2:2-4 in which *'emunato* is taken to refer to the reliability of the aforementioned *ḥazon*, the pronoun in *'emunato* thus continuing the reference of the pronoun subjects of the third-person verbs in v. 3, and of the pronoun in the phrase *lo* in that verse, see Janzen, "Habakkuk 2:2-4 in the Light of Recent Philological Advances."

8

"And Not We Ourselves"
Psalm 100:3 and the Eschatological Reign of God

For centuries, western Christians who sang, said, or read the Psalms in Latin understood Ps 100:3 to say, *ipse fecit nos, et non ipsi nos*. Those who used the *Book of Common Prayer* or the KJV similarly understood the psalm to say, "it is he that hath made us, and not we ourselves." And "Old Hundredth," William Kethe's familiar paraphrase of Psalm 100, beginning, "All people that on earth do dwell," has it that "without our aid he did us make."

But the Hebrew consonantal text, which reads *welo' 'anaḥnu*, "and not we ourselves," contains a marginal note indicating the judgment of the Masoretes that the text should read *welo 'anaḥnu*, "and we are his." (While the graphic difference in Hebrew script is obvious to the eye, involving the presence or absence of the letter *aleph*, the two Hebrew forms are homophones, sounding simply "*lo*," so that either could be explained as a secondary development from the other by a scribe producing a new manuscript under dictation.) While "not we ourselves" is supported by the Greek Septuagint (*kai ouch hemeis*) and the Latin Vulgate, "and we are his" is supported by a number of medieval Hebrew manuscripts, Aquila's revision of the LXX, the Aramaic Targum, Jerome's Latin translation, and the Qere.[1] Clearly, the manuscript evidence for this verse is divided, and judgments as to the original form of the text will turn on a variety of arguments.

A general consensus among modern interpreters prefers the reading, "we are his," and this is reflected both in modern translations of the Bible and in liturgical texts such as the English *Alternative Service Book* (1980),

---

1. See the apparatus in *BHS*.

PART TWO: Forays into a Biblical World

the American *Book of Common Prayer* (1977), the *Book of Alternate Services* (1985) of the Anglican Church of Canada, and William G. Heidt's edition of *A Short Breviary* (1962). One might conclude that if those who use "Old Hundredth" for hymnic purposes continue to say, "without our aid he did us make," they do so without the support of the Hebrew text as correctly interpreted.

In the present study, I shall review the major arguments for adopting this reading in the Hebrew Bible, but offer arguments in support of the reading, "not we ourselves." I shall base my argument not only on the context of this phrase within Psalm 100 itself, but also on the thematic development traceable within the small complex of psalms for which Psalm 100 serves as a conclusion. I note here that, in current Psalms studies, scholars are giving increasing attention to the structure of the Psalter as a whole as well as the structure of smaller groups within it. In this sort of analysis, such structuring is taken as evidence for theological reflection on the part of the Psalter's editors, and therefore as a basis for psalms-based theological reflection in our own day. One such focus of attention is the group in Psalms 93–100. As we shall see, some now take Psalm 100 as having been composed to conclude this group as a coherent unit. Within Psalm 100, v. 3 is then taken to achieve a breathtakingly bold theological step, in taking language previously distinctive of Israel's relation to God and extending it to embrace all humankind in an eschatological vision of the universal praise of God. The very center of that eschatological offering of praise consists in the affirmation,

> It is he that made us, and [not we ourselves / we are his];
> we are his people, and the sheep of his pasture.

If, then, Psalm 100 provides the theological vision in Psalms 93–100 with its theological capstone–or, as I would put it, provides the cosmic drama running through these psalms with their eschatological denouement–the precise nature of that capstone, the precise dramatic turn that ushers in that denouement, may rest in the disputed clause. In my view, we shall not come to the heart of the eschatological vision set forth in these psalms until we see how it turns on the affirmation, "not we ourselves." Conversely, this verse, so understood, puts its finger squarely on the fundamental challenge to the realization of that theological vision–the challenge posed by the pretension of the creature to self-creation.

But this raises the question of the palatability of the verse in our own day, when one does not need to be a Jean-Paul Sartre to understand oneself and humankind in general to be in some sense a "work [of one's

"And Not We Ourselves"

own] in progress." A theologian as firmly rooted in the tradition as Austin Farrer (who regularly, in the service of Morning Prayer, would have recited Psalm 100 in its traditional form) reiterated again and again in his theological writings his understanding of the mystery of human becoming in the formula, "God makes his creatures make themselves." At the end of this primarily exegetical study, then, I shall turn briefly to this theological issue and suggest one biblical resource for understanding Farrer's formulation in biblical terms.

Is "Not We Ourselves" Grammatical Hebrew?

We must begin with questions of Hebrew grammar. Commenting on the traditional understanding of the sentence common in his day—"He hath made us (chosen us to be His people), and not we ourselves"—J. J. S. Perowne writes, "but it is very doubtful if such a meaning would be thus expressed in Hebrew."[2] However, two passages in Job display precisely such a grammatical construction. In Job 15:6 Eliphaz says, *yaršiʿaka pika welo'-'ani*, "Your own mouth condemns you, *and not I;*" and in Job 34:33 Elihu says, *'atta tibḥar welo'-'ani*, "You must choose, *and not I.*" As in Ps 100:3, first an action is posited as performed by one party, and then that same action is negated as performed by another party; and in all three cases the grammatical construction is the same. Indeed, in Job 34:33 as in Ps 100:3 the subject of the verb is underscored by a preceding independent pronoun for emphasis.

One may also consider the construction in Gen 45:8, *lo'-'attem šelaḥtem 'oti hennah ki ha'elohim*, "It was not you who sent me here, *but God.*" Semantically, this sentence is equivalent to *ha'elohim šelaḥtem 'oti hennah welo'-'attem*, "it was God who sent me here, *and not you.*" In both the actual and the transformed version of Gen 45:8, a complete sentence is followed by its logical contrary in which only the alternate subject of the verb is given, with the rest of the sentence understood. Ellipsis of this sort is not uncommon. Another example occurs in 1 Kgs 18:18: *lo' 'akarti 'et-yišra'el ki 'im-'attah*, "I have not troubled Israel; *but you* [have]." Similarly, Isa 45:21, *mi hišmia' zo't miqqedem me'az higgidah halo' 'ani yhwh*, "Who told this long ago? Who declared it of old? [Was it] *not I, the* LORD?" In light of Isa 45:21, we may consider also 2 Kgs 18:22, *halo'-hu' 'ašer hesir ḥizqiahu 'et-bamotaw we'et-mizbeḥotaw*, "Is it *not he* [God] whose high places and altars Hezekiah has removed?" This is semantically equivalent

---

2. Perowne, *Psalms*, 212.

PART TWO: Forays into a Biblical World

to *mi 'ašer hesir mimmennu ḥizqiahu 'et-bamotaw we'et-mizbeḥotaw halo'-hu'*, "Who is it from whom Hezekiah has removed his high places and altars? Is it *not he* [YHWH]?"

In all these cases, the *negative particle + personal pronoun* stands alone, elliptically, over against a complete verbal sentence with whose expressed subject it either contrasts (Job 15:6; 34:33; Gen 45:8; 1 Kgs 18:18) or corresponds (2 Kgs 18:22; Isa 45:21). Thereby, the pronoun in the elliptical phrase identifies the (actual or negated) doer of the action indicated in the verb. There should be no doubt, then, that from a grammatical point of view the construction "not we ourselves" in Ps 100:3 is quite acceptable, indeed idiomatically pithy.

The Range of Variation in a Stock Expression

Interpreters who adopt the reading, "we are his," appeal to closely similar passages in Ps 79:13 and Ps 95:6. Norbert Lohfink lays out the parallels this way:[3]

Ps 79:13   *wa'anaḥnu 'ammeka weṣo'n mar'iteka*
           And we are your people / and the sheep of your pasture.
Ps 95:7   *wa'anaḥnu 'am mar'ito weṣo'n yado*
           And we are the people of his pasture / and the sheep of his hand.
Ps 100:3   *('anaḥnu) 'ammo weṣo'n mar'ito*
           (We are) / his people and the sheep of his pasture.

He comments, "The greatest agreement is between Ps 100:3 and Ps 79:13. The formulation in Ps 95:7 is similar, but somewhat different. It appears that we have to do here with a turn of phrase in psalmistic diction that within certain limits is variable."[4] Lohfink goes on to take Ps 100:3 as a version of the common covenant formula in which Yahweh is Israel's God and Israel is God's people–frequently imaged as the sheep of his pasture (as also in Jer 23:1; Ezek 34:31; Ps 74:1; and see Ps 77:21; 78:52, 71–72).

But if, as Lohfink says, this turn of phrase may vary slightly–presumably to fit the context–then his choice of "we are his" over "not we ourselves" is lame, and goes against the text-critical principle of *lectio difficilior*. According to this principle, where manuscripts disagree in their

---

3. The transliterations that follow have been changed from Lohfink's original into a simplified, general-purpose form.
4. Lohfink, "Die Universalisierung des 'Bundesformel in Ps 100:3,'" 175–76.

wording, that text-form is to be preferred that is harder to account for as a secondary scribal development. Given the frequency of the "he is our God, we are his people" formulation in the Bible, it is easy to imagine a scribe, in producing a new manuscript from another's dictation, hearing in the spoken *welo' 'anaḥnu*, "and not we ourselves," the more familiar *welo 'anaḥnu* (the spelled difference being indistinguishable to the ear). Or a scribe, on the basis of the more conventional formulation, could have undertaken to "correct" a manuscript reading *welo' 'anaḥnu*. It is more difficult to imagine a scribe familiar with the conventional formulation hearing a spoken *welo 'anaḥnu* as *welo' 'anaḥnu*. Such an inadvertent, or obtuse, scribal move is often observable in the manual transmission of texts; and it has the effect of "flattening" the text by replacing arresting and fresh turns of phrase with more conventional expressions and clichés. Of course, the text-critical principle of *lectio difficilior* is qualified by the requirement that the "more difficult" (that is to say, the more unusual, and less anticipated) text-form must make sense. One of the objects of the present paper is to show, not only that "and we ourselves" makes good grammatical sense, but that, in a fresh and arresting way, it enacts the resolution of the problem of sin and evil traced in its varied dimensions beginning with Psalm 93. Not only does the whole earth acknowledge God's mighty acts on behalf of Israel (as, for example, in Exod 15:14–16; Josh 5:1; Psalm 117), but this universal choir now confessionally celebrates God's creative/redemptive acts on its own behalf. When "and not we ourselves" is so understood, the scribal slip that produced "we are his" may be recognized as lamentably unfortunate. In his *Confessions*, Augustine refers to words as "those precious cups of meaning" (*Conf.* I, xvi, 26). In those terms, the scribal error turned a cup of wine into water. In short, as I shall argue, the confession "and not we ourselves" contributes in a unique way to the very thesis Lohfink advances as indicated in the title of his article: "The universalization of the covenant formula."

But Why the Need to Disavow Self-Creation?

This raises the question of whether the psalmist could reasonably be expected to give the rhetoric of the psalm such an unexpected turn. David M. Howard writes,

> this rendering ["and not we ourselves"] does violence to common sense and to the biblical world view. Human beings in general, and Israel specifically, are created by God; there is no

suggestion anywhere of self-creation. The traditional reading answers a question that few people, if any, were asking or even thought to ask. The contrast in the verse is not between YHWH's creative activity and humanity's (or Israel's) potential for self-creation. Rather, it is between YHWH and other gods.[5]

I have quoted Howard at length because he states so clearly and forcefully a view that may lie in the back of the minds of other interpreters who find the traditional translation unlikely. Yet he is palpably wrong. Not only is he wrong, but his comment overlooks, and obscures for his readers, a critical issue that crops up again and again in the Bible—an issue that, properly understood, identifies the root of human sinfulness. That issue is engaged in a variety of ways, some of which I shall trace in the following series of forays along several intersecting paths.

On Divine and Human "Making"

As many others note, the expression, "he has made us," need not refer to God's activity in primordial creation, but more likely refers to God's acts in history whereby Israel—and, in this psalm, all the peoples of the earth—owe their historical rise, national identity, and continuing existence to Yahweh.[6] Some earlier scholars supported Ps 100:3 as denying an implied assertion of self-creation by appeal to Ezek 29:3, where Pharaoh is reported, according to the Hebrew text, as saying, *li yeʾori waʾani ʿaśitini* "My Nile is my own; I have made myself." Nowadays, commentators and translators generally follow the translation in the Greek Bible, *emoi eisin hoi potamoi, kai ego epoiesa autous*, "The rivers are my own, and I have made them," a translation that presupposes as a Hebrew base text *ʿasitim*, "I have made them."[7] But evidence for the notion of "self-creation," in the

---

5. Howard, *The Structure of Psalms 93–100*, 92.

6. See, e.g., Perowne's expanded version of the traditional understanding, which he rejects only on grammatical grounds.

7. Walther Zimmerli, while agreeing with the current preference for the reading in the Greek Bible, draws attention to "statements made by Egyptian deities seeking to claim for themselves the power of creation or . . . who describe themselves, like Atum-Khepri, as 'self-created.'" But he doubts that Ezekiel could have "been aware of such sacral formulae of priestly theology" (Zimmerli, *Ezekiel* 2, 111). Zimmerli's doubts over the Hebrew text of Ezek 29:3 may be misplaced. From the characterization of Pharaoh in this passage as "the great dragon that lies in the midst of his streams," it is clear that Ezekiel is describing the hubris of Pharaoh in images derived from Mesopotamian and Canaanite religion, images that speak of human powers on analogy with,

sense of attributing one's success and standing in the world to one's own powers of action, is by no means limited to this passage. It can be detected in passages where the implication may not appear evident to us but would have been evident to persons living in the ancient Near East. For one summary exposition of the notion in ancient times of "humanity's (or Israel's) potential for self-creation," we may turn to Thorkild Jacobsen's discussion of the rise of the ruler metaphor for deity in ancient Babylonia. Of the rise of human kingship there as a permanent institution, Jacobsen writes,

> The impact of the new ruler concept on contemporary [that is, ancient Babylonian] thought can hardly be overestimated . . . [I]n literature a new form, the epic tale, took its place beside the myth. In the epic, man, represented by the ruler, is the hero, and the tale celebrates his prowess and his cleverness, *even to the point of challenging the authority of the gods.*[8]

Jacobsen gives several examples from the *Gilgamesh Epic*, in which the titular hero defies the gods, and he cites also the epic of *Enmerkar and the Lord of Aratta*, in which "the human ruler of Aratta bends all his cleverness to frustrate what he knows to be [the goddess] Inanna's will—and succeeds in no small measure." Jacobsen goes on to say,

> The trend to believe in man and his powers in defiance of the gods continued into the Period of Agade and found its most striking—and most cautionary—expression in the figure of the willful king, Naramsin, whose defiance of Enlil led to the destruction of Agade and the end of its rule.
>
> Religion met this trend toward hubris and lack of respect for the gods . . . with the assertion that in the long run the gods would not be flouted, that divine retribution was certain.

And again, later, Jacobsen writes,

> The Early Dynastic period at the beginning of the third millennium was . . . a "heroic" period in which war—the war of all against all—became the order of the day. In those parlous circumstances man tended to look to human martial prowess for salvation, and the idea of the human hero—subject of the

---

or as embodiments of, transcendent powers that other nations worship as deities. (Compare also Isa 27:1; 51:9–11; Ps 74:12–17.) Such imagery, we may note, occurs in Psalm 93 which launches in quasi-mythical terms—the arrogant self-assertion of the waters!—the conflictual drama finally resolved in Psalm 100. I shall return to Ezek 29:3 below.

8. Jacobsen, *Treasures of Darkness*, 79 (italics added). The next two quotations come from the same page.

epic and the heroic tale—tempted him to rely on man and human feats rather than on the gods for salvation. This trend is recognizable in much of the early Gilgamesh materials and finds its exemplar in Naramsin of Agade, of whom warning tales were told at a later date.[9]

The religious issue, as so described, is the issue not only of who is ultimately in charge of human affairs—the human king or the gods as divine rulers—but also of who ultimately shapes human destiny and provides effective security for human flourishing.

Given that Israel participated, in considerable degree, in the general climate of ancient Near Eastern culture, it would be odd if this issue of human hubris vis-à-vis heaven did not in some way manifest itself in Israel. It lies at the heart of the Eden story in Genesis 2–3 (see especially 3:5); it is implicit in a number of different ways in various biblical passages; and Ps 100:3 may be construed as gathering them all up in an explicit negation of such hubris. In the next section I shall draw on the analysis by Howard and others of the structure of Psalms 93–100, to argue that Ps 100:3 forms part of the punch with which this psalm brings this little collection to its climax. But first, I shall canvass a number of biblical passages where the issue of self-creation in the above sense is implicit.

Deuteronomy 8 opens with the well-known verse (v. 2) concerning the hunger and manna in the wilderness, as designed to teach Israel that they do not live by bread alone but by the word of their God. After a depiction of the promised land as a veritable cornucopia, we hear a warning "lest," when Israel has eaten and is full, "your heart be lifted up, and you forget the LORD your God . . . and you say in your heart, 'my power [*koah*] and the might of my hand have gotten [the verb is *'asah*, "made," as in Ps 100:3] me this wealth.'" For it is the LORD their God who gives them "power to get [again, *'asah*, "make"] wealth" (Deut. 8:12–18). As Patrick D. Miller puts the issue, "At bottom, it is the sin of assuming self-sufficiency in a world created and sustained by God's good care." Juxtaposed to the danger of attributing their prosperity in the land to their own doing, is the danger of going after other gods to serve them (8:19). As Miller says again, "the other danger when all is going well is intimated in Deuteronomy 6:14. It is the tendency to think that one's prosperity comes from the gods of the area, especially those associated with food and goods and prosperity."[10]

---

9. Ibid., 224.

10. Miller, *The Ten Commandments*, 24 and 25. In Deut 6:10–11, the sermonic emphasis in the phrases, "cities that you did not build, houses filled with . . . goods that

Worshipping other gods and glorying in the achievements of one's own "hand" are distinguishable dangers, but they are two sides of the same coin. This is implied in Psalm 115, which opens with the cry, "Not to us, O LORD, not to us, but to thy name give glory, for the sake of thy steadfast love and thy faithfulness!" Immediately there follows a contrast between Israel's God who is in the heavens (and therefore is invisible), and who "does [*'aśah*] whatever he pleases," and the objects of the nations' worship who are all too visible, for they are human artifacts, "the work of human hands [*ma'aśah yede 'adam*]" (v. 4). The contrast is between people who "make" [*'aśah*] gods for themselves and people whose God "does [*'aśah*] what he pleases." Though idol-worship has the appearance of worshipping one's creator, it in fact consists in worshipping the work of one's own hands and, in that sense, in self-deification.[11]

The issue engaged in Deuteronomy 8 crops up again in Judg 7:2, introduced similarly by the warning, "lest": "The LORD said to Gideon, 'The people with you are too many for me to give the Midianites into their hand, lest Israel vaunt themselves against me [*pen-yitpa'er 'alay*], saying, "My own hand has delivered me [*yadi hoši'ah li*]."'" There is no question that Gideon and his select band have an active role to play in the scenario that is to follow. But the temptation that lies before them will be to attribute their victory primarily to their own ability arising out of their superior numbers. To think this would be to "vaunt" themselves—as Psalm 115 says, to give the glory to themselves. The verb in Judg 7:2, *yitpa'er*, identifies the same attitude as that displayed by Assyria in Isa 10:15 (on which see below), concerning which God says, "Shall the axe vaunt itself [*yitpa'er*] over him who hews with it, or the saw magnify itself against him

---

you did not fill, hewn cisterns that you did not hew, vineyards and olive groves that you did not plant," presupposes a temptation and indeed tendency for Israelites to attribute these blessings to their own powers. The warning against idolatry in 6:14 again shows the close relation (as in Jer 9:14 vis-à-vis 25:6) between idolatry as worship of other so-called divine powers and self-sufficiency as implicit self-directed idolatry.

11. Isa 41:5–7 is instructive. Idol-makers fashion an idol and (in hubristic imitation of Genesis 1) declare of their work, "it is good." Then they fasten it with nails so that it "cannot be moved" (*lo' yimmoṭ*). This hubristic imitation of passages like Pss 93:1 and 96:10 (*tikkon tebel bal-timmoṭ*, "the world is established; it shall never be moved"), in which the stability of the cosmos is owing to Yahweh's creative action, is of course undercut in Deutero-Isaiah's rhetoric; for the very phrase that unmasks the implicit pretensions of the idol-makers to God-like creativity lampoons the createdness of the (unstable) objects on which they then rely as "gods" to deliver them. I shall return below to this theme of "not being moved."

who wields it? As if a rod should wield him who lifts it, or as if a staff should lift him who is not wood!"

A study of the phrase, "the work of X's hands" (*ma'aśah yad*), yields the following data. In the first place, the natural world (Pss 8:7; 19:2; 28:5; 92:4; 102:26; 111:7; 143:5), Israel (Isa 5:12; 19:25; 29:23; 60:21; 64:8), individuals (Ps 138:8; Job 14:15; 34:19; Song 7:1), and other nations like Egypt and Assyria (Isa 19:25), are spoken of as the work of God's hands. Secondly, the work of human hands in honest toil receives—or discloses—the blessing of God (Deut 2:7; 14:29; 16:15; 24:19; 28:12; 30:9; Isa 65:22; Ps 90:17; Job 1:10; Eccl 5:6). These latter texts reflect a religious sensibility within which a robust appreciation of the human capacity for productive work is grounded in the deeper awareness that that capacity derives from God, in such a way that productive human activity is the manifestation of the activity of God in blessing (my Mennonite forebears sometimes made wall samplers with the text, *von uns ist die Arbeit; von Gott ist der Segen*—from us, the work; from God, the blessing). But there is a work of human hands that is not blessed, for it consists in idolatry. Beginning with Deut 4:28, references to such "work" run through the Bible, occurring some two dozen times. Such idolatrous "work," as we have seen from Psalm 115, goes hand in hand with self-glorification, as when Jeremiah charges that the people "have stubbornly followed after *their own hearts* and after *the Baals*, as their fathers taught them" (Jer 9:14). The expression, "follow after," occurs over a score of times, especially in Jeremiah and Deuteronomistic passages. That "following after one's own heart" is likewise a form of (self-)idolatry, is implicit in the way "their own hearts" appears here in place of the customary "other gods" or the like; and that implication is reinforced with the following phrase, "after the Baals," as though the latter phrase unmasks the deeper religious significance of "after their own hearts." Jeremiah shortly follows with the exhortation, "Let not the wise man glory in his wisdom, let not the mighty man glory in his might, let not the rich man glory in his riches; but let him who glories glory in this, that he understands and knows me, that I am the LORD who practices steadfast love, justice, and righteousness in the earth; for in these things I delight, says the LORD" (Jer 9:23–24).

This theme of human celebration of its own power over against Yahweh comes to expression in another way in Isaiah 10, where Assyria, the rod of God's anger, attributes its successes to its own inherent abilities: "By the strength of my hand I have done it [*bekoaḥ yadi 'aśiti*], and by my wisdom, for I have understanding; I have removed the boundaries of peoples, and have plundered their treasures; like a bull I have brought down those who sat on thrones" (Isa 10:13). This is precisely the sort of human attitude that Jeremiah inveighs against. It is an attitude that is implicitly self-deifying, turning God into an instrument of human power and wisdom (Isa 10:15). The self-deification becomes all but explicit in Isaiah 47, where Assyria's later counterpart, Babylon, adopts for itself language of self-description that rightly belongs only to Yahweh (compare Isa 47:8, 10 with Isa 43:11; 44:6, 8; 45:5, 6, 21).

The Verb *ga'ah*

The theme of human self-glorification vis-à-vis God comes to expression also in occurrences of the verb *ga'ah* and its cognates. Having the basic meaning, "rise up" (like the swelling waters of a river, Ezek 47:5), the root in its extended uses typically carries connotations of triumph, exaltation, and pride. It can be used to describe nations in their wealth, power, magnificent architecture, and so on. But preeminently it celebrates the God of the Exodus, in the opening lines of the Song at the Sea: "I will sing to the LORD, for he has triumphed gloriously [*ga'oh ga'ah*]; the horse and his rider he has thrown into the sea." It is this triumph that establishes the incomparability of Yahweh among the gods (Exod 15:11) and results in the acclamation, "The LORD will reign forever and ever" (15:18). It is in relation to God's exaltation as divine king that the self-exaltation of humans (as Jacobsen has shown for Mesopotamia) can take on a rebellious and implicitly self-deifying character. In such a case, as Isaiah puts it, "The LORD of hosts has a day against all that is proud [*ge'eh*] and lofty [*ram*], against all that is lifted up [*niśśa'*] and high [*šapel*]" (Isa 2:12). Insofar as the terms *ram* and *niśśa'* are, in the Isaianic tradition, properly applied only to Yahweh and the servant of Yahweh (Isa 6:1; 33:10; 52:13; 57:15), "all that is lofty" in this sense (Isa 1:2; 2:11–14, 17; 10:12; 37:23;) repeats the *lèse-majesté* of Assyria in turning God into an instrument of human action and thereby implicitly committing self-deification. The theme arrives at its apotheosis in Isaiah 14, where the pretensions and fall of the king of Babylon are presented in terms of a motif familiar from Canaanite mythology: "How you are fallen from heaven, O Day Star, son of Dawn! How you are cut down to the ground, you who laid the nations low! You said in your heart, 'I will ascend to heaven; above the stars of God I will *set on high* [*'arim*] my throne; I will sit on the mount of assembly in the far north; I will ascend above the heights of the clouds, I will make myself like the Most High.'" (Isa 14:12–14) This passage is immediately preceded by the sentence, "Your pomp [*ga'on*] is brought down to Sheol" (14:11) A clearer indication could not be desired of the self-deifying implications of the attitude and aims of human autonomous *ga'on*.

Practical Atheism in Psalm(s) 9–10

With this in view, I turn next to Psalm 9–10,[12] with its portrayal of the wicked as exhibiting what commentators call "practical atheism." Such a characterization comes specifically in connection with 10:2–4:

> In arrogance [*ga'awah*] the wicked hotly pursue the poor;
> > let them be caught in the schemes [*mezimmot*] which they have devised.
> For the wicked boasts of the desires of his heart,
> > and the man greedy for gain curses and renounces [*ni'eṣ*] the Lord.
> In the pride [*gobah*] of his countenance the wicked does not seek him;
> > all his thoughts [*mezimmot*] are, "There is no God."

Commenting on the "arrogant" in 10:2, Peter Craigie writes, "They are arrogant . . . for they think they can pursue the afflicted in absolute immunity from any reaction by God."[13] And on 10:3c–5a Craigie writes, "The imagined words of the wicked indicate clearly their frame of mind and the structure of belief which permits their action. God 'does not call to account'; literally 'does not seek'; thus the wicked deny any reality to that epithet of God, that he is the 'Avenger (or Seeker) of Blood'. . . . The wicked are practical atheists, affirming that with respect to their actions, God is otiose, and thus they plan devices . . . against the afflicted with an unburdened conscience."[14]

Concerning atheism in the ancient world, Craigie observes that "the theoretical atheist was rare in the ancient (pre-Greek) world of the Eastern Mediterranean." Acknowledging that "true theoretical atheists" may be "very moral people," and that "some great religions (such as early

---

12. Since Psalms 9–10 are generally considered to have been composed as a single acrostic psalm, and since the thematics of the two currently separated parts are to be understood in integral relation to each other, for purposes of the present discussion I shall refer, in the singular, to "Psalm 9–10."

13. Craigie, *Psalms 1–50*, 123.

14. Ibid., 124.

Buddhism) involve a technical atheism, but at the same time are accompanied by a most profound and noble system of morality," he goes on to say,

> In contrast to the theoretical atheists, the practical or functional atheists, of whom the psalmist speaks, are a most dangerous species of human being. Ultimately, their character is determined not simply by dispensing with belief in God, but more specifically by dispensing with the concepts and precepts of morality and justice, which throughout the ancient world presupposed the existence of the gods or God. And it is the absence of morality which makes the functional atheist dangerous. The functional atheist, as portrayed by the psalmist, *is self-confident* ("I won't slip," v 6) and desires only such things as power and happiness of a sort.[15]

More on the Self-Confident Claim, "I Won't Slip"

Craigie's "I won't slip," in Ps 10:6 translates the expression *bal-'emmoṭ*. The biblical uses of this phrase (including its equivalent with the alternative negative particle *lo'*) bear tellingly on the thesis of this study. The phrase occurs primarily in the Psalter (13x), is echoed from there once in 1 Chronicles; and otherwise it occurs once in Job, twice in Proverbs, and three times in Deutero-Isaiah. Thematically, we may note the following uses:

(1) Psalms 93:1; 96:10 (echoed in 1 Chr 16:30); and 104:5 speak of God "establishing" the world so that it cannot be moved. As will be seen, this "immovability" is not intrinsic, but rests on the sustaining action of God. (2) Because God is in the midst of Zion, "she shall not be moved" (Pss 46:5; 125:1). Again, the immovability of Zion is dependent on the presence of God in its midst. (3) Those who trust in God (like immovable Mount Zion, Ps 125:1), who keep to God's paths, and whose life embodies God's claims, will not be moved (Pss 15:5; 16:8; 17:5; 21:7; 62:2, 6; compare, similarly, Prov 10:30; 12:3). (4) In contrast to the stable affirmations in the third usage, Ps 30:7 shows the psalmist retrospectively confessing some "slippage" in acknowledgement of God: "As for me, I said in my prosperity, 'I shall never be moved [*bal-'emmoṭ le'olam*].'" Here we see the psalmist (inadvertently?) beginning to anticipate his unending stability on the basis of his prosperity rather than on the God who gave it. It is a short

---

15. Ibid., 126–27.

PART TWO: Forays into a Biblical World

step from here to the claim in Deut 8:17, and to the portrayal of the arrogant in Ps 73:3–12. The momentary hiding of God's face (30:7), like the hunger in the wilderness of Deut 8:3, restores the psalmist to the sense of the divine source of one's security.

(5) But the wicked in Psalm 9–10, whose prosperity continues uninterrupted (10:3–5), takes the step that the speaker in Psalm 30 does not take. Thinking in his heart, "I shall not be moved; throughout all generations I shall not meet adversity" (10:6), all the wicked person's thoughts are, "there is no God" (10:4). (6) With this misplaced confidence we may compare two highly suggestive texts in Deutero-Isaiah: Isa 40:20, "He who is impoverished chooses for an offering wood that will not rot; he seeks out a skilful craftsman to set up *an image that will not move [lo' yimmoṭ]*"; and Isa 41:7, "The craftsman encourages the goldsmith, and he who smooths with the hammer him who strikes the anvil, saying of the soldering, 'It is good'; and they fasten it with nails *so that it cannot be moved [lo' yimmoṭ].*" As with the implied connection between Deut 8:17 and 8:19, the religious attitude expressed in Ps 10:6 and portrayed in Isa 40:20; 41:7 is at root one and the same: idolatry.

(7) Finally (as the eschatological counterpart to the foundational affirmations in Pss 93:1; 96:10; and 104:5), God promises exilic Israel, in Isa 54:10, "the mountains may depart and the hills be *removed [temuṭenah]*, but my steadfast love shall not depart from you, and my covenant of peace shall not be *removed [lo' tamuṭ]*, says the LORD, who has compassion on you." The language here may echo that in Psalm 46, where the people say (vv. 1–2), "God is our refuge and strength, a very present help in trouble. Therefore we will not fear though the earth should change, though the mountains *shake [moṭ]* in the heart of the sea."

The thematic movement traced above is, of course, a result of my organization of the passages, which for their own part are scattered throughout the Bible. But that this organization does not impose on the material a pattern foreign to the Bible is suggested by the same movement traced in Psalms 93–100 when these psalms are taken as a carefully edited group. I shall elaborate this point below.

Recurring to Craigie's Remarks on Psalm 9–10

To Craigie's remarks quoted above I would add the following comments on Ps 10:2–4: The verses display a chiastic structure in which v. 2 and v. 4 are the envelope and v. 3 the contents. The outside verses characterize the schemes/thoughts (*mezimmot*) of the wicked vis-à-vis all others, as arrogant and prideful. The inside verse identifies what drives the wicked: the desires of his *nepeš*—not his "heart" (KJV, RSV, NRSV, NIV), but his "appetite,"[16] greedy for gain, an appetite that leads him to "renounce [*ni'eṣ*] the LORD."

In Deut 31:20, we find a brief reiteration of the themes of Deut 8:11–20. The juxtaposition there of the three themes of "eating to the full," "despising [*ni'eṣ*] God" and "turning to other gods" implies, in effect, the deification of one's appetites as the ruling power in one's life. This becomes explicit in the Eden story in Genesis 2–3, where humankind is created, not simply as a "living being" (RSV, NRSV), or "living soul" (KJV), but dynamically, as a *nepeš ḥayyah*—one might say, in view of the general scenario in the Garden, a "lively appetite," surrounded with a garden of delights that are not only desirable (*neḥmad*) in appearance but good in the eating (2:9). These delights are freely offered and freely available within the terms of the one prohibition that brings human appetite under the rule of God the Creator. Under the seductive rhetoric of the snake, the primal couple comes to see the tree as not only "a *delight*[17] to the eyes" but also "desired [*neḥmad*] to make one wise" (Gen 3:6); that is, as enabling one to become like God (3:5). The "practical atheism" of Psalm 9–10 thus consists not simply in disregarding God but, in the process, elevating one's own desires to the throne that governs one's actions.[18] The opposing claimants for one's worship, then, are juxtaposed in the two lines of 10:3: "the appetites of one's *nepeš*"; and "the LORD."[19]

---

16. RSV translates 11x as "appetite" (e.g., Deut 14:26; Job 6:7; Isa 5:14; 56:11); and even KJV does so at Job 38:39; Prov 23:2; Eccl 6:7.

17. The word here is *ta'awah*, a strongly appetitive word, as may be seen in Deut 14:26 where "whatever you desire" (*te'awweh napšeka*) is paired with "whatever your appetite craves [literally, "asks for"—*tišaleka napšeka*]." Thus the slip-slide narrow line between appropriate and inappropriate exercise of one's appetites.

18. The most concrete and sustained portrayal of wickedness as unbridled appetite is given in Job 20, epitomized in vv. 12–13: "wickedness is sweet in his mouth . . . he hides it under his tongue . . . he is loath to let it go, and holds it in his mouth." The judgment intrinsic to such wickedness is most graphically portrayed cinematically in an aesthetically gross but thereby morally powerful concluding scene in the Monty Python Movie, *The Meaning of Life*, where the gourmand finally explodes from over-eating.

19. In this way of looking at the matter, the last commandment in the Decalogue implicitly returns to the theme of the first, as the prohibition of idols bears on one's relations to one's fellows in the context of the issue of legitimate and illegitimate enactments of one's desires.

PART TWO: Forays into a Biblical World

This sort of atheism is most pointedly set forth in the twin psalms 14 and 53, where saying in one's heart, "there is no God" goes hand in hand with "not . . . doing good," the latter an expression synonymous with "not doing *ḥesed*." Craigie writes of this psalm, "the fool is defined by the absence of lovingkindness, which in turn is the principal characteristic of the relationship of the covenant; he lives as if there were no covenant, and thus as if there were no God."[20] Intriguingly, Craigie writes of Ps 14:4, "These fools have 'eaten bread,' thinking that to be the staple of life, but have not called upon God; the wisdom of Moses, that 'man does not live by bread alone' (Deut 8:3), was entirely foreign to them."[21] Another way of putting this point is to say that those who, having eaten and are full, forget God, and attribute their prosperity to the power of their own hand (Deut 8:11–19), wind up as idolaters, a.k.a., practical atheists.

The upshot of all this is to propose that, contrary to Howard's view that "[t]he traditional reading [of Ps 100:3] answers a question that few people, if any, were asking or even thought to ask," the traditional reading addresses an issue that manifests itself in a variety of ways in different biblical genres and that comes to expression most pointedly in the self-exaltation of the practical atheist. That person—that group of persons—was a profound threat both to powerless individuals and to the social order, and could be spoken of in terms also used of the mythic forces of chaos. (Thus, if the sea can be called "proud" [*ga'on*, Job 38:11], and on that account serves as the mythic form and embodiment of the disorder that threatens God's cosmic order and rule [Job 9:8; 26:12], Job, who will shortly be accused of "doing away with the fear of God" [Job 15:4], complains in self-defense, "Am I the sea, or a sea monster, that thou settest a guard over me?" [Job 7:12].)[22]

Before leaving Psalm 9–10, I wish to take note of some recent comments on the status of this psalm in the Psalter. Working within the emerging view of the Psalter as a book organized around the thematic center of

20. Craigie, *Psalms 1–50*, 147.

21. Ibid., 148.

22. The first word in Job 10:16 is the verb *yig'eh*, in the third person singular. Scholars have emended it in various ways (compare NRSV). Habel takes it as "an impersonal, 'If one lifts oneself,' with Job as the implied subject" (*The Book of Job*, 184; see also 200). The rhetorical effect of this third person verb is to suggest that God typically puts down lion-like "uppity" human arrogance, before, in the next verse-half, applying God's general practice to himself (RSV, having both lines refer to Job himself, flattens the perspectival shift). Job is complaining that God sees him through the lens of another stock image of unbridled and unethical self-vaunting power. Compare Job 4:10–11.

"God's rule of the world, including Israel, the nations, and the whole of creation," Patrick D. Miller writes of Psalm 9–10,

> It may be that no other psalm so fully joins the basic themes of the Psalter—the rule of God, the representative rule of the king, the plea for help in time of trouble, the ways of the wicked and the righteous, and the justice of God on behalf of the weak and the poor. It is likely that the psalm was created precisely to bring all these notes into a single and powerful chord.[23]

In light of my own analysis above, I would add only that in Psalms 9–10 the "ways of the wicked" are traced to their ultimate source in their "practical atheism." Miller goes on to say that Psalms 9–10

> are almost a cardinal example of an Enthronement Psalm with the declaration of the Lord's kingship and the call for the Lord to rise up in judgment against the nations and peoples of the earth to manifest a righteous rule in the whole of the universe, as evidenced especially in the protection of the poor and needy from their oppressors.[24]

If my analysis of Ps 100:3 in its context has any validity, it is highly suggestive that the wicked oppressors of Psalms 9–10, as "practical atheists," are characterized in this "cardinal example of an Enthronement Psalm" as arrogant (*ga'awah*, 10:2), and that when we arrive at the cluster of Enthronement Psalms, the first one, Psalm 93, opens with a celebration of God's reign, as "robed in majesty [*ge'ut*]" (93:1); and the second, Psalm 94, calls on God as "judge of the earth," to "render to the proud [*ge'im*] their deserts" (Ps 94:2).

It remains, then, to consider how Ps 100:3 functions within the section of the psalter comprising Psalms 93–100. For, in arguing on behalf of the contemporary consensus in favor of the reading, "and we are his," Lohfink asserts that this reading "is necessary, above all, if it seems likely already for other reasons that Ps 100 as a whole was created to conclude

---

23. Miller, "The Ruler in Zion," 168–169. Concerning the central theme of the Psalter, Miller (ibid., 167 n. 1), draws attention particularly to the work of Mays, *The Lord Reigns*, for whom, as Miller says, "'The Lord Reigns' is the center of the Psalms, its root metaphor"; and to the work of Wilson, *The Editing of the Hebrew Psalter*, who, as Miller says, "has presented a cogent and persuasive argument that the climax of the Psalter is in the Enthronement Psalms in Book IV, announcing the enduring rule of the Lord of Israel in the face of the failure of the monarchy." The Enthronement Psalms in Book IV, we may note, comprise Psalms 93–99 and, as we shall see, the summarizing climax in Psalm 100.

24. Ibid., 172.

the preceding YHWH-is-king psalms."[25] As I propose to show, the thematics of Psalms 93–100, taken as an intentionally arranged group, in fact go to support the traditional text of Ps 100:3, in such a way as to bring to a fine point, and explicit resolution in the psalmist's act of praise, the fundamental issue on which the Kingship of God turns.

"And Not We Ourselves": Resolving the Conflict Drama in Psalms 93–99

As I noted earlier, one of the fruitful developments in recent Psalms studies is the attention that is being given to the setting of a given psalm within the context of neighboring psalms. The editorial activity by which psalms composed on disparate occasions for various purposes were combined into groups, and the groups into collections, and the collections into the Psalter as a whole, is increasingly viewed as itself a species of authorship, exhibiting theological reflection on the psalms so edited. The point may be illustrated from the way in which a poet like Robert Frost, having written individual poems at different times, each simply in and for itself, would select and arrange a group of them for publication as a coherent volume. Frost put the matter in these words when he spoke of

> looking backward over the accumulation of years to see how many poems I could find toward some one meaning it might seem absurd to have had in advance, but it would be all right to accept from fate after the fact. The interest, the pastime, was to learn if there had been any divinity shaping my ends and I had been building better than I knew. In other words could anything of larger design, even the roughest, any broken or dotted continuity, or any fragment of a figure be discovered among the apparently random lesser designs of the several poems?[26]

Martin Buber comments on the source-analysis of the Bible that was in full stride in his day and that summarized its "assured results" in such short-hand formulations as "JEDPR," where "R" represented the lowly Redactor; and he proposes that in fact we should view the "R" in such formulations as referring not only to the Redactor, but to Rabbenu–our Teacher.[27] Frost's words would provide a fitting rubric for such a Teacher.

---

25. Lohfink, "Die Universalisierung des 'Bundesformel,'" 175; cited in Hossfeld and Zenger, *Psalms 2*, 493.

26. Frost as quoted in Brower, *The Poetry of Robert Frost*, viii.

27. The quip was Franz Rosenzweig's originally (in his essay "The Unity of the

And we may take the current interest in the editorial shaping of the Psalter as going to school to such biblical Teachers. Work on Psalms 93–100 along such lines is documented, summarized and extended most recently in the Psalms commentary of Hossfeld and Zenger.[28] And it is most finely exemplified in the superb monograph of David M. Howard Jr., already referred to, on the structure of Psalms 93–100. As his analysis shows, one is not likely to over-read in tracing the most recondite of detailed verbal and thematic connections between these psalms. The tapestry woven through their combination provides a "constellation of intention" that vindicates the judgment of Terence Collins that the Psalter is "an integrated system, in which the final work 'has something to say quite independent of the intentions of the authors of individual psalms, the collectors of groups of psalms or the editors of the psalter'. . . . For [Collins] the Psalter's unity is at the implicit, subconscious level."[29] With these methodological perspectives in mind, I offer the following comments on those aspects of Psalms 93–100 that are especially relevant to the meaning and function of Ps 100:3.

Themes in Psalms 93–100 Bearing on Psalm 100:3

Psalm 93 opens with an affirmation that announces the central theme of Psalms 93–100: "The LORD reigns [*malak*, "is/has become king]." This is "seconded" by the affirmation that "he is robed in majesty [*ge'ut*]." These two affirmations synopsize Exod 15:1–18, which opens with the affirmation that Yahweh "has triumphed gloriously [*ga'oh ga'ah*]," and closes with

---

Bible"), but Buber echoed and affirmed it (see Avnon, *Martin Buber*, 50 and n. 6).

28. Two volumes of the English translation have now appeared: Hossfeld and Zenger, *Psalms 2*; and idem, *Psalms 3*.

29. Howard, *the Structure of Psalms 93–100*, 11, quoting Collins, "Decoding the Psalms: A Structural Approach to the Psalter," 41. On the challenge to recognize such "constellations of intention" within the Bible, one may compare some remarks of John Herschel, son of William Herschel, the great astronomer, and himself a noted astronomer. When a friend confessed himself unable to see "the dark lines seen in the solar spectrum by Fraunhofer," John Herschel said to him, "An object is frequently not seen, from *not knowing how to see it*, rather than from any deficit in the organ of vision . . . I will instruct you how to see them." Richard Holmes, who reports this incident, comments, "The point was that science must always be more than the simple observation of phenomena or data. It was simultaneously a subjective training in observational skills, self-criticism and interpretation: a complete education" (*The Age of Wonder*, 439–40; also, 116). Holmes' point applies also to biblical studies, as exemplified in the work of Howard and Hossfeld and Zenger.

the affirmation that "The LORD will reign [*yimlok*] forever and ever." But whereas the Song at the Sea celebrates God's kingship in Israel's deliverance from Egypt, Psalm 93 celebrates that kingship in God's establishment of the habitable world as immoveable (93:1b–2). In the Song at the Sea, the potential opposition to God's kingship is Pharaoh and Pharaoh's gods (Exod 12:12), and in defeating the power of Pharaoh God enlists the forces of earth, wind, and sea. But in Psalm 93 the sea makes its appearance as a potential threat to the stability of creation and thereby as a potential challenge to God's kingship. Hossfeld takes the sea here as ostensibly a reference to "the forces of nature" rather than the chaos-enemy Sea familiar from Canaanite mythology. But his comment is unnecessarily restrictive. Given the reverberations of this Canaanite mytheme in various biblical contexts, we are surely invited to hear such reverberations in, with and under this reference to "the forces of nature." Now, the way in which the motif of God as exalted King is counter-posed to Pharaoh in Exod 15:1–8 and to the Rivers/Seas who raise their voices in Psalm 93, raises in an intriguing way the possibility that the imagery in Ezek 29:3 may, after all, bear in some fashion on the thematics of Psalms 93–100 as a whole and on the connotations of the controverted clause in 100:3. If—to indulge our mythopoeic imaginations for a moment—we ask what the rivers and waves of the sea say in lifting up their voices against the divine king, according to Ezek 29:3 they say—or rather, the "great dragon that lies in the midst of his streams" says: "My streams are my own; I made them," which is tantamount, as we have seen, to saying, "I made myself."

If Psalm 93:5, "Thy decrees are very sure; holiness befits thy house, O LORD, for evermore," does not belong to the original form of the psalm,[30] it does belong to the psalm in its present setting as one of the apt touches of Rabbenu. (This reference to God's sanctuary vis-à-vis the rebellious pretensions of the "floods," is similar to another Psalmist's perspective, within the sanctuary, on the wicked in their prosperity vis-à-vis God's ultimate vindication. See Ps 73:17.) We are instructed thereby to associate the reign of God vis-à-vis the forces of disorder with the "decrees" of God which, synonymous with God's *torah*, form a central theme in Psalm 1, the introduction to the Psalter.

Psalm 94 follows 93 with an abrupt shift in genre. From hymnic celebration we move to an urgent petition to God to "shine forth" as judge of the earth and vindicate those who suffer at the hands of the "proud" (*ge'im*, v. 2), humans who implicitly assert their lofty pretensions in the face of

---

30. So Cross, "Notes on Psalm 93," 76.

the affirmation of God in Ps 93:1. These are the same figures encountered in Psalm 9–10; and like those figures they voice the attitude characteristic of practical atheists: "The LORD does not see; the God of Jacob does not perceive" (v. 7). Hossfeld comments on "the proud" in 94:2 that this term is "an image for the powerful, the rich, and those who have emancipated themselves from God." Of their attitude in v. 7 he writes, "the climax of their crime is the cynicism of practical atheism, which denies God's activity or even God's living presence."[31] If the sea in Psalm 93 is emblematic of the forces that would disrupt the habitable stability of God's creation, the proud in Psalm 94 embody those attitudes and actions in human society that similarly threaten the capacity of people to flourish in God's realm. Hence the urgent cry for the manifestation of God's rule in judgment.

Psalm 95 again shifts the tone from that heard in Psalm 94, opening with a joyous call to join in noisy thanksgiving and worship of One who is King above all gods, creator of the cosmos, and the worshipping community's maker ('ośenu)–this latter no doubt a reference to God's deliverance from Egypt and formation of the covenant at Sinai. We note that Ps 95:7 presents us with imagery encountered again in Psalm 100:3. Why is it, now, that this celebratory psalm suddenly shifts gears and, in the middle of v. 7, issues an urgent call, "O that today you would hearken to his voice!" followed by a strident warning, "Harden not your hearts," and a portrayal of the forty-years-long erring of the people in the wilderness? What, at root, did this "erring in heart" consist in? It consisted in testing God and putting God to the proof. That is to say, it consisted in subjecting God to the criteria of the people's own wisdom and expectations. But that was tantamount to making God answerable to, and an instrument of, their own concerns. The concluding verse, "therefore I swore in my anger that they should not enter my rest," may provide a more pointed clue. That divine oath is reported, as Hossfeld observes, in Numbers 13–14. In that incident, we may recall, the people as a whole drew back from entering the land because they feared the inhabitants and their fortified cities, and because "we seemed to ourselves like grasshoppers, and so we seemed to them." (Num 13:33). Their sense of the inhabitants' view of them made them feel as small as the inhabitants were mighty. Behind that sense lay the implicit sense that their God was too small to give them victory. But at the heart of their action lay an attitude that we have encountered in

---

31. Hossfeld in Hossfeld and Zenger, *Psalms 2*, 453–54. He goes on to say, "Such and similar quotations are found primarily in the psalms of the poor but also in the mouth of the heathen"; and in a footnote he says, "for the former see [Pss] 10:4, 11; 14:1; 53:2; cf. also 42:4, 11; for the latter, 79:10; 115:2."

PART TWO: Forays into a Biblical World

Deut 31:20 and Ps 10:3: they "despised" (niʾeṣ, Num 14:11, 23) their God. When Moses' intercession moved God to relent, to the point of sparing the people but sentencing them to live out their lives in the wilderness, so that only the next generation would enter the land, the people then committed an act of outright *lèse majesté*, in attempting to go up and take the land in their own right. When Moses warned them that such an attempt would now transgress the command of the LORD (14:41), "they presumed [*wayyaʿpilu*] to go up" (14:44). The verb *ʿapal* describing their action is of uncertain derivation, though some would connect it to a root meaning "to swell," and in the present instance, then, semantically similar to the root *gʾh*, "rise up, be boastful, proud." But the meaning is clear enough from context—the people presumed to attempt to accomplish by their own might what a short time earlier they had doubted that their God could do for them. In its own way, then, Psalm 95 identifies a potential threat to God's kingship and to the social order. If the threat in Psalm 93 lies in the direction of the "forces of nature" and what they connote, and the threat in Psalm 94 lies in the direction of human individuals presented as the enemies of the psalmist, Psalm 95 warns the worshipping congregation of a danger that lurks in its own bosom. Indeed, given that the language of Ps 95:7, as Lohfink shows, is that of the covenant relation, the move from that verse immediately into the appeal and warning of the following verses underscores that the danger lurks in the bosom of the very community that celebrates the covenant relation founded in God its maker. In its own way that danger is of a piece with those indicated in the two previous psalms: it consists in the masked desire and will to set oneself up against God.

Of Psalm 96 Howard writes, "This is the most purely praise-oriented of the eight psalms in the group." In tracing the links between this psalm and the others, he writes, "Psalm 93 shares more words with Psalm 96 than with any other psalm in the group, including the *yhwh mlk* clause and the entire poetic line *ʾp—tkwn tbl bl—tmwt*, which speaks of the unshakeable condition of the world that YHWH the king has established."[32]

One telling difference between Psalms 93 and 96 has yet to be noted—a difference that may account for the fact that Psalm 96 is the "most purely praise-oriented" psalm in the group. This difference consists in the fact that, whereas in Psalm 93 the floods and many waters of the sea lift up their voice in a display of might that challenges God's majesty [*geʾut*] and the stability of God's creation, in Psalm 96 the sea constitutes part of the congregation of creation called to the praise of the creator along with the

32. Howard, *The Structure of Psalms 93–100*, 176–77.

heavens, the field, and all the trees of the wood. The "roaring" of the floods (*dokyam*) in 93:3 now becomes the "thundering" (*yir'am*; RSV, "roar") of the sea in 96:11. Where even the sea joins in praise, the threat to God's reign and to the stability of the creation may be said to have vanished.

It may be noted that, whereas passages like Job 7:12; 9:8; Ps 74:13–14; Isa 27:1; and 51:9 echo the older Canaanite mytheme of victory by the defeat of enemy forces, here the victory consists in turning the voice of threat into the voice of praise. In that victory the sea loses none of its energy, none of its *élan*.[33] That energy, that *élan*, is simply (!) transformed into praise.[34] If Ps 22:3 is right in affirming that God is "enthroned on the praises of Israel," and if Alfred North Whitehead has captured the meaning of this affirmation in proposing that "the power of God is the worship he inspires,"[35] the shift from Psalm 93 to 96 traces what one may call the sublimatory path along which God's kingship is finally established. But, of course, the sea in Psalm 93 is described in its present activity, whereas in Psalm 96 it is as yet only addressed with a call to praise, and we are left to wonder whether it responds in the affirmative. For all that Psalm 96 is the "most purely praise-oriented" psalm in the group, the sobering fact is that this psalm does not conclude the group, and the next psalm shows why.

Of Psalm 97 Howard writes, "In v. 1, Psalm 97 continues the joyful exuberance of Psalm 96, but it quickly turns sober . . ." It is replete with references to God's "adversaries" (v. 3), "worshippers of images" who "make their boast in worthless idols" (v. 7), and "the wicked" (v. 10). That the fullness of God's coming envisaged in Psalm 96 has not yet fully arrived is signaled in 97:11, in the penultimate verse which RSV and NRSV translate, "Light *dawns* for the righteous, and joy for the upright in heart." This translation presupposes an emendation to the Hebrew verb *zaraḥ*, following several versions[36] and the latter verb's use in Ps 112:4 which reads, "Light *dawns* [*zaraḥ*] in the darkness for the upright." But the Hebrew

33. I know of no more evocative attempt to capture and re-present this *élan* than Claude Debussy's tone-poem, *La Mer*. This composition excels especially in giving musical voice to the sea's exuberant surging crash and thunder.

34. With this transformation of the chaotic energy of the sea into praise, we may compare Hans Loewald's profound analysis of the transformative work of sublimation, in which the primal energies of the id are not repressed, but channeled through symbolic processes, in such a way that the symbols, infused by and transforming those energies, can function as efficacious ideals for moral and spiritual aspiration and striving. See Loewald, *Sublimation*, esp. 476–77; and my discussion of Loewald elsewhere in this volume and in Janzen, "Sin and the Deception of Devout Desire."

35. Whitehead, *Science and the Modern World*, 192.

36. One Hebrew manuscript, and the Old Greek, Syriac, Targum, and Jerome.

PART TWO: Forays into a Biblical World

text in 97:11 reads *zarua'*, "is sown." Once again, on the principle of *lectio difficilior* it is easier to explain *zaraḥ* in 97:11 as a copyist's error under the influence of Psalm 112, than to explain the rise of *zarua'* in a context which would not conventionally suggest the notion of "sowing." Yet if the eschatological tenor of Psalm 126 may be expressed in the image of sowing in tears and reaping in joy, the proleptic praise called for in Psalms 93–100 may be said to arise from a light and a joy sown in the hearts of receptive hearers of the eschatological message while yet in the darkness of exile. For the new things of which that message speaks have not yet sprouted (Isa 42:9); and if they may be spoken of as now springing forth (Isa 43:19), that sprouting is in the form of the hope and praise kindled in receptive hearts by the message. The praise does not arise simply *in response to* God's eschatological action, but as the *manifestation in praise* of God's eschatological action beginning to sprout in the human heart amid the present ambiguous situation.

Of Psalm 98 Howard observes that it shares 25 lexemes with Psalm 96, "more than any other psalm-pair in the 93–100 section, and most of them are important in some way."[37] The "close correspondence in content and structure argues against entirely independent composition."[38] For present purposes, we may content ourselves with just two comments. First, Psalm 98 resumes the emphasis of Psalm 96 on praise following the "sober" turn in Psalm 97. Secondly, as Howard notes, "The use of *nhrwt* ["floods"] in both Psalms 93 and 98 is noteworthy . . . since they are the only two texts in the entire Bible where "floods" take any action vis-à-vis God: they rebel against him in Psalm 93 and they are told to praise him in 98."[39] This shift from 93 to 98 parallels the shift observed above from 93 to 96 in respect to the roaring of the sea, and supports my construal of the significance of that shift there.

Psalm 99 displays connections with Psalm 93 as well as shifts in the focus of generic themes found there. The precise expression, *yhwh malak*, "the LORD reigns," occurs for the first time in the Bible in Psalm 93 and for the last time in Psalm 99. But whereas in Psalm 93 God's throne is associated with God's creative actions in establishing the world (93:1–2), in Psalm 99 God's throne is the cherubim in the sanctuary in Zion, a theme evoking (as do vv. 6–8) Israel's story of redemption in Exodus. While Psalm 93 moves from cosmic creation and its reliable orders in v. 1 to

---

37. Howard, *The Structure of Psalms 93–100*, 144.
38. Ibid., 149.
39. Ibid., 179.

"And Not We Ourselves"

God's reliable decrees and holy house in v. 5, Psalm 99 begins and ends in that sanctuary, and the theme of holiness, which since Psalm 93 has reappeared in Ps 96:9; 97:12; and 98:1, becomes the leitmotif of Psalm 99 (vv. 3, 5, 9). What is it that challenges God's reign? In Psalm 93 the challenge comes from the "floods" and the "sea." In Psalm 94 it is posed by the "proud," who are practical atheists. In Psalm 95 the challenge comes in the form of covenant betrayal by the community of election and redemption. In Psalm 96–the member of this group most fully devoted to praise–the sea is called upon to join the community of praise. In Psalm 97 the challenge comes from God's "adversaries," "worshippers of images" and "the wicked." In Psalm 98 the threatening floods of Psalm 93 are called, like the sea in Psalm 96, to join the community of praise. Psalm 99 picks up the focus of Psalm 95 and elaborates the theme of the challenge to God's reign posed by the sins of the covenant community—sins against God's holiness, a holiness sponsoring God's "testimonies" (v. 7; compare Ps 93:5) and "statutes," at the heart of which is the call to forsake all other gods and to eschew the making of idols and images (compare Ps 97:7). The dramatic center of Psalm 99 lies in the interaction between Moses, Aaron, and Samuel who in various crises "cried to the LORD," and God "who answered them" as "a forgiving God." The crises? Moses intercedes most crucially in Exod 32:11–32 over the people's idolatry with the golden calf, and in Num 14:13–19 over the people's faithless fearfulness in face of the inhabitants of the promised land. In each instance the reign of God is challenged in the gravest degree; and in each instance God's threat of annihilating judgment (Exod 32:9–10; Num 14:11–12) gives way to a mercy that enables the community to survive a mitigated judgment. I take Samuel's intercession in connection with Israel's *lèse majesté* in desiring a king, and, in so doing, rejecting God from being king over them—an act tantamount to idolatry (1 Sam 8:4–8). The "forgiveness" implicit in this crisis consists in the fact that the divine king does not annihilate his rebellious subjects, but accedes to their request, and hears the continuing intercessions of Samuel on their behalf (1 Sam 12:23). Yet, Samuel exhorts the people to serve God faithfully (12:20–24) lest, if they continue to act wickedly, they and their king will be swept away (12:25).

Howard, following several other interpreters, finds "a great deal of difficulty" in the last line of 99:8, insofar as "the root *nqm*, commonly translated 'to avenge', occurs in the midst of a discussion about God's grace

PART TWO: Forays into a Biblical World

and forgiveness."⁴⁰ He solves the perceived problem by three interpretive moves:

1. He translates the line's opening conjunction *we-* (RSV "but") with "and." (The Hebrew conjunction can function either way.)

2. He takes the preposition '*al* (RSV "of") as here having a rare but attested meaning, "despite."

3. He takes the verb, *nqm*, to have here (as sometimes elsewhere) the meaning, "vindicate."

This enables him to conclude that "The entire verse thus portrays YHWH as acting on behalf of the faithful in Israel: he answered them (v. 8a), he forgives them (v. 8b), and he vindicates them (v. 8c)—all of this in spite of their sins. Indeed, this is the essence of forgiveness: God forgives and acts on behalf of sinners, who do not intrinsically merit his favor.⁴¹ At one time, in working on Psalm 99 solely in and for itself, I had independently worked out a similar construal of v. 8c—like Howard, because of the perceived difficulty of that line following the previous two. But in reconsidering this psalm in the context of Psalms 93–100, I have returned to the standard translation.

The issues here are exegetical, theological, and pastoral. *Exegetically*, the question is how 8c is to be taken in relation to the biblical portrayals of the prayers of Moses, Aaron and Samuel. It is clear from Exodus 32–34, Numbers 13–14, and 1 Samuel 8–12 (as also from the dialectic of grace and judgment in Exod 34:6–7), that God's forgiveness does not necessarily entail the suspension of the dynamic consequences of Israel's covenant betrayal; what God's forgiveness does is to render such consequences penultimate, and to encourage Israel to endure those consequences as a stringent, penitential path toward the realization of God's original and final purposes for the community. When Samuel's worst fears (1 Sam 12:25) are realized, and the armies of Babylon lay siege to Jerusalem, God responds to Jeremiah's intercessory overtures by saying that even the intercession of Moses and Samuel could not avert the looming disaster (Jer 15:1). Such hope as Israel may nevertheless cling to lies on the other side of the exile (Jer 31:31–34). If it is a fundamental exegetical principle that texts make sense in context, then the dialectic in Exod 34:6–7, and the thematic context of Moses' two intercessions, Samuel's promise of intercession (and Jeremiah's attempted intercession after their example), as well as the warning

40. Ibid., 86.
41. Ibid., 87.

exhortation in Psalm 95 with its allusion to the Moses generation, amply support the conventional reading of Ps 99:8c as beginning with a disjunctive "but."

The issue here is important for an adequate *theological* articulation of the relation between divine grace and judgment, as well as for a discerning distinction between forgiveness as a declaration and assurance of God's continuing and ultimate faithfulness to the promises that are the foundation of the people's existence, and judgment as the unavoidable working of the dynamics of action-and-consequence in the world as God has created and sustains it. To lose sight of the foundational priority, continuing reliability and ultimate realization of God's promises, is to be left with the sense that one's future turns on one's own ability to fulfill God's conditional claims—a sense that can lead to a religion of works-righteousness or to despair. To lose sight of the dynamics of act-and-consequence as in the nature of God's creation, in favor of an undialectical concentration solely on God's grace and forgiveness, is to ignore the warnings in Psalms 95 and 93, not to mention Exod 34:6–7; 1 Sam 12:25; and Jer 15:1.

The *pastoral* issue may be put in terms of Jer 6:14 and 8:11, which sum up God's charge against prophets and priests, scribes and sages (and today, we should add, exegetes, theologians, and pastors), that "they have healed the wound of my people lightly, saying, 'Peace, peace,' when there is no peace." The issue was aptly addressed by Archbishop Robert Leighton of Scotland (1611–1684) when, under the heading, "Superficial Reconciliations, and the self-deceit of forgiving," he wrote,

> When, after variances, men are brought to an agreement, they are much subject to this, rather to cover their remaining malices with superficial verbal forgiveness, than to dislodge them, and free the heart of them. This is a poor deceit. As the philosopher said to him, who being ashamed that he was espied by him in a tavern in the outer room, withdrew himself into the inner, he called after him, "That is not the way out, the more you go that way, you will be further in!" So when hatreds are upon admonition not thrown out, but retire inward to hide themselves, they grow deeper and stronger than before; and those constrained semblances of reconcilement are but a false healing, do but skin the wound over, and therefore it usually breaks forth worse again.[42]

---

42. Leighton as quoted in Coleridge, *Aids to Reflection*, 100–101.

PART TWO: Forays into a Biblical World

With this we may return to the function of Psalm 99 in the dramatic movement from Psalm 93 to Psalm 100. It is striking how the previous focus on the community's sin, in Psalm 95, is followed by an outburst of pure praise in Psalm 96; also, how the "sober" turn in Psalm 97 is followed by Psalm 98 which so closely resembles 96. If Psalm 99 returns, briefly in v. 8c, to the sobering theme of community sin as a threat to the reign of God, that theme is trumped by the celebration of how this particular divine king (in contrast to kings in general?) responds to such *lèse majesté* by sustaining the covenant relationship while holding them accountable to living out the consequences of their wrongdoing within the context of that ongoing relationship. This vision of God provides the immediate backdrop for Psalm 100, and in particular for the words in v. 3, "and not we ourselves."

The Conflict Drama Resolved in Psalm 100:3

Norbert Lohfink argues for the reading, "and we are his," for two major reasons: First, there is the likelihood "that Psalm 100 as a whole was created to conclude the preceding YHWH-is-king psalms."[43] This would mean, as others have also held, that v. 3 is to be construed "in light of Ps 95:6–7," which, it is held, implicitly favors the reading, "and we are his." And how does the connection with 95:6–7 favor the latter reading? This leads to Lohfink's second major reason: the fact that in both 95:6–7 and 100:3 we encounter the same version of the covenant formula, "Yahweh Israel's God; Israel, Yahweh's people," that recurs so often in a variety of specific formulations.[44]

But is the contested phrase a *sine qua non* of the covenant formula, such that the reading, "and not we ourselves," obscures it? By no means! Let us for the moment omit the problematic phrase and see how the passage reads: *yhwh hu' ʾelohim hu'-ʿaśanu / ʾanaḥnu ʿammo weṣoʾn marʿito*, "YHWH is God; it is he who has made us, / We are his people and the sheep of his pasture." In such a passage, the second line as a whole stands in apposition to the "us" that ends the first line, and the theme of the covenant formula is fully intact: YHWH is their maker; they are YHWH's people. What would "and we are his" add to this formula? Emphasis

---

43. Lohfink, "Die Universalisierung des 'Bundesformel,'" 175; cited in Hossfeld and Zenger, *Psalms 2*, 493.

44. For analysis of this covenant formula, see Lohfink, "Die Universalisierung des 'Bundesformel,'" 172–83.

through repetition, perhaps; but no new information. A distinctive feature of poetry in comparison with prose lies in the way it combines economy of expression with greater density of differentiated reference. In my view, "and we are his," while it may contribute to an emphasis achieved through its elaboration in the following clause, does not achieve the tensiveness of "and not we ourselves" as a phrase intruded into the very bosom of the covenant formula. How does "and not we ourselves" achieve that tensiveness?

As I have remarked in connection with Psalm 95, the threat to God's reign arises not only from transcendent powers symbolized by floods and sea in Psalm 93 or the arrogant practical atheists in Psalm 94, but also from within the covenant community itself. Even as it professes to belong to YHWH, that community, as Deuteronomy 8 "foresees," will tend toward the idolatry of self-reliance, the idolatry of attributing its history and its prosperity to its own doing or "making" (Deut 8:17, "My power [*koah*] and the might of my hand have made [*'asah*] me this wealth"). Psalm 99 has implicitly picked up the theme of Psalm 95 and, going beyond the latter's emphasis on the consequences of the people's rebellion, focuses primarily (though not exclusively) on God's forgiveness. This focus informs the conclusion to Psalm 100 (which celebrates God as epitomized in Exod 34:6): "Enter his gates with thanksgiving, / and his courts with praise! / Give thanks to him, bless his name! / For the LORD is good [*tob*]; / his steadfast love [*hesed*] endures forever, / and his faithfulness [*'emunah*] to all generations."

In Psalm 99:3 God's name is lauded as great, terrible, and holy. Here, God's name is blessed (v. 4) because Yahweh is "good." The connotation of the latter adjective may be gleaned from Exod 33:19, where God says to Moses, "I will make all my goodness [*tubi*] *pass before* you and will proclaim before you the name 'The LORD'; *and* I will be gracious to whom I will be gracious, and will show mercy on whom I will show mercy." If the conjunction "and" introducing last two clauses in this verse is explicative ("to wit, I will be gracious," etc.), then the two clauses that follow the conjunction, being formally similar to the "I will be who I will be" of Exod 3:14, present an exegesis, so to speak, of the divine name as relevant to the dilemma in which Israel finds itself. In that case, these two clauses form an *inclusio* with "my goodness," such that God's generic goodness is here specified as God's grace and mercy analytic in the divine name. This construal of Exod 33:19 is borne out by 34:6, where "the LORD passed before [Moses] and proclaimed, 'The LORD, the LORD, a God merciful and gracious, slow to anger and abounding in steadfast love and faithfulness [*ḥesed we'emet*]."[45] It is this fundamental revelation of God, in the face of the community's dire rebellions, that Ps 100:5 sums up in reference to God's goodness, mercy and truth.

The central point in Lohfink's essay is that, as the conclusion to the group of psalms celebrating God's universal reign, this psalm makes the bold move of drawing all nations within the compass of the covenant relation. How, then, may "and not we ourselves" serve that universalizing of the covenant formula? Let us review the theme of creation's address to God in Psalms 93–99.

In Psalm 93 the floods and the sea lift up their voice in challenge to God's reign by threatening the stability of God's creation. In Psalm 94 the "proud, practical atheists," say, "The LORD does not see; the God of Jacob does not perceive."[46] Psalm 95 presupposes the recollection of all the community's "murmurings" in the wilderness, in particular its fearfulness of going forward and its desire to return to Egypt. Psalm 96 calls on the sea to transform its roaring from challenge to praise. Psalm 97 calls on "the earth" to rejoice (v. 1), though the earth at present contains both "those who hate evil" and "the wicked" (v. 10). Psalm 98 calls on the sea, again, and the floods, to join the whole habitable world (*tebel*, vv. 7, 8, as in 93:1; 96:10, 13; 97:4) in praise of God's coming reign. Psalm 99:2 calls on "all the peoples" to praise God's "great and terrible name" for what it connotes: God's forgiveness and determination to continue with the community through the consequences of its rebellions, in response to the faithful intercessions of figures like Moses, Aaron, and Samuel. Now, in Psalm 100:3, on analogy with those three figures who interceded on behalf of their community, Israel on behalf of the whole earth utters the confession which will signal the cessation of all rebellions and eliminate all dissonances from the chorus of praise: "It is he who has made us, and not we ourselves."

---

45. With an eye to the reference to God's "great and terrible [*nora'*] name" in Ps 99:3, we may note that, in Exod 34:10, God tells Israel of the "marvels," and the "terrible thing" that God will do with them. In that context, I suggest, the reference is to God's forgiveness and reinstatement of Israel as "heir" of the promised land despite its radical idolatry in the golden calf. Such treatment of a rebellious vassal by its treaty overlord is so unlike the response of conventional overlords to such rebellion, as to fill those who behold it with awe and wonder, indeed with terror insofar as it undoes a worldview anchored/mired in the moral and political logics of zero-sum reward and punishment and leaves people at the mercy of ultimately unadulterated, non-negotiable mercy.

46. Implicitly rejecting the whole story that flows from Exod 2:23–25?

If the eschatological reign of God achieves its apotheosis *in* the conversion of rebellious powers *through* the act of praise, the community that uses the Psalter acts on behalf of the whole earth in uttering the central affirmation of v. 3. As such, this portion of the Psalter is a call to nations like Assyria in Isaiah 10 and Babylon in Isaiah 47 toward the day spoken of in Isa 19:25.[47] As a chorus that resolves the dissonant chords in creation's ambiguous responses to God, the confession, "and not we ourselves" is thus the climax of Psalms 93–100.

Is "Not We Ourselves" Palatable in Today's World?

I take the importance of the liturgical cry, "it is he that has made us, and not we ourselves," to lie in the way it both identifies and resolves the issue that lies at the heart of existence insofar as existence is rendered problematic and/or enigmatic by the presence and activity of evil in the cosmos. This issue is the problematic relation between the divine will as creating, sustaining, and redeeming the cosmos, and the (variable) freedom displayed in the behavior of the diverse creatures in the cosmos, whether that behavior be taken as embodying purposive freedom in pursuit of narrowly self- or group-serving ends in disregard of the well-being of all creation, or as displaying the interactions of the brute forces of inanimate nature and instinctual and appetitive drives of sentient non-human creatures. (In biblical passages such as Genesis 1 and Psalm 148, all creatures are viewed as coming into existence in response to divine call or command, and as capable of praising God in the sheer fact of their coming into being and "standing forth." One could also cite passages like Isa 40:26 and 48:13; and especially Job 38:35, where lightning, not unlike Abraham [Gen 22:1], Moses [Exod 3:4], Samuel [1 Sam 3:4], Isaiah [Isa 6:8], and Mary [Luke 1:38], calls out, "here we are!") The topic obviously is too vast and complex for meaningful exploration at this point in the present essay. What I want to say, rather, is that the group of Psalms 93–100, and especially Ps 100:3 as here interpreted, offers both unusually comprehensive and unusually focused resources for reflection on the biblical engagement of these issues. At the heart of these issues lies the core issue of the relation between the freedom of the creature and the achievement of the divine purposes in creation and redemption.

---

47. Cf. Paul's eschatological evocation of the biblical psalms to the same effect in Rom 15:8–13.

PART TWO: Forays into a Biblical World

At the outset, I referred to the Anglican theologian, Austin Farrer, and I want now to comment briefly on his work as it bears on these issues. Farrer summed up his understanding of the mystery of human freedom within the creative providence of God in the formulation, "God makes us make ourselves." He said this as one who, as an Anglican priest, would have repeated hundreds of times, in Morning Prayer, the line in the so-called *Jubilate Deo*, "It is he that hath made us, and not we ourselves."

Given his general approach to the behavior of *all* forms of creation vis-à-vis God's creative providence, Farrer's typical, more comprehensive statement is that God "makes things make themselves."[48] But that the eminent instance is human freedom is clear from his general argument, and it becomes explicit where—significantly enough—he connects the thought that God should make me make myself with "what happens when we pray ourselves, on the day when prayer comes really alive."[49]

Farrer derives this understanding from his study of the Gospel of John and, in particular, the presentation there of the relation of Jesus to "the Father." He approaches the figure of Jesus by summing up his argument in the following words:

> God makes his creatures make themselves, and they must truly make themselves by their own principle of action . . . [T]heir action is their existence, and if they did not act of themselves, they would not exist in themselves–in fact, God would have failed after all to create them.[50]

And, later:

> [I]n John's Gospel . . . Christ's Sonship is a continual dependence on his Father's Spirit. He neither speaks, nor acts of himself; his Father in him is the doer of his works. He has all the glory in the world, because his Father's glory is his; he has no glory in himself, but only what his Father bestows.[51]

Farrer's statement, that Jesus "neither speaks, nor acts of himself," while it does refer to a repeated refrain in John's Gospel, is not entirely accurate. In one fatefully significant place, John 18:17–18, Jesus says, "For

---

48. Farrer, *Saving Belief*, 72, 77; see also 51, 52, 74, 76, 82, 121, 124.

49. Ibid., 78.

50. Ibid., 124. I take Farrer to mean that, if creatures do not act "of themselves," but are merely passive puppets of a cosmic puppeteer, God will have failed to create creatures capable of response to God in any morally or spiritually meaningful sense.

51. Ibid., 126.

this reason the Father loves me, because I lay down my life, that I may take it again. No one takes it from me, but I lay it down *of my own accord*. I have power to lay it down, and I have power to take it again; this charge I have received from my Father."

I have explored this passage elsewhere—showing, in the process, the resonance between Jesus as here portrayed and Isaac in the Aqedah of Genesis 22.[52] The argument is too complex to sum up here, but I note that it bears in its own way on what is at issue in Psalm 100:3, that issue being the challenge so to act in responsible freedom that our actions become, not simply the assertions of our own "will to power," but covenanted responses to, and embodiments and disclosures of, God's activity in the world.

At this point I may also summarize my discussion, in an earlier essay in this collection,[53] of what I have come to call David's prayer of amazement in 1 Chronicles 29. That chapter shows David making provision for Solomon's construction of the temple after David's death. At the outset, David declares (29:2) how he has provided for the building of the temple out of all his state and personal resources "*so far as I was able.*" The italicized words translate the Hebrew phrase, *kol-koḥi*, literally, "according to all my power." Then he calls on his fellow Israelites also to make similar offerings "willingly." The Hebrew verb here has as cognate the noun, *nedabah*, a noun that older translations rendered, "freewill offering." As in Exodus 25–31 and 35–40, the sanctuary, as a place for God to dwell amid the people (Exod 25:8), is not to be built out of materials or money gathered as a tax. All materials and monies are to be offered freely and willingly. It is as though the material sanctuary, as built out of what is offered freely, is the outward and visible sign of the inward and spiritual reality wherein God seeks to dwell within the realm of human freedom—so to indwell humanity in the mystery of its freedom that human actions embody and disclose the dynamic presence and activity of the invisible, aniconic God.[54] Such a vision of the mystery of the divine-human relation comes to climactic expression when David turns from his address to his fellow Israelites and, on their behalf, addresses God in prayer.

David opens his prayer with an ascription of praise that first (1 Chr 29:11a) ascribes all greatness, power, glory, victory, and majesty to God.

---

52. Janzen, "(Not) of My Own Accord," 137–60.

53. See chapter 3.

54. I know of no exploration of this mystery that is as suggestive, theologically and existentially, as Coleridge's reflection on the interaction of the divine Spirit and the human spirit, in his *Aids to Reflection*, at "Moral and Spiritual Aphorisms: Aphorism VI," esp. 78–79.

Then (v. 11b) David acknowledges that all that exists in creation belongs to God, and that God is exalted as head over all. Then (v. 12) he acknowledges that riches and honor come from God as generously beneficent ruler over all, and that it lies in God's hand to bestow greatness and strength (*koaḥ* compare v. 2, referred to above). For all this, David (v. 13) gives thanks and praise.

Then, in contrast to the glory of God as so celebrated, David exclaims (vv. 14–15), "But who am I, and what is my people, that we should be able thus to *offer willingly* [or "freely"]? For all things come from thee, and of thy own have we given thee. For we are strangers before thee, and sojourners, as all our fathers were; our days on the earth are like a shadow, and there is no abiding." As I now hear David, his amazement is not simply over the "infinite qualitative difference" between God and humankind, such that God is cosmic landowner, possessor of all that is, and alone enjoying the power to dispense gifts of power and prestige, while humankind is a transient in the earth, having before God and in God's creation only the status of "strangers and sojourners." In ancient Israel, this last expression identifies those who did not enjoy an "inheritance" or parcel of the land meted out to Israel as presented in the book of Joshua; for these sorts of people were laborers and dependents of such a land-holder, who in turn could be called a *nadib*. A *nadib* was one who enjoyed the sort of freedom that could express itself in making a *nedabah*, a freewill offering, or, as the verbal form *hitnaddeb* expresses, "offer willingly." What amazes David is that, given the transient character of humanity as a "stranger and sojourner" on God's earth, and given that all things belong to God, *how is it*—and what does it signify concerning human being—that "we should *be able* thus to offer willingly?" The italicized words translate a Hebrew text that throws the amazing fact into sharper relief: "Who am I, and who are my people, *ki-naʿṣor koaḥ lehitnaddeb kazoʾt?*" The verb *ʿaṣar* means, "keep a firm hold on, hold back, restrain, arrest, lock up." Here, David is amazed beyond words at the realization that such transient mortals as ourselves are nevertheless given by God the gift and the power to hold as our own the power (*koaḥ*) to offer freely—which is to say, that we genuinely have the power to not offer unless we choose to do so. But when our sense of the generosity of God moves us to worship, then the mystery of the interaction of divine and human freedom finds its deepest expression in a prayer of amazement in the course of a genuine free act of self-offering, and that act has, as Farrer calls it, its "springing-point"[55] in God's gift to

---

55. This "springing-point," an image that recurs repeatedly in Farrer's writings in the sense of the "font" or "fountain" where the streams of free divine action issues

us of that freedom. In such an act, the worshipper is like the venturesome servants in Jesus' parable of the talents, for what is offered back to God is "value-added," and the "value" that is added is the divinely enabled free-will offering of ourselves in and through what we give.

I find many other passages in the Bible that bear on the issue as resolved in Psalm 100:3 as I construe that verse. But I know of no passage that presents the mystery so precisely and with such inexhaustible amazement as 1 Chronicles 29. Unless, that is, one eavesdrops on the scene in Gethsemane in Mark 14:36, and then, in light of that scene, re-reads the cosmic and eschatological vision that Paul is attempting to articulate—with prayer at its center—in Romans 8. For further comments on those two passages, see the "New Testament Afterword" in Part V of this volume.

---

in the stream of free human action, lies for Farrer, as for Coleridge, at an "unknown distance" from direct human conscious inspection (the phrase "unknown distance" is Coleridge's: *Aids to Reflection*, 79).

9

Standing on the Promises of God

On the Thematic Resonance of "No Foothold" in Psalm 69

PSALM 69:1–2 VOICES THE desperate cry, "Save me, O God! / For the waters have come up to my neck. / I sink in deep mire, where there is no foothold [*mo'mad*]; / I have come into deep waters, and the flood sweeps over me." At first glance, the expression, "there is no foothold," appears to function simply as an element in the familiar imagery for an existential crisis as a watery abyss. In this essay I hope to show that in fact this expression instantiates a frequent theme in the Psalms, and provides an apt means of connecting the quasi-mythical imagery of watery abyss with the social crisis that the psalmist is undergoing. I shall begin with a general comment on an emerging interpretive strategy in current Psalms studies, and then move in ever-narrowing concentric circles to Psalm 69 and the plaint that "there is no foothold."

Thematic Ligatures in the Psalms

For a long time, modern scholars studied the psalms as individual textual units to be interpreted, where possible, in relation to the historical and/or communal context in which they were composed. Where adjacent or neighboring psalms displayed common terms or images, these elements might be identified as "catchwords," means by which individual psalms might have been gathered into small groups; but no particular meaning was attached to them beyond this function. More recently, such "catchwords" have been studied as signs that the compilers and editors of the psalms intended such psalm groupings to display a thematic pattern that is itself as religiously and theologically interpretable as the individual psalms themselves. In the previous chapter, I have given an example of this new approach through an analysis and reflection on the group comprising Psalms 93–100.

In a related but distinguishable approach, Jerome F. D. Creach has argued that the Hebrew verb, *ḥasah*, "take refuge," and its cognate noun, *maḥseh*, "refuge," provide a clue to the editorial shaping of virtually the whole Psalter, and function to give expression to one of the most prominent existential concerns voiced in this book.[1] In setting forth his research methodology, Creach identifies a number of other verbs and nouns (such as words connoting "cover/shelter" and "place of escape") in the semantic "wordfield" of "refuge"; and he shows how these closely related words run through the Psalter like a many-stranded, tightly-woven thematic cord. This important thematic cord (or ligature, as I like to call it) arises first at the end of Psalm 2, "blessed are all those who *take refuge* [*ḥasah*] in him."

In the present study I shall undertake a probe into another such ligature-complex having at its heart the closely synonymous verbs, *'amad* and *qum* where they mean, "stand, take a stand." Other terms sharing in this complex ligature, or "wordfield," are verbs and nouns referring to a person's stability or instability, especially in reference to a person's feet. Tracing this "wordfield" even in outline exceeds the limits of the present study; but the terms are easily recognizable once the basic theme is identified. While Creach's core term, "refuge" (*ḥasah*), occurs in Psalm 2, I propose that Psalm 1 is marked by the core terms *'amad* and *qum*, and the related image of a transplanted tree. The centrality of these two verbs to Psalm 1, and their significance as launching a recurring theme through the Psalter, may be indicated by the following textual observations.

Psalm 1 begins with a contrast between two sorts of person: the (plural) "wicked / sinners / scoffers,"[2] and the (solitary) individual whose delight is in the *torah*. Whereas the former walk / stand (*'amad*) / sit—find their identity, and take up their social location—by being "in cahoots" with one another, the solitary individual who meditates on the *torah* is presented as a tree transplanted by streams of water whose life will be unceasingly fruitful. This is to say that the psalm begins by identifying two different existential locations. Now, in contrast to the rooted stability and

1. Creach, *Yahweh as Refuge*.
2. I note here, while reserving comment for later, that whereas the first two terms, "wicked" (*reša'im*) and "sinners" (*ḥaṭṭa'im*) are generic terms that may be applied to a variety of behaviors, the third, term, "scoffers" (*leṣim*) refers to a person's scornful and derisive attitude toward God (Prov 3:34), the claims of justice (Prov 19:28), and those who seek to serve God through embrace of the *torah* (Ps 119:51).

PART TWO: Forays into a Biblical World

fruitfulness of this individual, the gang of the wicked "are not so," for, like chaff, they have no rootage or stability, but will be carried away by the first wind that blows.[3] The psalm thus ends by reversing the "group/individual" situation that marked its beginning: the initially solitary individual now (by implication) stands (*qum*) in solidarity with the "congregation of the righteous," while the way of the wicked perishes as they are scattered to their solitary doom. Whereas Psalm 2 differentiates conflictual parties as political entities, Psalm 1 differentiates them as social groups in moral and religious opposition.

In light of this function in Psalm 1, one may note how the core image of "standing," or other terms in its "wordfield" (and, contrariwise, of falling, stumbling, being moved, or sinking), threads its way through the Psalter; and how the faithful individual (sometimes, explicitly, in the congregation or sanctuary) is imaged as a rooted and fruitful tree or vine. In what follows, I shall offer a few selected comments on the connotations of *'amad* as a context for my reflection on "there is no foothold [*mo'mad*]" in Ps 69:3.

On the Social Significance of "Standing"

The verb "stand" and its cognates (not to speak of other terms in this wordfield) refer to one of life's elemental aspects, such that they are indispensable in any vocabulary of 600 words. Compared with its various binary opposites, the word refers to an erect stance rather than sitting or lying down, stopping or remaining still rather than moving, and persisting or enduring rather than giving way or falling. All of these basic, one may say literal, meanings having to do with the human body in itself, also take on figurative connotations of the various relationships in which the individual exists, and the function or status of the individual within those relationships. These relationships may be horizontal, having to do with one's membership, status and/or function within a human community. (So, already, in Psalm 1.) But they may also be vertical, having to do with one's existence "before God." Individuals stand before God in postures of intercession or mediation for others (Gen 18:22; 19:27; Deut 5:5; 10:8; 17:12; 2 Kgs 5:11). Groups stand before God to receive covenant instruction

---

3. We may note that the word, *moṣ*, chaff, occurs frequently as an image for existential instability and transience (e.g., Isa 17:13; 29:5; 41:5; Hos 13:3; Zeph 2:2; Pss 1:4; 35:5; Job 21:18). Given its contrastive function in Psalm 1, it may be said to function as a dark contrasting thread in the complex ligature of the theme of "standing."

(Deut 4:10) and to offer praise (1 Chr 23:30). Beings heavenly as well as earthly can be depicted as standing before God in the heavenly court (1 Kgs 22:21; Jer 23:18). In these and other ways, the verb connotes what in our day we speak of as persons having "standing" in a community.

This connotation appears also in the cognate noun, 'omed, as well as the cognate noun, ma'mad (a virtual semantic twin of mo'mad in Psalm 69). The following examples are noteworthy: In 2 Kgs 23:2–3, the king reads in the people's hearing the words of the book of the covenant that had been found in the house of the LORD. Then the king "stood ['amad] by the pillar ['ammud] and made a covenant before the LORD . . . to perform the words of this covenant that were written in this book; and all the people joined in ['amad be-] the covenant." The verb here refers to a specific social/religious function and status, of the king on the one hand and of the people on the other. The parallel passage in 2 Chr 34:31–32 rings slight, but for our purposes significant, changes on the root 'md: "The king stood ['amad] in his place ['omed] and made a covenant before the LORD . . . to perform the words of the covenant that were written in this book. Then he *made stand* in it [he'emid] all who were present in Jerusalem and in Benjamin." Here the king's social status and function are associated with a physical place ('omed) in the sanctuary. To borrow sacramental language, the physical place is the outward and visible sign of his inward and social/spiritual "place" or status and function. In Neh 8:7; 9:3; 13:11; 2 Chr 30:16; 35:10, 'omed likewise refers to the "place" or "station" that priests, levites, and people respectively occupy in sacral assemblies. With this we may compare the noun, ma'mad which, in its five occurrences (1 Kgs 10:5// 2 Chr 9:4; Isa 22:19; 1 Chr 23:28; 2 Chr 35:15), carries similar connotations. In our idiom, all these occurrences indicate both one's physical location and one's "standing" in the community.

One's Standing in Others' Eyes

I want to draw attention now to a specific mode in which one can *experience* one's "standing" in relation to others, whether these others consist in one's political neighbors, one's community, or one's own family. This mode of experiencing one's standing has to do with the existential kernel of truth in the famous but largely discredited dictum of the philosopher, Bishop George Berkeley, when he posited that *esse est percipi*, "to be is to be perceived." The truth in this dictum lies in the formative or deformative, enhancing or diminishing, impact one's awareness of others' perceptions of

PART TWO: Forays into a Biblical World

one can have on one's psychological and even physical well-being.[4] (Belief in the power of the "evil eye," and the need to avoid its impact, is common in ancient Mediterranean cultures.) The biblical awareness of this is implicit in, among other things, occurrences of the verb, *qalah* ("be lightly esteemed, held of little account"), and its noun cognate, *qalon* ("dishonor, disgrace"); also in the related verb, *qalal*, "be slight, swift, trifling," which in a passive form can mean "be lightly esteemed" in someone's eyes (2 Sam 6:22). Especially instructive is Gen 16:4-5: "when [Hagar] saw that she had conceived, *she looked with contempt on her mistress.*" In Hebrew the italicized words literally read, "Sarah became little [of no account] in her [Hagar's] eyes." Then the scene shifts from Hagar's perspective to Sarai's, as she complains to Abram, "when [Hagar] saw that she had conceived, *she looked on me with contempt.*" Again, the Hebrew text reads, literally, "I became of no account in her eyes." Sarah's perception of Hagar's belittling perception of her leaves her feeling small. The "wrong" (*hamas*, literally, "violence") done to Sarai has to do with her sense of where she now stands in this multi-woman household, with its attendant dangers of being slighted in that household and by being neglected in practical affairs as well. Another instructive instance occurs in Num 13:32-33, where the

---

4. For a definitive brief psychological analysis, see Hans W. Loewald, "Superego and Time," in his *Papers on Psychoanalysis*, 43-52, republished in *The Essential Loewald*, 43-52. What Freud calls the superego—typically the parents' perceptions of and expectations for the growing child, Loewald terms the child's "ideal ego," an ideal still largely lying outside the child, and as present within its own psychodynamics as the shaping call of the parents, a call consisting in the child's perception of the parents' perception of it. But as the growing person internalizes this ideal ego, it comes to be an integral part of the person's own self-structure and self-perception, an "ego ideal," the ego's own aspiration for what it may become. The existential fatefulness at issue here may be appreciated, in biblical terms, by noting the almost incessant complaints of the solitary psalmist concerning others' false witness and undeserved reproach or taunting, and over against this the psalmist's sense of the accuracy of God's penetrating assessment of the psalmist's inmost being (Psalms 51 and 139), as also in the Priestly Blessing of Numbers 6 with its emphasis on the shining of God's face and the echoes of that blessing in Psalms like Ps 80:3, 7, 19. In the New Testament, the essence of Jesus' temptations in the wilderness lies in the competing perceptions of what his Messiahship should consist in, as offered by the Tempter on the one hand and the divine expectations that God places in Deuteronomy on Israel as his "son" (Deut 1:31; cf. 32:4-6), expectations especially placed on Israel's eventual king (Deut 17:18-20). The issue is engaged again at Caesarea Philippi, when Jesus asks his disciples about others' perception and estimation of him, and then asks them the same question. Peter's momentary accurate perception soon gives way to conventional expectations that Jesus rejects in the harshest way, identifying them as unwittingly echoing those of the Tempter in the Wilderness following Jesus' baptism.

spies sent ahead of the Hebrews to reconnoiter the promised land report that "all the people that we saw in it are men of great stature. And there we saw the Nephilim (the sons of Anak, who come from the Nephilim); and [reading the Hebrew literally] *we were in our own eyes like grasshoppers, and so we were in their eyes.*" They did not only "seem" so (RSV, NRSV), they "were" so (KJV)—such was the impact of the Nephilim's perception of them on their sense of themselves in their capacity to enter in and take the promised land.[5]

A positive version may be found in *"Jaqob's"* meeting with Esau following his night-long wrestling with the mysterious figure beside the *"Jabboq,"* during which he receives a new name, "Israel"—as if to suggest that the night-long wrestling, echoing the struggle between Jacob and Esau already in the womb (Gen 25:22), issues in a new birth of identity and character. Jacob had fled from home because of Esau's murderous hatred (Gen 27:41–45) over Jacob's repeated trickery. Now (33:1) Jacob sees Esau drawing near with four hundred men (!), so he takes steps to approach Esau in great humility and obeisance. Yet, finally face to face with Esau, he finds himself saying (33:10), "if I have found favor in your sight [Hebrew, "in your eyes"], then accept my present from my hand; for truly to see your face is like seeing the face of God, with such favor have you received me." With this we may compare God's address to Jacob's descendants who themselves have gone into exile and now are being encouraged to anticipate a return to their home (Isa 43:4): "you are precious in my eyes, and honored, and I love you."

Another positive instance: In Isa 40:11, the theme of God leading the exiles home is conveyed through the image of a shepherd leading a flock and gathering the lambs in his arms, carrying them in his bosom and gently leading those who are with young. The general resonance between this imagery and Jacob's words in Gen 33:13–14 is enhanced by the fact that the Hebrew word underlying "lambs" in Isa 40:11 (*tela'im*) actually refers to a variety of things in virtue of their surfaces as patched (sandals, in Josh 9:5), variegated in color (shrines, in Ezek 16:16, or spotted sheep, in Gen 30:32, 33, 35). The word *tala'*, then, in Isa 40:11, is not simply a plural form of the noun, *talah*, "lamb" (as in 1 Sam 7:9; Isa 65:25), but a deliberate allusion to the returning exiles as analogs to the variously marked sheep gained by Jacob in exile, while Jacob himself provides the basis for the

---

5. Compare the words of my six-year-old grandson, a *Star Wars* devotee, appealing from his three-year-old sister's harassment to his mother: "Mommy, Lulu is weakening me in the eyes of my enemies"; and then, not sure that his mother's intervention will suffice, adds, "Are gods alive?"

PART TWO: Forays into a Biblical World

portrayal of God as shepherd. In that case, one may be pardoned for reading Isa 43:1–7 as an address of this divine shepherd to (the descendants of) "Jacob/Israel," an address whose core message (resonating with the meeting between Esau and Jacob) is, "fear not . . . because you are precious in my eyes, and honored, and I love you."

To the examples of the effect on one's existential well-being of being negatively perceived, we may add the ubiquitous occurrences in the Old Testament of the theme of "reproach" or "taunting," signaled by the verb, *ḥarap* and its noun cognate, *ḥerpah* (generally translated in the LXX as *oneidizo/oneidismos*). To become the object of others' reproach is to have one's place among them, or relative to them, denigrated and destabilized, if not called into question altogether; and the effect on such objects of reproach is to threaten their very hold on life in this world. Here we may recall my observation above, in a footnote on the term, *leṣim*, "scorners," in Psalm 1. To be scorned or derided or scoffed at in regard to one's stance before God is—if the scoffing be effective—to have that stance undercut and subverted.

Standing in the Face of Reproach in Psalm 69

The presence of the theme of reproach is ubiquitous in the Hebrew Bible. With the case of Sarah and Hagar in mind, we may note that the first of 111 occurrences of the verb or noun comes in the context of the multi-wife family of Jacob. Because God "saw" that Jacob loved Rachel and "hated" Leah (that is to say, discounted her in comparison), God enabled Leah to bear numerous sons (Gen 29:31), beginning with Reuben (*re'uben*), so named because, as Leah said, "the LORD has looked [*ra'ah*] upon my affliction." In this cameo scene we have a graphic instance of someone enjoying one status (or standing) before God while suffering a disadvantaged emotional status within one's closest community. But Sarah's fecundity serves as an implicit reproach to Rachel's barrenness, until finally, with the birth to her of Joseph, Rachel is able to say, for her part, "God has taken away my reproach" (Gen 30:23).[6]

In the Psalter, the verb and noun for reproach occur thirty-one times. Its importance is indicated by its first occurrence, Ps 15:2–3, where it serves as the first mark of the sort of person who may sojourn in God's tent and dwell on God's holy hill: "who does not slander with his tongue, / and

---

6. For a fuller analysis of the complex Hagar-Sarah relationship, see Janzen, "Hagar in Paul's Eyes and in the Eyes of Yahweh."

does no evil to his friend, / nor takes up a reproach against his neighbor; / in whose eyes a reprobate is despised, / but who honors those who fear the LORD." Such a person (to extend the image of sojourning and dwelling) "shall never be moved" (15:5). This psalm, then, collocates terms fundamental to the issues in Psalm 69, where the Hebrew word, "reproach," occurs six times (69:7, 9 [2x], 10, 19, 20). Whatever it is that has drawn it on the psalmist's head, it is as though this reproach, and the effect it has on the psalmist's sense of belonging in the world, is the capstone to the psalmist's suffering. The effect of this reproach is to place the psalmist in a position of precarious solitariness, deprived of the life-nourishing support of his community. At the quasi-mythical (or poetically cosmological) level, this deprivation is experienced as taking the ground out from under him, leaving him to sink down into the waters of cosmic chaos "without a foothold." Just as the nouns ʿomed and maʿmad function to indicate both physical and social location, so here the noun moʿmad functions to connote loss of social standing and, more radically, loss of a foothold in the cosmos itself.[7] In this way of construing the phrase, "without a foothold" in Psalm 69 gives voice to a particularly dire point along the thematic ligature that begins in Psalm 1—a dire condition despite the assurances of "standing" given the faithful person in that first psalm.

"Standing" as a Ligature throughout the Psalter

At this point, we may pause for a brief synoptic survey of relevant occurrences of of the verb, ʿamad, "stand," and some of the other terms that belong to its "wordfield." (1) The core verb, ʿamad, occurs 32 times in the Psalter, of which the following are noteworthy (where necessary, I alter the RSV to render evident the use of the Hebrew verb, marking such with an asterisk): *18:33, "He made my feet like hinds' feet, and *caused me to stand* on the heights." *26:12, "My foot *stands* on level ground; in the great congregation I will bless the LORD." *31:8, "Thou hast not delivered me into the hand of the enemy; thou hast *caused* my feet *to stand* in a broad place." *69:2, "I sink in deep mire, where there is no *foothold*; I have come into deep waters, and the flood sweeps over me." *76:7, "Terrible

---

7. Erich Zenger (in Hossfeld and Zenger, *Psalms 2*, 172–173) identifies the language in Ps 69:1–4 as employing "the typical metaphorical world of the ancient Near East, in which chaos or death is described as a mighty force that arises in the midst of life and destroys it," whereas, in 69:5–13b "the imagery of water and chaos . . . is no longer in evidence; instead, the semantic field of mockery and contempt dominates," signaling the psalmist's "social ostracism."

PART TWO: Forays into a Biblical World

art thou! Who can *stand* before thee when once thy anger is roused?" *106:23, "Therefore he said he would destroy them—had not Moses, his chosen one, *stood* in the breach before him, to turn away his wrath from destroying them." *106:30, "Then Phinehas *stood* and interposed, and the plague was stayed." *122:2, "Our feet have been *standing* within your gates, O Jerusalem!" *130:3, "If thou, O LORD, shouldst mark iniquities, Lord, who could *stand*?" *134:1, "Come, bless the LORD, all you servants of the LORD, who *stand* by night in the house of the LORD!" *135:2, "you that *stand* in the house of the LORD, in the courts of the house of our God!"

I offer the following comments: (1a) In regard to the psalmist vis-a-vis human enemies, 18:33 comes after the "watery chaos/death" imagery in 18:4–5. With 18:33 we may compare 31:8. (1b) As I construe it, the plight of the solitary psalmist in 69:2 has its eschatological counterpart in 26:12. (1c) Among psalmic echoes of the tree image in Psalm 1 we may identify Ps 92:13–15 (in its use as a Sabbath psalm) as its eschatological apotheosis. (1d) In regard to human standing before God, we may note 76:7 and 130:3. The imagery in Psalm 130 is closely related to that in Psalm 69. For one thing, the word translated "depths" in 130:1 is the same as in 69:2, 14 (compare also Isa 51:10; Ezek 27:34). For another, 130:3 identifies the one thing that could render a person without a standpoint even before God: iniquity. But God's forgiveness of iniquity (expressed by the verb, *salaḥ*, and the noun, *'awon*, is no arbitrary or "one-off" affair. Its rootedness in God's character is clear from the classic texts in Exod 34:9 and Num 14:19, where (as also in Jer 31:34; 33:8; 36:3; 50:20) forgiveness concerns Israel as a whole. It is this theme of God's forgiveness of Israel's radical covenant infidelity that provides the ground both for the individual's hope in Ps 130:5 and for that individual's turn and exhorting encouragement of Israel in 130:7–8. The same ground for hope, I suggest, underlies the Psalmist's turn to others, and to Zion, in Ps 69:32–36. (1e) It is this ground for hope, then, that underlies also the affirmations of 106:23, 30; and, ultimately, 26:12; 122:2; 134:1; 135:2. Thus the final word on "standing" underscores the central theme of Psalm 1 with its imagery of rooted fruitfulness and standing (*qwm*) in the congregation of the righteous.

Of 51 occurrences of *qum* ("arise, stand"), the following are particularly relevant: *1:5, "Therefore the wicked will not *stand* in the judgment, nor sinners in the congregation of the righteous." *20:8, "They will collapse and fall; but we shall *rise* and stand upright." *24:3, "Who shall ascend the hill of the LORD? And who shall *stand* in his holy place?" *36:12, "There the evildoers lie prostrate, they are thrust down, unable to *rise*." *40:2, "He drew me up from the desolate pit, out of the miry bog, and *set* my feet upon a rock, making my steps secure." *41:8, "They say, 'A deadly thing has fastened upon him; he will not *rise* again from where he lies.'" *89:43, "Thou hast turned back the edge of his sword, and thou hast not *made him stand* in battle." *113:7, "He *raises* the poor from the dust, and lifts the needy from the ash heap." *119:28, "My soul melts away [*dalap*] for sorrow; *strengthen* [*qayyem*] me according to thy word!" 119:62, "At midnight I *rise* to praise thee, because of thy righteous ordinances."

A few comments: (2a) The connotations here are similar to those with *'amad*. In most instances, the action of "rising" has as its implicit outcome the regaining of one's ability to "stand." (2b) In 119:62 (as with *'amad* in 26:12; 122:2; 134:1; 135:2), the verb could as well be translated, "stand." (2c) In 119:28, RSV and NRSV "strengthen" is interpretive in light of the preceding expression, which is translated, "my soul melts." But in its other four occurrences, the verb *dalap* refers to the falling drip of tears or rain. And the factitive form, *qayyem*, literally, "make to stand," occurring eleven times in the Bible, occurs six times in Esther where it refers to issuing a permanent (we would say, "standing") ordinance governing religious practice. In view of the downward motion in *dalap* elsewhere, I take this verb to indicate downward motion also in Ps 119:28—a "sinking feeling," as we might put it—"My soul sinks for sorrow." The prayer, then, is "hold me up," or "sustain me." If the psalmist is able to stand under whatever weighs him down (repeatedly, in this psalm, the scorn and scoffing of the insolent), it is God's word that will revive him (119:25) and enable him to stand.

(3) There is no space here to peruse the 25 occurrences of the verb, *mot*, "totter, shake, slip, be moved," in the Psalter, many of them having to do with stability on one's feet.

(4) But we may note the verb, *taba'*, "sink," which forms part of the watery imagery in Ps 69:2, 14. Of its eight other occurrences in the Hebrew Bible, the following are especially noteworthy: (4a) Exod 15:4, "Pharaoh's chariots and his host he cast into the sea; and his picked officers are *sunk* in the Red Sea"; and (4b) Ps 9:15, "The nations have *sunk* in the pit which they made; in the net which they hid has their own foot been caught." Here the imagery is partly figurative and in part corresponds to the prose narrative in Exodus 14, where the similar imagery in Exod 15:4–10 is turned into a literal description.

(5) For our purposes, the two most noteworthy occurrences come in the book of Jeremiah.

PART TWO: Forays into a Biblical World

The Case of Jeremiah

I pause here to note once again how the texts in Scripture are inter-linked in myriad ways, which, when patiently traced and pondered, build up a sense of an existential world into which one is invited to enter and to stand. In the present instance, I am led to Jeremiah simply in following the inter-textual link afforded by the verb, *ṭabaʻ*, "sink" (as though clicking on a hyperlink in a website). But, as we shall see, Jeremiah is connected to Psalm 69 in more ways than this. For the present, we may note the following:

> So they took Jeremiah and cast him into the cistern of Malchiah, the king's son, which was in the court of the guard, letting Jeremiah down by ropes. And there was no water in the cistern, but only mire, and Jeremiah *sank* in the mire. (Jer 38:6)

As Zenger points out, this narrative and literary parallel to the figurative language in Psalm 69 is only the most striking of the connections between this psalm and this prophet. When Zedekiah later meets with Jeremiah in private, Jeremiah offers him what counsel he can, and warns that, if the king refuses to surrender, this is the vision Jeremiah has seen from God concerning this king's future:

> Behold, all the women left in the house of the king of Judah were being led out to the princes of the king of Babylon and were saying, "Your [i.e., Zedekiah's] trusted friends have deceived you and prevailed against you; now that your feet are *sunk* in the mire, they turn away from you." (Jer 38:22)

Here, the literal "sinking" scenario which Jeremiah had himself just lived through becomes a figure for Zedekiah's social and political fate. He has lost his former standing as a king on a throne among his people, and, after his sons and nobles are killed before his eyes, he is blinded and thrown into prison for the rest of his life (Jeremiah 39 and 52).

Among the connections between Jeremiah and Psalm 69, Zenger takes Ps 69:7, "it is for thy sake that I have borne reproach," as a word-for-word quotation of Jer 15:15.[8] Following other interpreters at Ps 69:20 in taking the form *ʾanuša* as an adjective, "incurable," rather than (as in RSV, NRSV) a verb, "I am in despair," he sees in it an allusion to Jer 15:18, "Why is my pain unceasing, my wound *incurable* [*ʾanuša*], refusing to be healed?" We may note that, like Psalm 69, Jer 15:18 voices the prophet's

---

8. Ibid., 180.

Standing on the Promises of God

complaint to God in the face of enemies and even his own family who reproach and taunt him, and that it is this reproach that constitutes his deepest wound.

To these specific connections between Psalm 69 and Jeremiah we may now add the thematics of "standing." Those thematics are presented in Jeremiah with particular richness:

1. The verb 'amad refers to those who stand in the king's presence as his servants (Jer 36:21; 52:12, where RSV "served" translates Hebrew "stood before").

2. It refers to those assembled people who stand in the sanctuary (Jer 7:10; 28:5), on the assumption (in these cases, gratuitous) that they are, so to speak, in good standing with God.

3. It refers, in Jer 44:15, to the Jewish refugees in Egypt who, gathered together, men and women, in a "great assembly," hear Jeremiah charge them with idolatry. Again, this is no *ad hoc* crowd, but a body that considers itself knit together by, and addressable concerning, its formal practices in worship.

4. Of the political powers of Edom and Babylon and their kings, God says, in announcing the judgment about to fall on them, "What shepherd can stand before me?" (Jer 49:19; 50:44)

5. Over against all these groups and their dubious claims to standing, Jeremiah announces the divine word that, because the house of the Rechabites has remained faithful to God, "Jonadab the son of Rechab shall never lack a man to stand before me."

6. As for Jeremiah, his "standing" is bi-directional:

    a. he stands in the court of Yahweh's house, or elsewhere, to announce God's word to the people (Jer 7:2; 17:19; 19:14); and

    b. he stands before God to intercede on behalf of the people (Jer 15:1; 18:20).

Of course, his credentials, his standing, as a prophet is implicitly challenged by the "false prophets" who proclaim peace over against his word of judgment, and especially by the "prophet" Hananiah (Jer 28:11) and Pashhur the priest (Jer 20:1–2). But over against the human estimation of who does and who doesn't have standing with God and therefore as a member of the people of God, Jeremiah sets another claim, when he says,

PART TWO: Forays into a Biblical World

> Who among them has *stood* in the council of the LORD to perceive and to hear his word, or who has given heed to his word and listened?
>
> ...
>
> But if they had *stood* in my council, then they would have proclaimed my words to my people, and they would have turned them from their evil way, and from the evil of their doings. (Jer 23:18, 22)

Here we have a standing of another order. Over against the rootage of human identity and status and existential efficacy as derived from one's participation in a human community—a community whose determinative power vis-à-vis its members ranges from the dimensions of politics and economics to the dimension of approbation and disapprobation, honor and reproach—over against this, we see Jeremiah claiming a standing that is not derived *therefrom*, and that is not adjudicable *thereby*, for it is not located *therein*. This is not to claim that the "locus" of the divine council floats "above" the human realm in a "heaven" that exists on the same four-dimensional co-ordinates that mark the physical contexts and dimensions of human existence. But it is to suggest that the term, "stand," here—and in Psalm 69:2, not to mention the other figural uses of the verb in the Psalms—is more than a "mere" figure of speech. I want to propose that the claim to "stand" in the divine council expresses an interior sense that is the experiential ground of Jeremiah's conviction of having been called as a prophet. More broadly, I shall argue that this image of standing in the divine council over against a community in which one's standing is called radically into question, to the point where one is both literally and figuratively "in the pits," goes to the heart of the biblical presentation of the reality of God in human experience and conviction.

In exploring this issue further, I follow Abraham Heschel in his description of the prophetic consciousness as he lays it out in the introduction to his monumental study, *The Prophets*. As he says at the outset, his aim "is to attain an understanding of the prophet through an analysis and description of his *consciousness*, to relate what came to pass in his life—facing man, being faced by God—as reflected and affirmed in his mind."[9]

In taking this approach, Heschel seeks to avoid the "absolute objective and supernatural" approach of dogmatic theology and at the same time the tendency of some psychologizing approaches to "reducing [prophecy] to a subjective personal phenomenon" and thereby disregarding "the prophet's

---

9. Heschel, *The Prophets*, xiii.

awareness of his confrontation with facts not derived from his own mind." For Heschel, prophecy "is composed of revelation and response, of receptivity and spontaneity, of event and experience."[10]

In a sentence, Heschel's reading of the prophetic experience "from within" is that it consists, at its core, in the prophet's *sympathetic* participation in the divine *pathos*, as that pathos arises in God with variable tone and texture in relation to Israel's varying situations in the world and responses to or defections from the covenant. What the prophet experiences is a feeling of powerful pathos *within* himself that is not simply derived *from* himself nor simply from the situation around him; and, for the prophet, that pathos is the pathos of God.

To my mind, perhaps the best example of this in the Hebrew Bible is, not surprisingly, the experience of Moses at the burning bush (see chapter 1). There, God reports to Moses what God has seen, heard, and now knows of the Hebrews' affliction in Egypt (Exod 3:7), moving God to "come down to deliver them" (3:8). This is the dawning awareness in Moses of what the narrator had reported already in 2:23–25; and the shift in wording between these two passages is suggestive. In 2:23–25, God hears, sees, remembers the ancestral covenant, and "knows" the enslaved Hebrews' condition; in 3:7–8, the temporal reference turns from past to future, from "remembering," to "coming down to deliver." But immediately, Moses' sense of *God's* coming down to deliver is transformed into the sense that God is sending *him* to deliver the people (3:10).

In Heschel's terms, Moses' awareness of the pathos of God for Israel becomes Moses' own sympathetic participation in that pathos, such that he finds that (sym)pathos urging him in a vocational direction with an insistence that overcomes his initial resistance. I have deliberately said, "(sym)pathos," to draw attention to the mystery at the heart of this experience. To a radical humanist psychoanalyst, or, for that matter, to a Jungian analyst, all that is presented as a datum for interpretation is the prophet's consciousness as suffused with a depth of feeling that impels toward action, together with the prophet's report that that depth of feeling arises within the prophet from "another place" (so to translate *maqom*, in Esth 4:14, the Hebrew word there being a noun cognate with the verb *qum*, "to stand").

This "place" is not plottable on the four-dimensional coordinates of the created order. But this way of putting the matter is not quite accurate. Whatever else may be said about it, by way of poetic imagery to

10. Ibid.

PART TWO: Forays into a Biblical World

convey what is not literally sayable (for example, the imagery of a "divine council"), this "place" is, so to speak, "ingredient" in Moses' or Jeremiah's consciousness. By "ingredient," I mean "entering into, so as to become a part of." If there is any truth to Moses' or Jeremiah's testimony to the experience of their call, it means that, however transcendent God is, Moses and Jeremiah, as creatures with bodies, are plottable on this world's four-dimensional grid, and if God is in any slight degree ingredient in their experience, then the invisible, infinite, and eternal God is in that degree ingredient in this world. And whatever the standpoint of these two persons vis-à-vis the Pharaohs and Babylons of this world, or vis-à-vis the variable perceptions and attitudes of their own families and communities, these two, at the same time, occupy another invisible but existentially experienceable standpoint—which Jeremiah refers to as the divine council—*right where they are*. There is no way to verify or falsify claims to such an experience other than to act on the claims it makes on one's own response. This is in accord with the strange word of assurance that God gives Moses at the burning bush: "This shall be the sign for you that I have sent you: when you have brought forth the people out of Egypt, you shall serve God upon this mountain" (Exod 3:12).

Before I return to the case of Jeremiah, I want to return to two texts quoted above in tracing the theme of "standing as a ligature throughout the Psalter." The texts are these: Ps 76:7, "Terrible art thou! Who can *stand* before thee when once thy anger is roused?"; and Ps 106:23, "Therefore he said he would destroy them—had not Moses, his chosen one, *stood* in the breach before him, to turn away his wrath from destroying them." The Hebrew word translated "breach" can refer to an opening broken into a city wall allowing enemy forces to enter the city (e.g., 2 Kgs 25:4). In this literal context, to "stand in the breach," is to stand in that opening and fend off attackers with one's weapon(s) and one's body. The word can also be used of a social breach (Judg 21:15); and figuratively, it can also be used in contexts where Israel has fundamentally broken the covenant, as in Ezek 22:30: "I sought for a man among them who should build up the wall and stand [*'amad*] in the breach before me for the land, that I should not destroy it; but I found none." The implication is that the covenant is in several senses a kind of "wall." It functions as a "boundary-marker" distinguishing, and separating, between Israel and other peoples (Deut 4:5–8). Also, insofar as the Sinai covenant is conditional, with sanctions for disobedience, the keeping of the covenant serves as a protection against God's wrath—Israel becoming an enemy of God through such breach of covenant. In the context of the present essay, how are we to understand the standpoint of Moses, who *stands in the breach* between a people who have lost their standing before God and a God who is bent on destroying the city and scattering its inhabitants among the nations? What is it that Moses stands on, in that "no-man's-land"? Does he simply stand "on his own two existential feet"? In the ancient world that would itself constitute rebellion. If Moses was successful before God in "standing in the breach," and if Exod 32:13, and Num 14:13–19 are any indication of what it was that gave him a standpoint, Moses "took his stand" upon the promises of God, and upon God's steadfast love and forgiveness. Insofar as those promises are founded in the mystery of God's generosity and compassion—and insofar as Moses must appeal past God's wrath to that compassion and generosity—Moses may be said to takes his stand on what is utterly hidden from a conscience occluded by such deep rebellion, a stand on what only a Moses, faithful in all matters pertaining to the covenant people (Num 12:7), dares to believe is still there.[11]

To resume, then, the case of Jeremiah. What does he present as evidence of his claim to be—unlike the false prophets—a true prophet of Yahweh? All he can offer, finally, is the implicit claim that he has "stood in the council of Yahweh" (23:18, 22). But what does this *feel* like? By "feel," I do not refer to Jeremiah's emotional state, but to something more basic, for which the "ordinary" flesh-and-blood analogy is the bodily awareness that one enjoys in standing amid a company to which one rightfully belongs, and among whom one's credentials for participation cannot be called into question. One may think of the way a toddler stands pressed close to a mother's thigh in the presence of strangers, or of a young child snuggled against its father's chest as he reads a bedtime story. Or one may imagine a scenario in which one has crashed a coming-out party for a wealthy debutante. One may have rented a tux and faked an invitation card, and now mingles among the other guests. Suddenly one finds oneself in the presence of the young woman and her parents. What is that feeling in one's knees? Such is the look on their faces that it would be a mercy for the floor to open beneath one, and for one to descend into a watery abyss, foothold or no foothold. But think of someone else who, though not of that social circle but earlier the young woman's high school classmate, has received from her a genuine invitation. Now, in mingling with the other guests, this

---

11. Compare the lines from William Cowper's hymn, "God Moves in Mysterious Ways": "Behind a frowning providence / God hides a smiling face." In Exod 32:11, "Moses *besought* the LORD," the italicized verb translates a Hebrew idiom that, taken literally, is "sweetened the face."

PART TWO: Forays into a Biblical World

young man finds himself the object of raised eyebrows, pointed fingers, and decidedly unpleasant whispered jibes and slanders. He begins to feel distinctly uncomfortable, and is on the verge of leaving when, at the door, he runs into the young woman and her parents. To a relief that verges on disbelief, he hears her say, "Mom! Dad! This is the fellow I've been telling you about," and, turning, "I'm so glad you could come." What is his feeling then? From his time in a church choir, he hears in his ears—but for the first time with a depth of existential understanding—the words of the hymn that begins, "Glorious things of thee are spoken, / Zion city of our God," and especially the verse, "On the rock of ages founded, / Naught can shake thy sure repose; / With salvation's walls surrounded, / Thou mayest smile at all thy foes." That feeling is of a secure footing beneath one and strong walls around one. Just so, Jer 1:17–19 indicates the prophet's awareness that, though Jerusalem and the cities of Judah have become lumped together with all the nations as God's enemies (1:13–16), the capital city thereby losing its status before God as "inviolable Zion," Jeremiah himself will continue to enjoy the standing and assurances that once belonged to it. But this awareness is not a passive matter. The "I will make you this day a fortified city" is the divine side of what on Jeremiah's side is experienced as the exhortation to "gird up your loins; arise [*qum*!], and say to them everything that I command you" (Jer 1:17). Here, I suggest, the verb *qum* is synonymous with the word *'amad* that occurs several times where Jeremiah is commanded to go and deliver a specific word. What is the subjective feeling in that verb, especially alongside the exhortation to "gird up your loins"?

What I mean to be getting at is simply this: That the divine pathos is distinguishable from the prophetic (sym)pathos; but it is not experienced apart from the prophet's consciousness of a sympathy with that pathos.[12] Similarly, God's promise to make Jeremiah a fortified city is distinguishable from, but not experienced apart from, the call to him to gird up his loins and to *rise-and-stand* for God amid the people. Indeed, as God says, "do not be dismayed before them, lest I dismay you before them" (1:17). Taken simply at a conceptual level, this is a terrifyingly conditional assurance! But it does convey the way in which the divine assurance is not simply infused into a passive recipient, but is to be met by a response that (unlike that of Israel's spies in the face of the gigantic inhabitants of the

---

12. Compare God's repeated, "in my anger a *fire* is kindled," or the like (Jer 4:4; 11:16; 15:14; 17:4, 27), with Jeremiah's "If I say, 'I will not mention him, or speak any more in his name,' there is in my heart as it were a burning *fire* shut up in my bones, and I am weary with holding it in, and I cannot" (20:9; cf. 5:14).

land, in Numbers 13) appropriates that assurance and internalizes it as the ground of one's stance within the community.

Standing before God: The Case of Daniel

In support of such a reading of the experience of Moses and Jeremiah, I offer here a brief reading of Daniel's visionary experience as presented in Daniel 8–10. In chapter 8 a vision appears to him (8:1), consisting of animal images representing human political powers, and "holy ones" speaking first to one another. One of the latter, in the figure of Gabriel, addresses Daniel "where I stood [*'eṣel 'omdi*]" (8:17). In view of my comments earlier on the social connotations of the noun, *'omed*, I take Daniel here as referring, not simply in an ad hoc way to where he happens to be standing, but to the *post* or *position* in which he finds himself; for, clearly, he is standing in, or before, what Jeremiah would call "the council of Yahweh." Moreover, the echoes of Hab 2:3 in Dan 8:17, 19; and 10:14 place Daniel squarely within the trajectory along which prophetic vision transforms into apocalyptic vision.[13] When Gabriel comes near where Daniel is stationed, he states, "I was frightened and fell upon my face." As Gabriel speaks to him, he reports (8:18), "I fell into a deep sleep with my face to the ground; but he touched me and *set me on my feet* [*wayya'mideni 'al 'omdi*; literally, "caused me to stand at my station]." When Gabriel has finished interpreting the vision, he instructs Daniel to "seal up the vision, for it pertains to many days hence" (8:26; cf. 7:28) At this, Daniel reports, "I, Daniel, was overcome and lay sick for some days; then I rose and went about the king's business; but I was appalled by the vision and did not understand it."

Then, "in the first year of Darius," Daniel reads in the book of Jeremiah of the seventy years that must pass "before the end of the desolations of Jerusalem" (9:1–2). So he turns his face to God, with fasting and sackcloth and ashes, praying and supplicating "because thy city and thy people are called by thy name" (9:3–19). Again Gabriel appears to him to give him wisdom and understanding (9:20–27). Yet again, "in the third year of Cyrus king of Persia," a word is revealed to Daniel. He has been in mourning and fasting for three weeks, when a figure of overwhelming aspect appears before him, whose words are "like the noise of a multitude"

---

13. The language in Hab 2:3, *'od ḥazon lammo'ed / we-yapeaḥ laqqeṣ*, "still the vision awaits its time / and testifies to the end," is echoed in Scripture only in Dan 8:17, *le'et qeṣ heḥazon*, "the vision is for the time of the end"; 8:19, *lemo'ed qeṣ*, "the appointed time of the end"; and 10:14, *'od ḥazon layyamim*, "the vision is for days yet to come."

(10:1–6). At this, those who are with Daniel flee to hide themselves, and, he reports:

> I was left alone and saw this great vision, and no strength [*koaḥ*] was left in me; my radiant appearance was fearfully changed, and I retained no strength [*lo' 'aṣarti koaḥ*].[14] Then I heard the sound of his words; and when I heard the sound of his words, I fell on my face in a deep sleep [*nirdam*] with my face to the ground. (10:8–9)

Presumably the "deep sleep" (as already in 8:18) is of the sort we see in passages like Gen 15:12 and Job 4:13, where the Hebrew word indicates the state of mind of one undergoing a visionary experience (we might call it an altered state of consciousness). But in Ps 76:6 it describes the state of the formerly stouthearted in the face of God's rebuke. In either case, it appears to refer to a state of consciousness in which one's normal faculties and bearings are "undone."

With this, a hand touches Daniel and sets him *trembling* on his hands and knees. (The Hebrew verb here occurs in Isa 19:1 in parallel with the melting of the heart.) At this the figure speaks more reassuringly:

> O Daniel, man greatly beloved, give heed to the words that I speak to you, and stand upright [*'amad 'al 'omed*, literally, "stand at your post," or, "stand your ground"], for now I have been sent to you." While he was speaking this word to me, I stood up [*'amadti*] trembling. (Dan 10:11)

The figure then says to him, "fear not" (v. 12) and reassures Daniel that he has come in answer to Daniel's prayers, in order to help him understand "what is to befall your people in the latter days. For the vision is yet [*'od ḥazon*] for days to come" (v. 14)

These words render Daniel dumb and, again, face to the ground. This time the figure touches his lips, and he is able to say:

> O my lord, by reason of the vision pains have come upon me, and I retain no strength [*lo' 'aṣarti koaḥ*]. How can my lord's servant talk with my lord? For now no strength remains in me [*lo' ya'amod bi koaḥ*], and no breath [*nešamah*] is left in me. (10:16–17)

---

14. We may note that the idiom here is the same as that occurring in 1 Chr 29:14. For a fuller treatment, see chapter 3. In that scene, David is filled with amazement that he and his people—like all humans, only "strangers and sojourners," that is, persons of no independent standing vis-à-vis their benefactors and employers—have the power to make freewill offerings to build a sanctuary for God to dwell in their midst.

Again Daniel uses the idiom, "retain strength." But now he augments that idiom, in two related ways. For the devout Hebrew, one's *nešamah* is given one at birth (Gen 2:7); and every breath continues to be God's ongoing gift, so that, for example, even when Job swears an oath of innocence before the God who has embittered him through his terrible losses, he swears that oath with the breath and the spirit (*ruaḥ*) that God gives him (Job 27:3). But before this figure, even after being reassuringly addressed as "greatly beloved" (Dan 10:11), and told that he is to receive understanding in a vision of things yet to come, Daniel has no *nešamah* left in him. Moreover, no strength (*koaḥ*) remains in Daniel. Given the references in this passage to Daniel as falling on his face or trembling on hands and knees, and the call to him to "stand at your post" (Dan 10:11), I suspect that the expression, *lo' ya'amod bi koaḥ*, in Dan 10:17, like the similar expression in Josh 2:11,[15] touches on his continuing inability to stand erect through the weakness in his limbs. But of course, even though Daniel says he has no breath in him with which to speak to this figure, he does have enough breath to say *that*!

On hearing this figure's reassuring address, Daniel reports that again this figure touches him, "and he strengthened [*hizzeq*] me" (Dan 10:18). The verb, as in Isa 35:3, is in the active voice, indicating that the figure before Daniel *does* something to him. As in Isaiah 35, this figure strengthens Daniel by *saying* something to him. And what he says, in essence similar to Isaiah 35, is, "O man greatly beloved, fear not, peace be with you; *be strong* and *be of good courage*" (Dan 10:19a). Here the RSV's two italicized phrases represent what in Hebrew is a repetition of the same verb, "*ḥazaq*, and *ḥazaq*." Following this, the RSV reads: "And when he spoke to me, I was strengthened [*hithazzaq*], and said, 'Let my lord speak, for you have strengthened [*hizzeq*] me'" (10:19b).

The RSV here, following the KJV and followed in turn by the NRSV, translates the form of the verb, *hithazzeq*, in the passive: "I was strengthened." This is no doubt because of the fact that the verb *ḥzk*, in a different verbal stem indicating the active voice, occurs immediately afterward in Daniel's response to God: "You have strengthened me." But this is to obscure the subtle shift in nuance conveyed by the shift in verbal forms, from "be strong" to "I strengthened myself" to "you have strengthened me," a shift in nuance going to the heart of what Heschel would call the dynamics

---

15. Josh 2:11: "And as soon as we heard it, *our hearts melted*, and *there was no courage remaining* [*lo' qamah 'od ruaḥ*] in anyone." The verb *qum* here carries the connotation of standing.

of the prophetic consciousness. To appreciate this, we need to take a brief detour to consider some typical uses of the verb, *ḥazaq*, together with its close synonym, *'ameṣ*. (This detour also serves to introduce a Hebrew expression that I shall study more closely in Chapter 10.)

On Some Hebrew Expressions Involving the Verbs *ḥazaq* and *'ameṣ*

The basic form of the verb, *ḥazaq*, means, "be strong." In the jargon of grammarians, when a verb of this kind occurs in this form, it can carry what is called an "ingressive" connotation, a connotation of "entering into a state," in the present instance a state of being strong. When the middle consonant is doubled (in this instance, *ḥizzeq*), a verb like this takes on what is called a "factitive" connotation, in which the subject of the verb (here, the strange figure) "makes" the object of the verb (here, Daniel), "strong." Or, one could say, the figure enables Daniel to become strong. But the form *hitḥazzeq* carries yet a third connotation. When the middle root letter is doubled, and the verb is prefixed with the syllable, *hit-*," this Hithpael stem of the verb carries the connotation of reciprocity, reiteration, or reflexive movement.[16] The verb *ḥazaq*, in this Hithpael form, occurs 26 times in other passages in the Hebrew Bible, and always elsewhere it carries a reflexive meaning, "to strengthen oneself." A good example is Gen 48:2, where Jacob is on his deathbed. When Joseph is told that his father is failing, he comes to him hoping for a blessing for his two sons. When someone tells Jacob of Joseph's arrival, "then Israel [Jacob's other name] *summoned his strength* [*hitḥazzeq*] and sat up in bed." In many of these passages, the RSV translates *hitḥazzeq* with "be of good courage," or "strengthen oneself."

Given that the Hithpael stem of *ḥazaq* everywhere else carries a reflexive, and not a passive, connotation, we may explore the implications of its having the same connotation in Dan 10:19. The whole passage may be read like this:

> And he said, "O man greatly beloved, fear not, peace be with you; *be strong* and *be of good courage* [*ḥazaq we-ḥazaq*]." And when he spoke to me, I summoned my strength [*hitḥazzaq*],

---

16. So, e.g., the verb for blessing occurs, in its basic stem, only in the passive, *baruk*, "blessed." In the Piel ("factitive") stem it means, "to make blessed, to bless." In the Hithpael stem (as in Gen 22:18; 26:4; Ps 72:17; Jer 4:2) it functions reflexively, "to bless oneself."

and said, "Let my lord speak, for you have strengthened [*ḥizzeq*] me."

This combination—of exhortation and the shifts in connotation of this verb—lies at the heart of the mystery in which the presence and activity of God is experienced in its effects, those effects consisting in the subject's own sense of, as it were, being called upon to call upon one's own inner resources and to take one's stand on them.[17]

With this we may compare the more than a dozen occurrences of the paired verbs, *ḥazaq* and *'ameṣ*. In Deut 3:28, God instructs Moses to "charge" (or command) Joshua, and in so doing, to "encourage [*ḥazaq*] and strengthen [*'ameṣ*] him." In Deut 31:6, Moses' similar words to Israel are followed by the exhortation (similar to God's words to Jeremiah in Jer 1:17b), "do not fear or be in dread of them." This theme crops up again in Deut 31:23; Josh 1:6, 7, 9, 18; 10:25; Isa 35:3 (with reference to weak hands and feeble knees); Nah 2:1; 1 Chr 22:13; 28:20; 32:7. In all these passages, persons are called on to call on and muster their strength.

Standing before God: The Case of Habakkuk

Given the connections noted above between Daniel and Habakkuk through Daniel's use of Habakkuk's language, we may note that the latter book begins with the prophet—like Daniel after him—interceding with God for Israel. On hearing God's first answer, Habakkuk intercedes again, then (in Hab 2:1) "takes his stand" (*'amad*), to "watch," and "stations himself"—all three verbs indicating a formal posture consistent with a specific function before God on behalf of the community. Following the divine response in chapter 2, Habakkuk responds (in the present form of the text) with a poem in classical form and with classical portrayals of God's action in judgment and redemption. This picture produces an initial effect in Habakkuk similar to the one in Daniel:

> I hear, and my body trembles,
>   my lips quiver at the sound;
> rottenness enters into my bones,
>   *my steps totter* beneath me.

Then his total aspect changes, as he says,

---

17. On this complex way of reporting an experience of God's strengthening through one's summoning one's strength, see the remarkable report discussed in Chapter 10.

> I will quietly wait for the day of trouble
> 	to come upon people who invade us. (Hab 3:16)

and he concludes,

> Though the fig tree does not blossom,
> 	nor fruit be on the vines,
> the produce of the olive fail
> 	and the fields yield no food,
> the flock be cut off from the fold
> 	and there be no herd in the stalls,
> yet I will rejoice in the LORD,
> 	I will joy in the God of my salvation.
> GOD, the Lord, is my strength;
> 	he makes *my feet* like hinds' feet,
> he *makes me tread* upon my high places. (Hab 3:17–19)

Back to Psalm 69

I want now to show how the preceding forays into dimensions of the theme of "standing" bear on Psalm 69 in particular, and in the Psalter generally. I shall begin by comparing the treatment of the "watery abyss" theme in Psalms 18, 40, and 69. In Psalm 18, the opening verses celebrate God as the psalmist's "strength"—*ḥezeq*, the noun cognate with the verb *ḥazaq* that we encountered at the heart of Daniel's experience in Daniel 10. This "strength" is then characterized through the images of "rock, fortress, deliverer, rock, refuge, shield, horn of salvation, and stronghold," all connoting stability and ability to withstand attack. (This list, we may note, contains a number of terms in Creach's "refuge" wordfield.) Following this laudatory opening, the psalmist reports his plight in terms of the watery abyss (Ps 18:4–5). From these depths, the psalmist cried to God for help (18:6), and that help is then portrayed in theophanic imagery generically similar to some of the imagery in Habakkuk 3. The result of God's deliverance is that "he brought me forth into a broad place" (Ps 18:19). We may recall that in Ps 31:8 the psalmist rejoices because "thou hast caused my feet to stand [*ha'amid*] in a broad place." That the same connotation is implicit here is suggested by the recapitulation in Ps 18:31–33: "Who is God, but the LORD? / And who is a rock, except our God?—/ the God who girded me with strength, / and made my way safe. / He made my feet like hinds' feet, / and set me secure [*ha'amid*, literally, "caused me to stand"] on the heights." Again two verses later, the psalmist says (18:35–36), "Thou hast given me the shield of thy salvation, / and thy right hand supported me, / and thy help made me great. / Thou didst give a wide place for my steps under me, / and my feet did not slip [ma'adu]."[18] Finally, in 18:49, the psalmist declares, "For this I will extol thee, O LORD, among the nations, and sing praises to thy name." Here the second verb ('azammerah) is marked with a cohortative suffix, indicating that here the psalmist calls upon himself to sing praises in response to God's deliverance.

Psalm 40 presents the same general picture, of which I note the following elements:

(1) The psalmist begins with a brief reference to his plight, his prayer, and God's "hearing" of that cry. Then the psalmist says, "He drew me up from the desolate pit, / out of the miry bog, / and *set* [*heqim*] my feet upon a rock, / making my steps secure. / He put a new song in my mouth, / a song of praise to our God." (Note that "set" translates *heqim*, a causative form of the verb, *qum*, "to stand".) What is striking here is that the psalmist includes within God's saving actions the putting of a song of praise in the psalmist's mouth. The act of praise is the psalmist's own, he has uttered it unstintingly and unreservedly (40:9–10); and yet even as he sings it he experiences its uprising in his throat as a prolongation and new form of God's saving action for and in him; for the uprising of praise in the psalmist's throat is the voicing of an uprising of existential energy in the psalmist's heart and soul.

(2) However, the psalm contains no explicit parallel to the *self*-exhortation to praise and extol God, such as we saw in Ps 18:49. Rather, the psalm ends on a note of renewed prayer for deliverance (Ps 40:11–17). Yet there is an indirect echo, in that Ps 40:16 ends in an *other*-directed exhortation: "May all who seek thee / rejoice and be glad in thee; / may those who love thy salvation say continually, / 'Great is the LORD!'"

What, then, of Psalm 69? Strikingly, there is no explicit indication of God's deliverance from the watery chaos that is devoid of any foothold. As late as v. 29 (and compare vv. 14–15), the psalmist seems still to be in that plight, as he cries, "I am afflicted and in pain; let thy salvation, O God, set me on high [*teśaggebeni*]!" (The last verb is cognate with the noun, *miśgab*, "lofty stronghold," which occurs at the end of Ps 18:3.) Yet, with no indication that God has acted, or even "heard" the psalmist's plea (as in Ps 40:1),

---

18. We may note the exquisite way in which the "reversal of fortunes," from the danger of slipping to being enabled to stand secure, is mirrored in the reversal of consonantal letters in the verbs at the end of v. 33 ('*amad*, "stand") and v. 36 (*ma'ad*, "slip").

the psalmist suddenly shifts from a minor to a major key, and announces (Ps 69:30), "I will praise [*ʾahallelah*] the name of God with a song; I will magnify him with thanksgiving." As the cohortative suffix on the first verb shows, what we have here is the psalmist calling on himself to praise God. If in Ps 134:1 the psalmist can exhort, "Come, bless the Lord, all you servants of the Lord, who *stand* by night in the house of the Lord," here, in an existential darkness, the psalmist calls upon himself to engage in such an act of praise—an act, I suggest, that is the expression of a sense within the psalmist—despite the absence of any foothold in the space-time social and material world—of a standing before God from which he can utter such anticipatory praise. If this association seems too far a stretch, we may consider how, in Psalm 42–43, the psalmist repeatedly addresses himself with words of comfort, encouragement, and resolve to praise (42:5, 11; 43:5), and the ground of this resolve is his memory of having participated in such a joyous singing procession in the past (42:4).

Psalm 77 provides an unusually informative example of the dynamic that I am tracing here–so unusual that its precise force, as captured in the KJV, is obscured in the RSV and NRSV. With RSV and NRSV, "I commune with my heart in the night; / I meditate and search my spirit," compare the following translations:

- Ich denke des Nachts an mein Saitenspiel,
  und rede mit meinem Herzen;
  mein Geist muss forschen. (Martin Luther)

- I call to remembrance my song in the night:
  I commune with mine own heart:
  and my spirit made diligent search. (KJV)

- I will call to mind my song in the night,
  I will muse in my heart,
  and my spirit maketh anxious search. (Delitzsch)

- I call to remembrance my song in the night:
  I commune with mine own heart;
  And my spirit maketh diligent search. (ASV)

- I remembered my songs in the night.
  My heart mused and my spirit inquired. (NIV)

- I will remember my harp,
  in the night with my heart,
  I will consider, so that my spirit broods. (Hossfeld)

Standing on the Promises of God

The first line consists of the following Hebrew units: (1) *'ezkerah*, a cohortative form of the verb *zakar*, "remember," with the nuance of calling upon oneself: "let me remember/call to mind;" (2) *neginati*, a first person possessive suffix, "my" attached to the noun, *neginah* which is derived from the verb, *nagan*, "touch (strings), play a stringed instrument," and can refer here to the instrument or the music played on it or the song sung to that music. (3) *balaylah*, "in the night." The second line reads, literally, "I will meditate with my heart" (compare Luther, KJV above), while the third says, "I will search my spirit." What we have here is a graphic portrayal of what I have called in chapter 5 "prayer as self-address," and what we have just seen exemplified in Pss 42:5, 11; 43:5. Here, we see how it is undertaken. One engages one's heart in an internal dialogue and searches one's spirit, through the instrumentality (!) of the songs one has been accustomed to sing.[19] More specifically, as the cohortative mood of the verbs indicates, one calls upon oneself to do this.

We may compare Ps 108:1–2, "My heart is steadfast, O God, my heart is steadfast! / I will sing and make melody! Awake, my soul [*kabodi*, literally, "my glory"]! Awake, O harp and lyre! / I will awake the dawn!" (The verbs, "sing," and "make melody," are in the cohortative, the self-address being underscored by the following imperative, "awake," addressed to one's soul or "glory.") What is happening here? Elsewhere in the Bible, the imperative "awake" is several times addressed to God, with the plea that God may come to deliver the plaintiff from a dire situation. (See Isa 51:9–10; Pss 7:6; 35:23; 44:23; 59:5.) In Isa 52:1 God responds in kind to the people's call in 51:9–10, saying in 52:1, "Awake, awake, put on your strength, O Zion." Here we find explicit expression of the implicit connotation that the call to "awake" includes a call to the addressee to arouse one's strength. One could not ask for a better example of what is involved in the reflexive verb-form, *hithazzeq*, "to muster one's strength." In Psalm 77, as in Psalm 108, the muster involves the calling to mind of deep felt memories encoded in the pulsating vibrations of music and the words they accompany. Those memories may lie so deep in the soul that current afflictions and travails may obscure them; so it is by means of the "sonar probes" of instrumentally accompanied song that one searches one's spirit for those moving emotions.[20]

19. Compare the case of the infant Jessica, cited in Chapter 5.

20. Loewald, in his essay, "Superego and Time," presents a view of the structure of the self not simply as three-tiered (id, ego, superego), but as temporally dimensioned, such that the id or the unconscious is the presence of the past subliminally in the present, the superego is the presence to the ego of the claim or lure of the future, and

PART TWO: Forays into a Biblical World

In Psalm 77, the initial self-exhortation in v. 6 is then elaborated in vv. 11–20. Resuming the cohortative mood, the psalmist says, "I will call to mind the deeds of the LORD; / yea, I will remember thy wonders of old. / I will meditate on all thy work, / and muse on thy mighty deeds." Then vv. 13–20 presents the content of that remembering and meditation and musing. The mighty deeds, of course, are those celebrated, in the first instance, by the song in Exod 15:1–18. They are deeds that manifest God's might (*'oz*, Ps 77:14). But their transformative effect on the psalmist is evident in the tone and mood that steals over the psalmist's words as he moves toward and then settles into the imagery of the last verse: "Thou didst lead thy people like a flock by the hand of Moses and Aaron." If the song here recalled celebrates God's *'oz*, "might, strength," the act of recalling and celebrating that strength has the effect of revivifying the psalmist's fainting spirit (77:3). In this light, we may propose that songs such as Israel's foundational hymn in Exodus 15, when learned and internalized, become a sort of stored potential energy, available whenever "awakened" through being called upon. But this energy is distinguishable from the primal energy of Freud's id, insofar as that raw, relatively formless primal energy, channeled into meaningful form through the symbolism of language and music, is capable of service also in shaping the superego in meaningful hope, which is to say, shaping one's sense of what the future may hold for one and claim of one.[21]

A Last Word, then, on Psalm 69

The psalmist, finding no foothold in the watery abyss as such, somehow finds a foothold in the ideal company of those who truly have Zion's well-being at heart; and he experiences this foothold in his self-exhortation to sing and praise God in that company. Thus the psalmist, alone and a social outcast at the beginning, finds himself in the end where Psalm 1 tells us

---

the ego is the self in its present activity of engaging each of the other temporal dimensions in the light of the other. In Psalm 77, I suggest, we see the psalmist searching in his deep self for the presence within him of the softly but still vibrating tones of past celebration, in order to regain a sense of hope.

21. Here I am drawing on Loewald's profound discussion of the dynamics of genuine sublimation, in which the primal energies of the id are not repressed, but channeled through symbolic processes, in such a way that the symbols, infused by an transforming those energies, can function as efficacious ideals for moral and spiritual aspiration and striving. See Loewald, *Sublimation*, reprinted in *The Essential Loewald*, 439–527.

Standing on the Promises of God

that those who meditate in the *torah* finally will stand–in the congregation of the righteous.

It is not uncommon nowadays to hear and read criticisms of religion in America as being excessively focused on the individual, and for individuals as being excessively focused on their own present status and future prospects before God, with the result that the here-and-now social and communal dimensions of religious concern are neglected. Such criticism has its valid aspects. But an emphasis on the communal dimension of religious concern can itself become excessive and distorting, not to say unjust, where the standpoint of the individual comes to be viewed merely as a function of the community as a whole. It is salutary to be reminded that, among the types of psalms that make up the Psalter, the single most frequent type of psalm is the complaint of the individual. And again and again, at the heart, or as a prominent feature, of the psalmist's plight lie the attitudes, words and actions of others against the individual, all of which have the effect—as so vividly expressed in Psalm 69—of depriving the individual of his or her standing in the community. In such cases—if this psalm is any guidance—the individual is on solid ground in appealing to God not only as a place of refuge but as one's standpoint before God and therefore also in the community.

It remains to be observed, again, that Psalm 69 is second only to Psalm 22 in frequency of quotation or echo in the New Testament, in reference to Jesus in his crucifixion. As a concrete "place" in the space-time-material world, the cross is also the sign that one who is impaled on it is thereby bereft of any legal, moral, and spiritual standing in the world of human relations. The call, then, that Jesus issues to his followers to take up their cross and follow him, may be viewed as a call to forsake the temptation to seek one's *ultimate* existential standpoint in any of the structures or institutions of creation. However positive a role they may play as part of the ordering of God's creation and as mediations of God's blessing for creation, the promise they hold out to the individual of providing an *ultimate* standpoint is a temptation to idolatry. The call of Jesus, like the call of Psalm 69, is to stand on the promises of God.

The call of Psalm 69, so construed, calls to mind the quip that an atheist is one who has no invisible means of support. What, then, to make of the strange resonance between Psalm 69 and the stoutly secular credo of the prominent twentieth-century philosopher Bertrand Russell? In his credo, published under the title, "A Free Man's Worship,"[22] he follows his

22. In Russell, *Why I Am Not a Christian*, 104–16.

summary description of the cosmos (that all things originated and will end in a 'hot nebula whirl[ing] aimlessly through space")[23] by saying

> Such, in outline, but even more purposeless, more void of meaning, is the world which Science presents for our belief. Amid such a world, if anywhere, our ideals henceforward must find a home. That man is the product of causes which had no prevision of the end they were achieving; that his origin, his growth, his hopes and fears, his loves and his beliefs, are but the outcome of accidental collocations of atoms; that no fire, no heroism, no intensity of thought and feeling, can preserve an individual life beyond the grave; that all the labours of the ages, all the devotion, all the inspiration, all the noonday brightness of human genius, are destined to extinction in the vast death of the solar system, and that the whole temple of Man's achievement must inevitably be buried beneath the debris of a universe in ruins—all these things, if not quite beyond dispute, are yet so nearly certain, that no philosophy which rejects them can hope to stand. Only within the scaffolding of these truths, only on the firm foundation of unyielding despair, can the soul's habitation henceforth be safely built.[24]

Such a cosmos surely offers as powerful a challenge to human aspirations as any watery abyss confronting our psalmist. Amid such a world Russell seeks a *home* for his ideals—the ideals he does cherish, which are noble and admirable. I note his imagery: Only within a *scaffolding* erected on such a ceaseless "flux," only on the *firm foundation of unyielding despair* evoked by such an accidental flux, can the soul's *habitation* be *safely built*. He goes on to characterize the threat to the ideals he espouses, and to a life built upon their foundation, as coming both from the mindless brute material forces at work in this cosmos and from those humans who would act out their base passions and achieve their ignoble desires through a similar display of brute, coercive power. In the face of these two forces, natural and social, he opposes simply his ideals—ideals that he takes to be purely human constructs. "Shall we worship force, or shall we worship goodness? Shall our God exist and be evil, or shall he be recognized as the creation of our own conscience?"[25]

The "God" that exists independent of humankind consists, for Russell, in the brute forces of nature around one and within one in the form

---

23. Ibid., 105.
24. Ibid., 106–7.
25. Ibid., 109.

of one's animal self, forces that he calls "evil," while the Good that he seeks to serve is a good that arises purely as a human project. "If Power is bad, as it seems to be, let us reject it from our hearts. In this lies Man's true freedom: in determination to worship only the God created by our own love of the good, to respect only the heaven which inspires the insight of our best moments."[26] Continuing in that vein, Russell writes, "Let us learn, then, that *energy of faith* which enables us to live constantly in the vision of the good; and let us descend, in action, into the world of fact, with that vision always before us."[27] And, resonating with another biblical trope to be considered in Chapter 10 on Psalm 27, he writes of the help that we owe to one another: "Be it ours to shed sunshine on their path, to lighten their sorrows by the balm of sympathy, to give them the pure joy of a never-tiring affection, to *strengthen failing courage*, to instil *faith* in hours of despair."[28]

Ontologically, the psalmists would disagree with Russell. But would they not resonate with his description of the existential texture of faith, explored with respect to the phenomenology of consciousness? Does the faith of the psalmists, amid all that challenges it, not at times have the feel of having no foothold other than the sheer determination of the psalmist to affirm the reality of all that the psalmist has affirmed aforetime (e.g., Psalm 42–43)? Does the cry to God not at times feel like self-address (as in the case of Hannah)? If the confession in Ps 63:8, "My soul clings to thee; thy right hand upholds me," may, on one reading, offer two perspectives on one "thick" conjoint act, the human act of clinging arising out of and in turn establishing its foothold in the deeper reality of the divine act of upholding; in another reading, the two affirmations may be taken in temporal succession, such that only in retrospect does the act of naked human clinging disclose the deeper fact of a divine upholding. If, nevertheless, both perspectives affirm God as independent of the projective outreach of the soul's clinging, in contrast to Russell's stubborn clinging to ideals taken as nothing more than human constructs, one may be pardoned for imaging Jesus say to Russell, as to the scribe of the law, "you are not far from the kingdom of God" (Mark 12:34).

---

26. Ibid.
27. Ibid., 110 (italics added).
28. Ibid., 115 (italics added).

PART TWO: Forays into a Biblical World

A Belated Confession

In the interest of full disclosure, I offer here a final word on how I came to undertake this analysis of Psalm 69 centering in verse 2. It began when I pondered, for the umpteenth time, what Heb 11:1 means when it says that "faith is the *hypostasis* of things hoped for, the *elenchos* of things unseen." Finding the discussion of the two Greek words in recent commentaries instructive but not conclusive, I undertook a study of the Hebrew texts translated by these words. In the case of *hypostasis*, I came in due course to Ps 69:2, where the Greek translation of *'en mo'mad*, "no foothold," is *ouk estin hypostasis*.

When I considered that Psalm 69 is one of the most frequently quoted and echoed Psalms in the New Testament,[29] and that the Letter to the Hebrews has as one of its chief aims to exhort and encourage followers of Jesus not to become disheartened under various afflictions—especially such things as reproach[30]—I entertained the strong suspicion that the word *hypostasis* in Heb 11:1 means something like what this whole paper has been exploring: That, when surrounded by a Jewish and Gentile world that systematically subjects followers of Jesus as the Christ to *oneidismos*, "reproach," and so deprives them of the social standpoint that they would claim for their lives as followers of Jesus as Messiah, it is their *faith* (elsewhere in Hebrews, their *hope*) that provides them with that standpoint. Needless to say, that faith is itself grounded in God's faithfulness. But, as Abraham Heschel might say, the experience of God as ground is manifest to consciousness as an inner sense of having a standpoint and an inner strength to endure. But an analysis of the plausibility of this construal of *hypostasis* in Heb 11:1 is a project for another essay (see chapter 16).

---

29. Zenger gives the following listing (I convert their psalm versification into what appears in English Bibles): 69:4 (John 15:25); 69:9 (John 2:17; Rom 15:13; Heb 11:26); 69:21 (Mark 15:23, 36 and parallels in Matthew and Luke; John 19:29); 69:22–23 (Rom 11:9–10); 69:24 (Rev 16:1); 69:25 (Acts 1:20); 69:28 (Phil 4:3; Rev 3:5; 13:8; 17:8; 20:12; 15:21–27). In addition to the echo of Psalm 69 in Heb 11:26, we may note the congruent connotations of the quotation of Isa 35:3 in Heb 12:12: "Strengthen [*hazzequ*] the weak hands, and make firm [*'ammeṣu*] the feeble knees."

30. *Oneidismos*, six times in Psalm 69 LXX; and in Hebrews at 10:23; 11:26; 13:13 ("abuse" in RSV and NRSV.)

10

The Verb *ya'ameṣ* in Psalm 27:14
Who Is Strengthening Whom?

MY FORAY INTO PSALM 69, in the preceding chapter, eventually led across the trail of issues encountered already in chapter 5. But the way in which elements in Psalm 69 intersected with those issues led to a consideration of additional motifs, as expressed through the use of characteristic expressions. Among these expressions I identified the verb-pair, *ḥazaq—'ameṣ*, and gave a brief survey of their connotations when used in tandem. That survey was sufficient for my discussion of the complex dynamics involved in the question of Daniel's standing before God. But it made no mention of Ps 27:14, which employs these two verb in tandem but in a highly unusual way. So unusual is this usage that the translation in KJV, "be of good courage, and he shall strengthen thine heart," which follows the Hebrew precisely, is modified in RSV and NRSV, "be strong, and let your heart take courage," a result that brings it into conformity with the usual meaning of these verbs when paired.

In the present essay, I want to examine with a fine-tooth comb the texts that bear on what is at issue in Ps 27:14. That issue is among the most complex and elusive—if not, indeed, the most complex and elusive—in religious experience within a biblical world and in theological reflection on that experience. The issue concerns the form and nature of divine-human interaction in the depths of the human spirit. It concerns questions of divine and human initiative, divine and human receptivity to each other's initiatives, patience and hope as gifts of the divine Spirit and as practices of the human spirit, and so on. In such matters, experience itself shades off from a bright center of consciousness to vague, penumbral awareness, and language that attempts to represent this complexity must itself resort to unusual means of indirection and intimation, sometimes taking stock expressions and bending them to its own probing purposes, in such a way that what is attempted by way of intimation, of "inkling," depends on the torque that is set up between the conventional expression and its new, bent form. I take the use of the verbs *ḥazaq* and *'ameṣ* to be a case in point. In order for us to appreciate what the psalmist is attempting to say in this verse, we must attend carefully to what these verbs generally mean elsewhere. What follows, then, is a painstaking tracking of their usage elsewhere in the Bible, leading up to a full appreciation of their novel impact at the end of Psalm 27. At the end of this foray, I will present a short twenty-first-century statement that strikes me as catching the force of this last verse exactly.

A Preliminary Review of the Hebrew Text in Psalm 27:14

Psalm 27:14 reads as follows: *qawweh 'el-yhwh / ḥazaq weya'ameṣ libbeka / weqawweh 'el-yhwh*. The LXX translators rendered the middle line, *andrizou, kai krataioustho e kardia sou*, "be brave, let your heart become strong," thus construing *libbeka* as the subject of the verb, *ya'ameṣ*. In the Latin Vulgate, the line was similarly rendered, *viriliter age; et confortetur cor tuum*. But Miles Coverdale rendered the line, "be strong, and he shall comfort thine heart," and the translators who produced the KJV followed him with "be of good courage, and he shall strengthen thine heart." More recent English translations have reverted to the Greek and Latin versions, RSV and NRSV reading, "be strong, and let your heart take courage"; JB reading, "be strong, let your heart be bold"; and NIV reading, "Be strong and take heart." The JPS (1917) likewise rendered the line, "Let thy heart take courage," and its successor, the NJPSV (Tanakh), reads, "be . . . of good courage." These recent translations reflect a general consensus among modern commentators. The comments of Franz Delitzsch are representative:

> Instead of *we'emaṣ* (Deut xxxi. 7), it is said, as in [Ps] xxxi. 25, *weya'ameṣ libbeka*, let thy heart show itself strong. The translation: may he (Jahve) strengthen thine heart . . . would require *ye'ammeṣ*; but *he'emiṣ*, like *hirḥib*, [Ps] xxv. 17, belongs to the class of intensive denominatives, in which Hebrew is by no means poor, and in which Arabic is especially rich.[1]

---

1. Delitzsch, *Psalms*, 1:440. In the Deuteronomy passage, Moses exhorts Joshua in the words of the stock expression (occurring eleven times in the MT), *ḥazaq we'emaṣ*,

With what Delitzsch terms "the class of intensive denominatives," we may compare the following comment on the Hiphil stem in GKC §53d.

> Among the ideas expressed by the *causative* and *transitive* are included . . . a series of actions and ideas, which we have to express by periphrasis, in order to their being represented by the Hiph'il-form. To these *inwardly transitive* or *intensive* Hiph'ils belong . . . Hiph'il stems which express the obtaining or receiving of a concrete or abstract quality.[2]

This passage goes on to include *'mṣ* among the verbs so functioning in the Hiphil. Similarly, BDB gives the meaning of this verb's Hiphil stem as "exhibit strength, be strong," while *HALOT* gives the meaning, "show strength, prove to be strong."

The purpose of the present study is to re-open the question as to whether *ya'ameṣ* is to be construed as having *leb* or God for its grammatical subject. The interest here is not simply precision in translation, but precise nuance in the portrayal of the nature of human experience, as expressed in this verse, in the move from inner weakness and anxiety to inner strength and encouragement. As I shall argue, Coverdale and the translators of the KJV had it right, and the ancient and modern versions have obscured the three-dimensional perspective on the existential dynamic in question. I shall present my argument through a detailed examination of all the occurrences of the verb, *'mṣ*.

The Verb *'mṣ* in the Qal Stem

The verb *'mṣ* occurs forty-one times in the MT.[3] In the Qal stem it appears sixteen times; in the Piel, nineteen times; in the Hiphil, two times (Pss 27:14; 31:25); in the Hithpael, four times. The verb occurs in the Qal

---

"be strong, and take courage." Delitzsch's comparison of *he'emiṣ* to *hirḥib* backfires; see further below.

2. GKC §53d (italics in the original). The part of the quotation following the last ellipsis comes in subsection (a) in the text. Following subsections identify uses of the Hiphil, (§53e) to express "the entering into a certain condition and, further, the being in the same"; (§53f) "stems which express action in some particular direction"; and (§53g) *denominatives* which express the *bringing out*, the *producing* of a thing, and so are properly regarded as causatives" (all italics original). It would appear that Delitzsch's construal of the Hiphil in Ps 27:14 conforms to GKC §53g.

3. The following analysis, including statistics, has been made with the use of the *Accordance Bible Software* (version 8.1.1, November 2008) produced by *Gramcord*.

stem most often (eleven times) in the stock expression, *ḥazaq weʾemas*, "be strong, and take courage." This expression is accompanied six times by the mirroring stock expression, "do not fear or be dismayed."[4] All eleven occurrences come in direct speech, in which one party exhorts and encourages another party. A moment's reflection on the *inter*personal and *intra*personal dynamics here will set the stage for further comments.

On the one hand, when Moses summons Joshua, in Deut 31:7 (to take Delitzsch's example), and in the sight of all Israel says, "be strong, and take courage," these words, as imperatives, function to elicit a response in Joshua. The verb here functions in an *inter*personal speech-act. The response will consist in Joshua mustering his inner strength and morale. Yet the *intra*personal act arises in response to words that are heard as conveying what they call for. One's ability to "take heart" is due in part to the "heartening" encouragement given in the exhortation.[5] One can say that these words, as "spoken-by-Moses/appropriated-by-Joshua," constitute a single complex word-event in two dimensions.

In five of the six instances in which the stock expression is accompanied by the mirroring counterpart, "do not fear or be dismayed," these combined stock expressions are followed by a clause that begins with *ki*, "for." The *ki* clause refers either to an accompanying and enabling divine action (Deut 31:7; Josh 10:25) or reassures the hearer(s) that God will go or be "with you" (*ʿimmak* or the like; Deut 31:6, 23; Josh 1:9; 1 Chr 28:20; 2 Chr 32:7). In Deut 31:6 and 1 Chr 28:20, the latter prepositional phrase, *ʿimmak*, is accompanied by yet another stock expression, *loʾ yarpeka weloʾ yaʿazbeka*, "he will not fail you or forsake you." (The latter expression, together with the prepositional phrase, *ʿimmak*, occurs also in Deut 31:8 and Josh 1:5.) The implication is that this single complex "word-event" displays not just two dimensions involving the human speaker and the human actor, but also a third "accompanying" divine dimension. The reference in these passages to God's "withness" suggests that God is present to both speaker and hearer in and through the spoken and received word. The word, then, is, in such contexts, the nexus of a three-dimensional event. As such, the word is "super-saturated" with meaning and power. Moreover, insofar as these words do not simply "go in one ear and out the other," but sink down into the hearer's memory, where they may be

---

4. The 11 occurrences (with the 6 "accompanied" instances with asterisks) are: Deut *31:6, 7, 23; Josh 1:6, 7, *9, 18; *10:25; 1 Chr *22:13; *28:20; 2 Chr *32:7.

5. In J. L. Austin's terminology, verbs used in this way function "performatively," performing what they mean (*How to Do Things with Words*).

recalled (re-called) whenever needed, those words deep in the hearer's memory become one mode of God's continuing to be with the hearer.

(Two analogues on the human plane may illustrate the dynamics that I am identifying here. Little Jonathan Coalson came home from kindergarten and said to his mother, "We learned a new song in school today! I can sing it for you—it's still in my mouth." In Midland, Texas, in 1987, an eighteen-month-old child, Jessica McClure, fell down an old abandoned well. Because of the danger of a cave-in, the rescue dragged on for hours. In the meantime, rescuers lowered a two-way microphone into the well to offer reassurance to the presumably terrified child. To their amazement, instead of a terrified child, they heard a reasonably calm little girl singing to herself songs that her mother had often sung to her.)

In addition to its occurrence in these 11 instances, the Qal stem occurs four times (Gen 25:23; 2 Sam 22:18 // Ps 18:17; Ps 142:7) in the comparative construction, *'ms min*, "stronger than." (In Ps 142:7 the psalmist cries to God for help because persecutors are "*too strong* for me"; and in 2 Sam 22:18 // Ps 18:17 David testifies to God's deliverance from "my strong [*'az*] enemy" when those who hated him were *too mighty* for me." As David goes on to say, however (2 Sam 22:38), God's deliverance comes in the form of David's gaining such strength that he is able to turn and pursue his enemies to defeat them. The dynamic here, then, is implicitly what I have sketched above.) Once, the Qal stem occurs absolutely, where, in contrast to the men of Israel who were "subdued" (*wayyikkane'u*) the men of Judah *prevailed*, "because they relied [*niš'anu*] upon the LORD, the God of their fathers."

The Verb *'ms* in the Piel stem

For purposes of the present study, we may note the following distribution of usages. (1) The Piel can describe God's action in "making firm the skies" (Prov 8:28), or human actions in restoring and "strengthening" the temple (2 Chr 24:13). Here the grammatical objects of these actions are inorganic, presented simply as being acted upon. (2) In Isa 44:14, an idolater singles out a growing tree and apparently nurtures it ("has it grow strong") for idol purposes; while in Ps 80:15, 17 God tends a stock that God's right hand has planted—the stock being the "son of man" or Davidic king. Vegetation is not simply a passive bit of inorganic matter, but displays an organic life of its own.[6] One may recall here Thorkild Jacobsen's characterization of one

---

6. Such "life of its own" is reflected in the use of the word *automatos* in Mark 4:28

PART TWO: Forays into a Biblical World

prominent metaphor for deity in fourth millennium Mesoportamia: "*élan vital*, the spiritual cores in phenomena, indwelling wills and powers for them to be and thrive in their characteristic forms and manners." Again, he writes of these as "forces in nature . . . intuited as the life principle in observed phenomena, their will to be in this particular form."[7] Anyone who has observed the stages by which a germinated bean seed thrusts its two leaves up through a crack in the soil and grows of its own accord (*automatos*) will appreciate that whatever one may do to assist such a plant in its growth, by way of watering it and staking it up, one is still reliant on the plant's own internal power of growth. Thus, the action involved in the Piel stem of *'mṣ* in relation to vegetation is both semantically and existentially on the border between "strengthening" action expended on inanimate objects and such action expended on, or reflexively within, humans.

(3) In relation to humans, the Piel verb in the first place can characterize a person as simply possessing or displaying strength: Prov 24:5, where "one who has strength" translates the expression, *me'ammeṣ koaḥ*. In Amos 2:14 and Nah 2:1, the latter expression refers to the act, or the attempt, to call on and muster one's strength in the face of a sudden threat. (In these two instances, *koaḥ* is the grammatical object of *'immeṣ*, where the verb is "inwardly transitive,"[8] though one can also identify the noun as a component with the verb in one verbal idea, an idea conveyed absolutely by the Hithpael stem.) Closely related to this idiom is the occurrence in Prov 31:17, "She girds her loins with strength [*'oz*] and makes her arms strong [*watte'ammeṣ zero'oteha*]." With these closely related expressions we may compare the three instances where the object of the verb is *leb*. In Deut 15:7 each Israelite is enjoined, "you shall not harden your heart

---

as well as in classical Greek texts to refer to vegetation growing with no visible help.

7. Jacobsen, *The Treasures of Darkness*, 20, 73.

8. A child's bedtime prayer provides a good example of an "inwardly transitive" verbal construction. While one would normally say, "I lie down to sleep," where the verb is intransitive, in the child's prayer, "now I lay me down to sleep," the child is both subject (I) and object (me) of the verb. To say that the total verbal action, "I lay me down" is (like, "I lie down") intransitive is to say that the action does not work an effect on anything outside the actor, while to say that it is "inwardly transitive" is to say that the subject's action works an effect on that same person as object of the action. Another word for this is "reflexive." If we want to situate this kind of verbal construction in relation to the two most common forms in English, the active and the passive voice, we could call it (after its counterpart in Hebrew and Greek) the middle voice. For my argument that the middle voice is logically the fullest form of action, while the active and the passive voice are complementary abstractions from it, and for the existential and theological implications of such an analysis, see Janzen, "Hagar in Paul's Eyes and in the Eyes of Yahweh."

The Verb ya'ameṣ in Psalm 27:14

[lo' te'ammeṣ 'et-lebabeka] or shut your hand against your poor brother." In 2 Chr 36:13 it is said that Zedekiah "rebelled against King Nebuchadnezzar, who had made him swear by God; he stiffened his neck and hardened his heart [waye'ammeṣ 'et-lebabo] against turning to the LORD, the God of Israel." Here again we have what one might call an "inwardly transitive" use of the verb. But in Deut 2:30 the latter expression involves an "outwardly transitive" use of the Piel, when Moses says of Sihon, "the LORD your God hardened his spirit and made his heart obstinate [we'immeṣ 'et-lebabo]." Though the respective uses of this expression are grammatically identical, semantically they must be distinguished because the first two instances function as "inwardly transitive," while the third instance involving Sihon functions as a causative, with the sense, "God caused Sihon to harden his heart."[9] This leads us to the remaining group of occurrences.

The following occurrences involve an action of one party toward a second party in order to effect or elicit a change in the latter party. In Deut 3:28, (a) God instructs (b) Moses to charge (c) Joshua, and "*encourage and strengthen* him." Here we have the three dimensions of the one word-event that I discussed above in connection with the eleven occurrences of this stock expression using the Qal stem. But it is worth underscoring that the encouraging-and-strengthening occurs through words as spoken and appropriated. Similarly, in Isa 35:3 messengers are instructed to "*Strengthen* [ḥazzequ] the weak hands, and *make firm* ['ammeṣu] the feeble knees," and that they are to do this by speaking is clear from 35:4: "Say to those who are of a fearful heart [nimhare-leb], 'Be strong [ḥizqu], fear not!'" With this we may compare the opening words in Isa 40:1-2, "Comfort, comfort my people, says your God [dabberu 'al-leb yerušalam]," literally, "speak *upon the heart* of Jerusalem." This idiom, occurring seven times in the MT where one party addresses another, occurs once in self-address (1 Sam 1:13).[10] As indicated by the parallel verb naḥem in three instances (Gen 15:21; Isa 40:1-2; Ruth 2:13), the effect to be sought in "speaking upon the heart" is that of consoling and assuaging a sorrow or a fear. In Isa 35:3-4, with its reference to hands and knees, and in the eleven instances of the Qal stock expression examined above, the sought-for effect is that of nerving and emboldening the hearers to steadfastness in the face of an enemy.

In Ps 89:21, God says to the Davidic king in a vision, "my hand shall ever abide with him, my arm also *shall strengthen him*." Since the

---

9. For my analysis of the complex picture that emerges concerning the (comparable) interaction between God and Pharaoh in the book of Exodus, see Janzen, *Exodus*, esp. 69-79.

10. For my analysis of the idiom in reference to Hannah, see chapter 5.

PART TWO: Forays into a Biblical World

passages already canvassed in this group, and also those to come, all have to do with strengthening through speech, one may wonder whether the reference to God's hand and arm is a figurative way of characterizing the sustaining help this visionary message will continue to provide as lodged in the memory and resolve of the king. For, in Job 4:4 (where Job retorts sarcastically, "Your words have upheld him who was stumbling, and you have *made firm* the feeble knees") as well as Job 16:5 ("I could *strengthen* you with my mouth, and the solace of my lips would assuage your pain"), the strengthening comes through encouraging words.

There remains one more occurrence of the Piel, where the nature of the action is unclear. When Rehoboam comes to the throne of his father, Solomon, the Chronicler tells us (2 Chr 11:16–17) that

> those who had set their hearts to seek the LORD God of Israel came after them from all the tribes of Israel to Jerusalem to sacrifice to the LORD, the God of their fathers. They strengthened [*wayeḥazzequ*] the kingdom of Judah, and for three years they *made secure* [*wayeʾammeṣu*] Rehoboam the son of Solomon, for they walked for three years in the way of David and Solomon.

Given how the stock expression, occurring eleven times with Qal stem verbs and once or twice with the Piel stem, appears as a spoken encouragement in 1 Chr 22:13; 28:20; 2 Chr 32:7, one wonders whether in the present instance the echo of that expression does not imply some sort of moral support—perhaps their assurances that, unlike Jeroboam (2 Chr 11:14b–15), they would loyally support him on the throne; which is to say that, in "walking for three years in the way of David and Solomon," they continued their support of the dynastic cult in Jerusalem instead of following Jeroboam in his cult of the calves.

The Verb *'mṣ* in the Hithpael Stem

In Ruth 1:18, "Naomi saw that [Ruth] *was determined* to go with her." The intensive-reflexive (or "inwardly transitive") connotation of the verb can be expressed in English idiom as, "Ruth had fixed it in her mind." In 1 Kgs 12:18 // 2 Chr 10:18, when "all Israel" stoned Adoram whom the king had sent to oversee the forced labor, Rehoboam "*made haste* to mount his chariot, to flee to Jerusalem." That the connotation of haste here is conjectural is indicated in *HALOT*'s alternative, "he *managed* to mount," an option chosen in the NIV. I suggest that the verb here refers to *an intensely concerted action* in which, as we might say, he "put everything he had into getting back to town."

Finally, in 2 Chr 13:1–7 we have a scene in which Abijah of Judah and Jeroboam of Israel array their forces opposite one another, at which point Abijah addresses Jeroboam and "all Israel" across the intervening space. In that speech, he recalls how, when Jeroblam rebelled against the Judean overlordship, "certain worthless scoundrels gathered about [Jeroboam] and defied [*wayyit'ammeṣu*] Rehoboam the son of Solomon, when Rehoboam was young and irresolute [*rak-lebab*] and could not withstand them [*hithazzaq lipnehem*]." Several aspects of this usage call for comment. (1) The stock expression encountered so often with Qal stem verbs is here pried apart so that one verb is applied to one party in the confrontation and the other verb is applied to the other party. It cannot be doubted that that stock expression reverberates softly in the background; but precisely because of this the present transformation of the expression achieves an arresting freshness of force.[11] (2) As with the Qal and Piel forms of the expression, both verbs appear in the same stem. (3) The reasons Rehoboam is no match for them are that he is young and he is *rak-lebab*. The connotation of the latter expression is evident from its occurrence in Deut 20:8, in the context of rules for engaging in warfare. As the mustered force draws near to battle, the priest addresses it, beginning with the words (20:3–4),

> Hear, O Israel, you draw near this day to battle against your enemies: let not your heart faint [*'al-yerak lebabkem*]; do not fear, or tremble, or be in dread of them; for the LORD your God is he that goes with you [*'immakem*], to fight for you against your enemies, to give you the victory.

I suggest that one signal way in which God goes "with" them is in and through this word spoken authoritatively to them on God's behalf, and appropriated faithfully by—one may say—"all whose hearts make them willing [*kol-'iš 'ašer yiddebennu libbo*]." (The expression here is found,

---

11. Given how the conventional appearance of this stock expression has the two verbs describing a self-referential action within a given individual—an action, moreover, that has the effect of marshalling all that individual's resources in a single concerted stance and action—one might be tempted to propose that, in distributing the two members of this verbal hendiadys between two opposing factions in what is, after all, still one body of descendants of Abraham and Sarah as people of the LORD, and then intensifying the verbal idea of the Qal stem by throwing the verbs into the Hithpael stem, the writer is representing in a marvelously compact way the tragic irony of the scene, in which a people turns against itself energies originally meant only to be aimed against a common foe.

of course, in Moses' call for voluntary contributions to the construction of the desert sanctuary [Exod 25:2]; but its applicability to a summons to war, and the nature of the sought-for response, is suggested by Judg 5:2, "When locks are long in Israel, / when the people offer themselves willingly [*hitnaddeb*]—/ bless the LORD!" [NRSV].) But then, among the various types of those who are discharged from military service, anyone who is "fearful and fainthearted [*rak hallebab*]" is to be sent home, "lest the heart of his fellows melt as his heart." Perhaps we are to imagine that the majority of troops, on hearing the priest's call to "fear not," take it to heart in such a way that their stalwart, voluntary readiness is evident from their posture and their faces, while some of those being sent home show signs of fright and faint-heartedness already at the mere prospect of battle. Such an attitude with its accompanying behavior would "send a message" contrary to the priest's words, a message that would "get inside the heads" of others and subvert their inner resolve. We could note here that when it came to getting out of town (2 Chr 10:18), Rehoboam displayed no lack of ability to *marshall his energies* and act with dispatch. But when it came to facing the enemy, he lacked the ability to hear in that call anything that could move him to issue a clarion call to those same inner resources.

The Verb *'mṣ* in the Hiphil Stem

Finally we come to the two occurrences of the verb in the Hiphil stem; in Ps 27:14, "Wait on the LORD: *ḥazaq weya'ameṣ libbeka*: wait, I say, on the LORD"; and Ps 31:24, *ḥizqu weya'ameṣ lebabkem*, all you who hope in the LORD." The question that I make bold to pose, over against BDB, *HALOT*, and the grammatical comments of Delitzsch and GKC, is whether the related data, as marshaled and deployed above, provide any clues that would tip the balance of probability as to how we should construe these Hiphil stems.

Weighing the Pros and the Cons

I shall begin by reintroducing the comment quoted above from GKC §53*d*, this time including words with which the passage continues:

> Among the ideas expressed by the *causative* and *transitive* are included . . . a series of actions and ideas, which we have to express by periphrasis, in order to their being represented by the

Hiph'il-form. To these *inwardly transitive* or *intensive* Hiph'ils belong . . . Hiph'il stems which express the obtaining or receiving of a concrete or abstract quality. (In the following examples the *Qal* stems are given, for the sake of brevity, with the addition of the meaning which—often together with other meanings—belongs to the *Hiph'il*).[12]

My point, now, is to draw especial attention to the qualifying words that appear between the em-dashes: "often together with other meanings." What GKC acknowledges here is that, among the twenty-seven root words there listed as displaying an "*inwardly transitive* or *intensive*" meaning, we "often" find such a root also displaying "other meanings." I offer the following examples where a Hiphil stem displays an *inwardly transitive* meaning in one place and a *causative* meaning in another:

1. *gbr*, Hiphil: (a) "with our tongue *we will prevail*" (Ps 12:4); but also, (b) "he shall *make strong* a covenant with many" (Dan 9:27).

2. *ḥzq*, Hiphil: (a) "display strength" (2 Chr 26:2); but also, (b) "make strong, strengthen" (Ezek 30:25, "I will strengthen the arms of the king of Babylon").

3. *prḥ*, Hiphil: (a) "flourish" (Ps 92:13; Job 14:9); but also, (b) "cause to flourish" (Isa 17:11; Ezek 17:24).

4. *ṣlḥ*, Hiphil: (a) "prosper" (Gen 39:2; Deut 28:29; Ps 1:3); but also, (b) "cause to prosper" (Gen 39:3; Ps 118:25).

5. *'rk*, Hiphil: (a) "to become long"; but also, (b) "to cause to become long."

In cases 2, 3, and 4, usage (a) refers to a human or other creaturely grammatical subject, whereas in usage (b) the grammatical subject is God and the grammatical object is creaturely. This by itself should indicate that, prescinding from any other considerations, the Hiphil stem of this sort of verb (having to do with "a concrete or abstract quality," as GKC puts it) may, in a given instance and without prejudicing its connotation in other instances, convey either an (a) or a (b) sort of connotation. The fifth case, that of *'rk* in the Hiphil stem, calls for fuller comment. (This comment will also lay the groundwork for aspects of the discussion of the last line in Psalm 23, in the next chapter.)

---

12. GKC §53d (italics original).

PART TWO: Forays into a Biblical World

The Special Case of the Verb *'rk* in the Hiphil Stem

Of the seventy-one occurrences of *'rk*, thirty-one are Hiphil, of which twenty-six indicate temporal duration. Of these twenty-six, eighteen occur in some sort of conjunction with the noun, *yamim*, "days." With this we may compare the nominal phrase, *'orek yamim*, which occurs nine times, including Deut 30:20 where it functions in place of this book's usual verbal construction (eleven times). The verbal construction displays the following variations: (a) In four instances, the noun *yamim* is the subject of the Hiphil verb functioning as an "inward transitive," and is suffixed with a personal pronoun of the one whose "days may be long" (e.g., Exod 20:12, *ya'arikun yameka*). (b) In fourteen instances, the unsuffixed noun *yamim* is the direct object of the Hiphil verb functioning causatively, while the personal subject of the verb is the one who is "prolonging" his days (e.g., Deut 4:40, *ta'arik yamim*).[13]

This verb, then, dramatically emphasizes the point gleaned from the previous 4 examples—that verbs of the sort discussed in GKC §53*d* can function at the grammatical level either as "inwardly transitive" or as causative. But it must be acknowledged that the difference is only *grammatical*; whether the person is indicated in the pronoun suffix in type (a) or as subject of the verb in type (b), in both cases that person is the beneficiary of the verbal action, so that at the *semantic* level we have here, again, an "inwardly transitive" construction. To this point, then, it may remain unclear how the occurrences of this verb bear on one's construal of the Hiphil form, *ya'ameṣ*, in Pss 27:14 and 31:24.

But there remains one other occurrence of type (b) that brings the incidence of this tight but variable set of stock expressions into direct bearing on one's construal of *ya'ameṣ*. I refer, of course, to 1 Kgs 3:14:

> we'im yelek bedarke yhwh lišmor 'et-ḥuqqaw we'et-miṣwotaw ka'ašer halak dawid 'abiw wehe'erik yamim

> And if he will walk in the ways of the LORD, keeping his statutes and his commandments, as his father David walked, then he will lengthen his days.

Here we have almost exactly the grammatical context of *ya'ameṣ* as in the two psalms, and we are posed with the same question: Does the final "he" in 1 Kgs 3:14 refer to David's dynastic heir, or to "the LORD"? Posed this

---

13. Type (a) in Exod 20:12; Deut 5:16; 6:2; 25:15; type (b) in Deut 4:26, 40; 5:33; 11:9; 17:20; 22:7; 30:18; 32:47; Josh 24:31; Judg 2:7; Isa 53:10; Prov 28:16; Eccl 8:13.

The Verb ya'ameṣ in Psalm 27:14

way, the usage in all the other instances of this stock expression makes it an open and shut case: Whether in type (a) or type (b), the expression never occurs with "the Lord" as the subject of the Hiphil verb. By analogy, then, one could argue concerning *ya'ameṣ* in the two psalms, that the stock expression involving the paired occurrence of the verbs *ḥzq* and *'mṣ* always has the same party as the subject of both verbs, and so we must assume the same to be the case in these two psalms, despite the shift from the usual Qal stem of *'mṣ* to the Hiphil stem, a shift presumably made (as Delitzsch implies) for purposes of intensification of the verbal idea.[14]

Except that in presenting the text of 1 Kgs 3:14 as it appears above, I "cooked the books." In fact, and against all precedent, this text reads,

> we'im telek bidrakay lišmor huqqay umiṣwotay ka'ašer halak dawid 'abika weha'arakti 'et-yameka
>
> And if you will walk in my ways, keeping my statutes and my commandments, as your father David walked, then I will lengthen your days.

This usage is astonishing. Literarily, it occurs in the same Deuteronomic tradition as thirteen of the eighteen occurrences of the expression involving *'rk* and the noun *yamim*. Given that almost knee-jerk compositional practice, one would fully expect that the result clause in 1 Kgs 3:14 would read either *weha'arakta yamim*, "that you may prolong your days," or *ya'ariku yameka*, "that your days may be long." Yet in full view (or memory) of such repeated usage, the writer departs from it and uses the Hiphil form with God as causative subject.

The issue here, then, as in Psalms 27 and 31, is not merely *grammatical* and *literary*, having to do with the possibility of novel usage of a stock expression; rather, the issue is *theological*—which is to say, it is *existential*.

In the case of the paired verbs *ḥzq* and *'mṣ* the grammatical subject, in the Qal stem, is the human recipient of the exhortation presented by another human. But insofar as Moses is instructed by God to "encourage and strengthen" Joshua in Deut 3:28, both the human issuing of the exhortation and the human reception of it are grounded in God's spoken

---

14. But if the aim was simply intensification of the meaning of the Qal, why resort to a Hiphil form in a use for which there is no biblical parallel? (Ps 31:24 is not such a parallel, insofar as Pss 27:14 and 31:24 are so closely related that they have all the separate evidentiary weight of the testimony of two witnesses in court whose accounts of an incident are word-for-word identical.) The psalmist could have avoided all confusion by using the Hithpael stem that does exhibit such a meaning in its other occurrences.

word of instruction. The theological underpinning of human existential engagement that is implicit in Deut 3:28 may be taken to underlie all the occurrences of this paired expression. But this implicit theological underpinning (rendered explicit only in echoes of the paired expression in Pss 80:17; 89:21; Isa 41:10) is muted in favor of the emphasis on the human actions in the exhortation and reception—in order, as Martin Buber puts the matter in another context, to touch "the power of fate-deciding"—in order, in this instance, to penetrate past a person's or a group's understandable emotional reaction of fear and dread in the face of dire threat, so as to touch and move the will or "power of fate-deciding" to a posture of steadfastness.[15]

It is the same with the issue of "prolonging days." The relative distribution of the neutral expression, "that your days may be prolonged" (four times) and the rhetorically more loaded, "that you may prolong your days" (fourteen times), highlights a similar rhetorical aim in these latter passages—by throwing the rhetorical emphasis on the hearers as subjects of the verb 'rk, to penetrate the person's or the group's bent for self-determination toward self-serving aims at the expense of others, and to move them to *torah*-obedience.[16] What the writer in 1 Kgs 3:14 achieves, in throwing the final focus on God, is—precisely through a departure from customary usage (which is to say, through a departure from what the habituated hearers or readers expect in such a context)—to underscore that,

---

15. God is represented as grammatical subject of 'mṣ in the Piel stem with positive connotation in Isa 41:10; Pss 80:17; 89:21. The emphasis here on God's part in the multi-dimensional dynamics of "strengthening" accords with the dire communal situation reflected in all three contexts. But note the sobering implications of the words in Jer 1:7–8, 17: "The LORD said to me, 'Do not say, "I am only a youth"; for to all to whom I send you you shall go, and whatever I command you you shall speak. Be not afraid of them, for I am with you to deliver you, says the LORD . . . gird up your loins; arise, and say to them everything that I command you. Do not be dismayed by them, *lest I dismay you before them*.'"

16. Buber, *Prophetic Faith*, 103–4, where he says, concerning true prophets, "The true prophet does not announce an immutable decree. He speaks into the power of decision lying in the moment . . . The power and ability are given to every man at any definite moment really to take his choice, and by this he shares in deciding about the fate of the moment after this, and this sharing of his occurs in a sphere of possibility which cannot be figured either in manner or scale. It is to this personal decision of man with its part in the power of fate-deciding that the prophetic announcement of disaster calls. The alternative standing behind it is not taken up into it; only so can the prophet's speech touch the innermost soul, and also be able to evoke the extreme act: the turning to God." To the degree that the present study is found to be persuasive, it may serve to broaden Buber's point beyond the sphere of prophecy to all speech uttered or received in God's name.

given the concern for human response and responsibility, the issue does not lie with humans alone, nor even primarily there. I suggest that this novel turn in a stock expression—and the novel turn in Psalms 27 and 31—delivers the concerned person from the burden of supposing that all depends on him or her, and at the same time prevents an incipient pride (as in Deuteronomy 8 and Psalm 30:6) from attributing one's actions and well-being to one's own powers of moral striving.[17]

All this does not prove that the Hiphil verb, *ya'ameṣ*, in Psalms 27 and 31 is to be taken as referring to *God's* action working in and through the Psalmist's faithfulness (27:13). But it does provide a linguistic analogy for so construing *ya'ameṣ*, as well as a way, evidenced in other text-forms, of interpreting this novel usage in existential and theological terms. But there are other aspects of usage that, in my view, tilt the probability in the direction that I am proposing.

Final Assessment

I recur here to other occurrences of *'mṣ* in the Piel stem outside of the paired stock expression, and wish to make two points:

(A) Where a person is the grammatical subject, the "self-strengthening" that would nerve and empower that person to resolute stance and action is expressed in the following ways: Amos 2:14, "the strong shall not retain his strength [*lo'-ye'ammeṣ koḥo*]"; Nah 2:1, "gird your loins; collect all your strength [*'ammeṣ koaḥ me'od*]"; Prov 24:5, "a man of knowledge [is mightier] than he who has strength [*me'ammeṣ-koaḥ*]"; and Prov 31:17, "She girds her loins with strength [*be'oz*] and makes her arms strong [*watte'ammeṣ zero'oteha*]."

(B) But where *'mṣ* in the Piel stem involves the *heart*, the idiom has to do, not with a person finding the morale and strength to *do* something, but with a person stubbornly *refusing* to act in response to a call for compassionate or theologically appropriate action. The instances, once again, are these:

1. Deut 2:30, "But Sihon the king of Heshbon would not let us pass by him; for the LORD your God hardened his spirit and made his heart obstinate [*we'immeṣ 'et-lebabo*], that he might give him into your hand, as at this day."

---

17. Recall the analysis and argument in chapter 8.

2. Deut 15:7, "If there is among you a poor man, one of your brethren, in any of your towns within your land which the LORD your God gives you, you shall not harden your heart [*lo' te'ammeṣ 'et-lebabeka*] or shut your hand against your poor brother."

3. 2 Chr 36:13, "He also rebelled against King Nebuchadnezzar, who had made him swear by God; he stiffened his neck and hardened his heart [*waye'ammeṣ 'et-lebabo*] against turning to the LORD, the God of Israel."

We may recall Delitzsch's objection that "[t]he translation: may he (Jahve) strengthen thine heart ... would require *ye'ammeṣ*." To this objection, and to his argument on behalf of the general consensus, I enter three counter-objections. First, in following the Qal imperatives in Pss 27:14 and 31:1 with *ya'ameṣ libbeka*, the author would have left his words open to construction on analogy with the other occurrences of this expression. Thus, the weight of the usage in Deut 2:30; 15:7; and 2 Chr 36:13 militates against Delitzsch's proposal.

Second, Delitzsch's comment that *he'emiṣ*, like *hirḥib* in Ps 25:17, belongs to the class of intensive denominatives, is inconsistent with his comment on Ps 25:17, where he corrects MT's *ṣarot lebabi hirḥibu mimmeṣuqotay hoṣi'eni* to *ṣarot lebabi hirḥib umimmeṣuqotay hoṣi'eni*, and, citing Ps 119:32 (*ki tarḥib libbi*) translates, "enlarge the straits of my heart."[18] That is to say, his emendation of Ps 25:17 employs *hirḥib* with a connotation that is *not* intensive denominative. Psalm 119:32 is no analogue to Ps 25:17, for it deals with an increase in understanding. Psalm 25:17 deals with a condition of anxiety, or *angst*, a constriction of the chest familiar to those, for example, who, month after month, find it difficult to stretch the paycheck to pay all the bills. In any case, the translation Delitzsch offers in his comment, "enlarge the straits of my heart," shows that God is the subject of *hirḥib* and *ṣarot lebabi* is the verb's direct object. So if this verse is a guide to the construal of *weya'ameṣ libbeka*, it supports my construal of the latter clause.[19]

18. Delitzsch, *Psalms*,1: 424.

19. Both RSV and NRSV render Ps 25:17 as, "Relieve the troubles of my heart, and bring me out of my distresses," but a marginal note reads, "Or *The troubles of my heart are enlarged; bring me.*" That the preferred reading is correct is suggested by the collocation elsewhere of the root words, *rḥb* and *ṣrr*. Thus, Ps 4:2 reads, "Thou hast *given me room* [*hirḥabta li*] when I was in *straits* [*baṣṣar*]"; Ps 31:9–10 reads, "thou hast not delivered me into the hand of the enemy; / thou hast set my feet *in a broad place* [*bammerḥab*]. / Be gracious to me, O LORD, for I am in *straits* [*ṣar-li*]"; in Ps 118:5, the psalmist says, "*Out of my distress* [*min-hammeṣar*] I called on the LORD; the

Third, if the authors (or editors) of Ps 31:25 had wanted to follow the Qal imperative *ḥizqu* with an intensified form of its pair-verb, *'mṣ*, why resort to a Hiphil form not exemplified outside these two passages, when such an intensification could easily be achieved by casting the verb into the Hithpael? Before I offer my counter-interpretation in favor of the construal of Hiphil *'mṣ*, I want to make a brief excursion to Ps 27:8, to suggest how one current translation of that difficult verse may shed light on v. 14, and how, conversely, my reading of v. 14 supports that translation of v. 8.

A Brief Excursion to Psalm 27:8

On the face of it, the first line in the Hebrew text of 27:8 is, to say the least, odd. It reads, *leka 'amar libbi baqqešu panay*. The clause in the last two words is clear: a call in the imperative plural, "seek ye my face." But the first three units are unclear. The prepositional phrase, *leka*, normally means, "to thee," or "for thee," or "concerning thee"; the following verb, *'amar*, means "he said"; and the third unit, *libbi*, "means, "my heart." The most natural translation would be, "to/for you my heart (has) said." Most English translations follow the KJV in the general sense, "Thou hast said, 'Seek ye my face.'" RSV offers, "My heart says of you." But NIV renders it, "My heart says of you, 'Seek his face!'" with a marginal footnote, "or *To you, O my heart, he has said.*" JPS, followed by the NJPSV (Tanakh), renders the whole verse, "In Thy behalf my heart hath said: 'Seek ye my face.' Thy face, LORD will I seek."

I propose to follow the grammatically straightforward reading of the verse presented in JPS and NJPSV (Tanakh), and to take its sense in light of the above analysis of three-dimensional "word-events," especially their ability to continue to resonate in the deep memory where they can be recalled as and when needed.[20]

LORD answered me and *set me in a broad place [bammerḥab]*"; in 2 Sam 22:7 // Ps 18:7, David says, "In my distress [*baṣṣar-li*] I called upon the LORD," and in 2 Sam 22:37 // Ps 18:37 , he says, "Thou didst *give a wide place [tarḥib]* for my steps under me"; and in Job 36:16 Elihu says, "He also allured you out of *straits [mippi-ṣar]* into a *broad place [raḥab]* where there was no cramping." The universality of the concrete imagery here, at least in regard to one's plight, is indicated in the existential connotations, arising in part out of bodily feelings accompanying the emotional states, of *ṣarar / ṣar* in Hebrew, *thlibo / thlipsis* in Greek, and *anguere / angustia* in Latin—not to mention the adoption of the Latin terms to refer to angst, anxiety, and angina. In view of the frequent occurrence of this trope in the Hebrew Bible, the meaning of the expression in Ps 25:17 should be beyond debate.

20. Bearing in mind Gen 37:9–11; 1 Chr 29:16–18; Prov 4:20–23; 22:17–18 (and

PART TWO: Forays into a Biblical World

The psalmist is in dire straits. Over against the surrounding and besieging presence of his enemies, the psalmist voices his desire to dwell in God's house and to "behold the beauty" of God (Ps 27:4). In the recesses of his heart we may imagine a soft resonance of such sayings in the sanctuary as the one in Ps 105:4//1 Chr 16:11, "Seek the LORD and his strength; / seek his face continually." Now, encouraging and exhorting words such as these rise to the surface, not as a memory but as transformed into a present address.

Here, a true anecdote. Bobby, a toddler, returns home with his nanny to a fourth-floor walkup apartment. They mount the stoop, open the outer door, enter the vestibule, press the buzzer, and wait for the door to be buzzed open. While they wait, Bobby, tuckered out after a long day in the park, is overheard to say to himself, "That's all right, Bobby, that's all right; why you fussin?" Is Bobby encouraging himself for the climb? Or is he repeating to his heart, on behalf of his nanny, words that she has presumably said to him in similar occasions? Just so, the psalmist, uttering on behalf of God the verb in the plural, "seek ye" (a clear sign that words like those in Ps 105:4 originally were heard as addressed to a congregation), then responds, "thy face, LORD, will I seek."

Having in mind such a reading of Ps 27:8, I offer the following interpretation in favor of the construal of Hiphil *'mṣ* at the psalm's end. As I have argued, the various usages explored in this study may be plotted on a three-dimensional semantic field that includes three agents: (1) one person who encourages a second person; (2) the latter person appropriating the encouragement through an "inwardly intransitive" act of self-exhortation and encouragement;[21] and (3) God who implicitly is present and active in and through each of these persons. In Ps 27:8, the psalmist's self-address reflects a congregational setting such as Ps 105:4 in which the words of God were originally spoken by a priest on behalf of God.[22]

What the author in Ps 27:8 achieves, then, in a condensed manner typical of poetry, is to represent all three of these agential dimensions in

---

Luke 2:19, 51); recollect also Ps 119:11: "I have laid up [or, "hid, squirreled away," *ṣapan*] thy word [*'imra*] in my heart [*libbi*], that I might not sin against thee."

21. As, e.g., repeatedly in the refrain of Psalm 42–43.

22. Psalms scholarship has long since arrived at the conclusion that certain sentences in the Psalms are best heard not as those of the psalmist, but as spoken by a priest or other cultic functionary. Some recent commentators account for 27:14 by attributing that verse to a priest who responds to the foregoing verses on behalf of God. In my view, such priestly words need not be heard as directly intervening here, but—as in v. 8—lying behind the psalmist's self-exhortation.

The Verb ya'ameṣ in Psalm 27:14

one sentence: a priestly voice as "channeled" by the psalmist; the psalmist as encouraged and exhorted in the opening verb; and God, now invoked, not simply as the ultimate agent behind the priestly exhortation, but directly as the third-person subject in the Hiphil verb, *ya'ameṣ*.[23] The rhetorical emphasis on God as subject and actor in this verb is underscored by the unexpectedness of this verb-form as following *ḥizqu*. Given the familiarity of the word-pair Qal *ḥizqu* + Qal *'imṣu* (as in Deut 31:6 and often), the reader expects the same sequence here. Instead, the psalmist makes an unprecedented shift to *ye'ammeṣ*, to signal that the transforming action, however mediated by a human mediator, present or as recalled from the past, is to be attributed directly to the divine agent. Such a reading of Ps 27:14 also adds support to the translations of Ps 27:8 in JPS, NJPSV (Tanakh), and NIV as cited above.

Earlier, I proposed that the repeated emphasis on God as "with you," in this general rhetorical context, may be taken to suggest that the divine presence and activity resides, in significant part, in the human word faithfully uttered and faithfully appropriated, relied on and acted upon. What I propose, now, is that, by the novel shift to *ye'ammeṣ*, the "real presence"[24] of God is brought home to the hearer with even greater attractive force. In such a reading, what J. L. Austin calls "performative language" becomes "transformative."

If my argument does nothing more than to re-open the issue in 27:14, and to establish that the verb in the intervening clause may be taken either way, then future translations might do well to indicate in a marginal note the reading they do not adopt in the body of the text. The principle here would be that a translator should not make interpretive decisions for the non-Hebraist reader if the ambiguities in the Hebrew text are such as can be re-created in the target language. I myself consider that the strong probability lies with Coverdale and the KJV. Whatever their grammatical reasoning, their exegetical instincts (perhaps as steeped in the theo-logic of Scriptural passages squirreled away in their own hearts) were sound.

---

23. Compare the multi-dimensional character of the Priestly Blessing in Num 6:22–27, where "you shall bless the people" becomes "I will bless them." See further chapter 3.

24. To invoke George Steiner's book *Real Presences*.

PART TWO: Forays into a Biblical World

Two Modern Afterwords to Psalm 27

It is customary in our time, in looking for light on the meaning of Scripture, to turn to those whose education and experience qualifies them as Professors of Scripture in the academy and as writers of commentaries and other authoritative tomes. In this addendum to the present essay, I want to quote two modern writers, not for their commentary on Psalm 27, but for their testimonial uses of it in their own personal situations.

1. The first is Samuel Taylor Coleridge, known to the general public as a romantic poet, but also a man of profound philosophical abilities, and a student of the Bible in Hebrew and Greek who, in the last several years of his life, devoted a large portion of his notebooks to running commentary on his systematic readings in the Bible. In March of 1828, he entered a series of reflections, some of them almost short essays, on matters of fundamental philosophical and theological import. One such extended reflection, beginning with Note 5814 and continuing through at least Note 5816, concerns the bearing of the *Logos* doctrine of John's Gospel on some long-standing philosophical debates. Note 5814 ends with the words, "this borne in mind . . . I may now proceed to the necessity and purpose of the Incarnation."[25] But instead of so proceeding, Coleridge breaks off and enters words now labeled Note 5815:

> I feel as if my heart sank or shrunk within me, as the several no less abstruse than momentous themes pass rapidly before me / . First, to submit to a new Trial the long controverted and falsely or most superficially decided Cause of the Realists versus Nominalists—to exhibit the vast importance and deep interest of this question—and the necessity of reversing the judgment . . . and to demonstrate the dynamic Objectivity of the ideal, Kind, and Nature. The Son of God submitted to be of one *Kind* with Men.[26]

The sinking feeling thus arises within him and is noted, then gives way to the onward flow of abstruse reflection that continues into Note 5816; and the latter goes on for two whole printed pages, ending with the intention to explore the possibility of

> Insight into the possibility and the necessity of *theanthropism*, i.e., the sinking of the Logos into Man in order that Man might

---

25. Coleridge, *Notebooks*, 5: no. 5814.
26. Ibid. 5: no. 5815.

rise into the Logos, & with the Logos ascend into the Glory of the Father—[27]

At this point, as the long dash shows, Coleridge puts his reflection on pause (to use TV language) and interjects with what the editor calls Note 5817, in which the momentary feeling expressed in Note 5815 sweeps over him again:

> How is it possible, I repeat, to . . . cast the eye on the road that stretches before me, and to glance in succession at the several rocks to be scaled and defiles to be threaded and obscure Jungles to be cut thro'—and not quail and . . . inwardly confess the utter disproportion of my strength to . . . my Task-work. Verily, "I had fainted, unless I had believed to see the goodness of the Lord, for to be strong in my weakness, Wait on the Lord. Be of good courage, and he shall strengthen thine Heart. Wait, I say, on the Lord. Through God we shall do valiantly."[28]

With this Coleridge returns to the theme of the last sentence in Note 5816, reiterating the need "of obtaining insight into the necessity of such and such truths," and continues in that abstruse mode for another half page.

We may note how Coleridge's first digression, in Note 5815, has the appearance of an interior "aside" in which he is communing with himself; but the second "aside," in Note 5817, beginning as such a communing with himself, imperceptibly modulates, with the aid of remembered Scripture, into "prayer as self-address," in which the psalmist's call to "Wait on the Lord" becomes Coleridge's call to himself in the same spirit. Such is the dynamic of that prayer, that entry into the place—the "secret place"—where the prayers of successive generations join one another in a single chorus, that by the time he has participated in the words of Ps 27:13–14, he goes on to appropriate to himself the words of Ps 60:12 // 108:13 (in, of course, the KJV), "Through God we shall do valiantly." Though he doesn't quote the latter text's following line, "for he it is that shall tread down our enemies," the reference in Note 5815 to his struggle in alliance with philosophical Realists versus the Nominalists of his day suggests the resonance of that line for him. Indeed, Ps 60:12 itself, with its back-and-forth shift—(a) "Through God," (b) "we shall do," (a') "for he it is that shall"—identifies in its own way the complex dynamics at the heart of Ps 27:14. The power, for Coleridge, of those biblical words of prayer is such that, in entering for

---

27. Ibid., 5: no. 5816.
28. Ibid., 5: no. 5817.

PART TWO: Forays into a Biblical World

a moment into the place where they voice his own concern, he is strengthened to go on with his abstruse researches.

2. At times, the testimony of those who read the Bible only in their own language, and without access to biblical scholarship, but who have read the Bible as though their lives depended on it, offers an interpretation the illuminative power of which is breathtaking for its freshness and its perspicuity. I offer in evidence for the themes explored in this essay, a document written by my father in August or early September of 1936. He lay in an infirmary in a distant city, sick also at heart over affairs at home where household economics, already strained by the Great Depression, were further strained by his own debilitating chronic ill-health, and anxious for our mother who now, three months into another pregnancy, had to manage for herself and care for four young boys including me at age four.

Recently, in rummaging through a box of old documents, I came across a letter written by our mother to our father and dated August 30, 1936. On its back I identified our dad's handwriting which covered the whole page except for a space at the top. Looking more closely, I was able to make out four lines of his script, scratched with perhaps a safety pin along the top of the page, and then a capital letter P on the fifth line, before a pencil took over with the rest of the word "[P]salm." I shall simply quote the relevant part of what he wrote, letting it speak for itself, except for voicing my hunch that the "noonday angel" was my mother's letter arriving at his sick-bed moments earlier. I think the psalmist—and Coleridge, himself chronically unwell—would recognize a kindred spirit and a kindred experience in the following words (the four lines plus one letter scratched onto the page are indicated by italics).

> *We have cried from beneath the crushing load of burdens too*
> *heavy to be borne and anxieties too grinding to endure.*
> *We cried long and earnestly but no answer came, and we thought*
> *to have fainted beneath the added weight of the Lord's silence.*

Psalm 27. But somehow we endured; and looking back upon it all, we wonder at the reserve strength which we found within ourselves when the day was evil & our prayers went unheard. But our prayers were heard. God answered us by the coming of the noonday angel whose name is "strength of the day." The needed strength was not dropped upon us as a miracle from heaven; it is God's gracious and kindly way to give us his best gifts in so unobtrusive and interior fashion that they seem to be our own, part of our natural equipment. In the day when I cried,

Thou answeredst me; we only knew that where we expected weakness, there we found strength.

. . . We may follow Job through the furnace of affliction . . .

I can't think of a more succinct expression of the *experience* reflected in Ps 27:14, than the words, "the needed strength was not dropped upon us as a miracle from heaven; it is God's gracious and kindly way to give us his best gifts in so unobtrusive and interior fashion that they seem to be our own, part of our natural equipment."

11

Revisiting "Forever" in Psalm 23:6

IN ENGLISH TRANSLATIONS GENERALLY, up through JPS (1917) and RSV (1952) as well as NIV (1973), Psalm 23 has been rendered as ending with the psalmist's hope to dwell in the house of the LORD "forever." But the translations in JB (1966, "as long as I live"), NJPSV (Tanakh) (1985, "for many long years") and NRSV (1989, "my whole life long"; text note, "Heb *for length of days*") signal a wide consensus among contemporary biblical scholars that the psalmist's hope is more limited in scope, being confined to this present life. To be sure, Mitchell Dahood in his Anchor Bible Commentary on this verse argues for an unambiguous reference to "eternal life." But his claim to find references to such a hope in several psalms has not persuaded many. W. Sibley Towner reflects the consensus when, in an essay on the rendition of various psalms in English paraphrases and hymns, he writes,

> Had the psalmist wanted to close the poem with the word "forever," a perfectly good Hebrew word, *lĕʿôlām*, was available. Instead, the choice was made to use the phrase *lĕʾōrek yāmîm*, "for length of days," or even, perhaps, "for a long time." NRSV renders the phrase paraphrastically, "my whole life long" (guided by the synonymous expression in v. 6a, *kōl yĕmê ḥayyāy*, "all the days of my life"), but avoids introducing the theme of eternity. No evidence suggests that ancient Israel had any concept of life after death until the very end of the canonical period (see Isa 26:19; Dan 12:3). Some might wish to argue that Psalm 23 provides such evidence; however, the plain meaning of the Hebrew phrase with which it closes carries us no further than this life at most. The psalmist simply hopes to be near God always.[1]

---

1. Towner, "Without Our Aid," 24–25. On the Scottish paraphrase, "And in God's house forevermore / My dwelling place shall be," Towner comments, "While one may argue that 'forevermore' is simply a logical and necessary extension of the trust that

There is no question that "forever" is an interpretive rendering of *le'orek yamim*, and that "for length of days" or the like is more accurate if perhaps literally lame. But it is less clear to me than it is to Towner that "the Hebrew phrase with which [Psalm 23] closes carries us no further than this life at most." The question is how tightly religious aspiration expressed in poetry is to be tethered to the "plain meaning" of its terms. It is the purpose of this essay to propose that when the psalmist "hopes to be near God always," the extent of the "always" is more open-ended—its aspirational reach less clearly delimited—than Towner's "plain meaning" would admit. My argument will rest on a variety of intertextual claims, including, indeed, a possible reason for the choice of the phrase *le'orek yamim* rather than the phrase *le'olam*.

Aspects of Experience "in God's House"

When tending sheep, the shepherd engages in at least three types of activity: leading, providing, and protecting. The shepherd leads the sheep to pastures and watering-places, sees to their nourishment, and in both activities guards them against danger and hostile enemies. Insofar as the sheep are pictured in Psalm 23 as following the shepherd along various paths, and then lying down, the psalm is governed by the imagery of motion and rest, of journey and arrival. James Luther Mays comments that this psalm "is at points connected with the language of Israel's testimony to its salvation in the exodus," and provides half a dozen connections, one of which is the occurrence in Ps 78:19 of the statement that "God 'prepared a table' for Israel in the wilderness."[2] The narrative sequence in Exodus–Joshua and in Psalm 78 displays the same alternation of motion and rest, journey, and arrival, as we find in Psalm 23. In Exod 15:1–18, God's leading of the people ends with God bringing them to God's "holy abode"

'Thou art with me,' that would be an argument more typical of a Christian reader than of an ancient Hebrew writer" (ibid., 25). One wonders how Towner might have worded things in light of Jon D. Levenson's brilliant analysis in his *Resurrection and the Restoration of Israel*. One could, that is, easily trace Towner's "extension" within Jewish contexts that entertained the hope of resurrection (e.g., the Pharisees) or immortality (e.g., Wisdom of Solomon). See further below (esp. the addendum).

2. Mays, *The Lord Reigns*, 117–18. See also Towner, "Without Our Aid," 26 n. 17, who references David Noel Freedman as arguing "forcefully that the language of Psalm 23 is deeply rooted in the Exodus imagery of Yahweh as shepherd of the flock, Israel, in the wilderness." The divine shepherd imagery appears in Pss 77:24; 78:52; and, indirectly, in 78:70–72.

PART TWO: Forays into a Biblical World

(v. 13), God's "sanctuary" (v. 18). Psalm 78 echoes the "abode" language of Exod 15:17 preliminarily in v. 54, but then goes on to end at the sanctuary in Zion (v. 69) with David as shepherd-king over Israel (vv. 70-72).

This language of motion and rest, of journey and arrival, bears on the construal of the last verse in Psalm 23. The first line resumes the imagery of motion (cf. vv. 3b-5), while the second resumes the imagery of rest (cf. vv. 2-3a, 5). The last line involves an image that appears in other psalms. The most obvious is Ps 27:4 which includes the desire "that I may dwell in the house of the LORD all the days of my life." The fact that the first part of this line is identical to the first part of Ps 23:6b, while the second part, "all the days of my life," is identical to the temporal phrase at the end of Ps 23:6a, would appear to confirm Towner's construal of *kol-yeme ḥayyay* ("all the days of my life") in v. 6a and *le'orek yamim* ("for length of days") in v. 6b as synonymous. But the shift from motion in 6a to rest in 6b invites another construal, in which the connotations of *le'orek yamim* in v. 6b are not strictly determined by *kol-yeme ḥayyay*. What those connotations might be is suggested by the following psalms passages.

Psalm 30 celebrates deliverance from the threat of death. If that threat evoked the psalmist's night-long "weeping," the deliverance evokes the psalmist's "joy" (v. 5), and the psalm itself is the expression of that joy. But while the threat of death evokes weeping, death itself issues in silence (v. 12). The contrast between weeping and joy—both of which are signs of life—is deeper than the contrast between silence and praise. As Mays puts the matter in another context, "praise is the sound of life," whereas "the special silence of which the psalms speak" is "the symptom of death." He goes on to say, "You can see where this essential correlation of praise and life is leading. Because praise is so much the purpose of God, and so essentially a dimension of life, death itself is on the way to being transcended."[3] It is noteworthy, then, that Psalm 30 ends with the words, "O LORD my God, I will give thanks to thee forever [*le'olam*]." What does the psalmist mean by this last word? Are we to take it in its "plain meaning"?

Psalm 52 celebrates God as refuge in the face of a wicked enemy. Because that enemy loves evil more than good, and lying more than truth, God will "break you down forever [*laneṣaḥ*]" and "uproot you from the land of the living." In contrast, says the psalmist (vv. 9-10a),

---

3. Mays, *The Lord Reigns*, 71.

> I am like a green olive tree[4] in the house of God.
> I trust in the steadfast love of God forever and ever [ʿolam waʿed].
> I will thank thee forever [leʿolam], because thou hast done it.

Here the "plain meaning" of the text (as in Ps 30:12) is that the psalmist intends to give thanks to God "forever." But we "know" that the text is not to be read that way, for we "know" that, except in the latest stages of Israelite piety as attested in the Bible, Israelites entertained no clear hopes for post-mortem existence. Why then these terms?

Psalm 61 is a cry "from the end of the earth," that God would lead the psalmist to "the rock that is higher than I." The psalmist prays, "Let me dwell [ʾagurah] in thy tent forever [ʿolamim]!" Here again we see the shift from movement to rest, from journey to arrival. The psalm concludes with the resolve, "So will I ever [laʿad] sing praises to thy name, as I pay my vows day after day."

Psalm 84 celebrates God's sanctuary as a hoped-for destination of often-tearful pilgrims, and affirms that even a day in God's courts is superior to a thousand of any other days. How happy, then, the sparrow who finds in it a home and the swallow a place for her nest; how enviable "are those who dwell in thy house, ever [ʿod] singing thy praise!" Erich Zenger speaks of v. 5 4 as a beatitude

> for all those who "dwell" there, whether because they belong to the cultic and service personnel of the Temple; or because, as Jerusalemites, they have the privilege of being able to go whenever they wish to the Temple and its liturgy; or they are existentially "at home" there because they are "rooted" in that place.[5]

The third alternative implies an existential internalization of one's identification with those (Levites?) whose vocation is to minister in the sanctuary.

In none of these passages, of course, do we find a "plain" assertion or assumption of what will later be referred to as "eternal life." Yet it is striking that in the context of praise and joy within God's sanctuary, the psalmists employ language that elsewhere, especially in reference to God, unambiguously affirms everlastingness. Is this simply a matter of "oriental hyperbole," or "poetic license"? But what is it that moves one to such license, such hyperbole? In "fish stories," hyperbole adds to the narrative color and

---

4. The contrast here between the fate of the wicked and the destiny of the faithful is imaged in similar terms in Psalm 1 and Psalm 92. Trees often signify fruitfulness and longevity; compare Gen 21:33; 23:17; Ps 92:12–14.

5. Zenger in Hossfeld and Zenger, *Psalms 2*, 355.

heightens the dramatic interest of what, if reported in plain terms, would not strike the listeners as all that remarkable. I do not take the psalmists to be telling fish stories. In these contexts I take hyperbole to be the attempt to convey a sense of an experience that is out of the ordinary, or rather a sense of experiencing what is out of the ordinary. However rare such an experience, such an awareness, may be in our own day, whenever it happens it beggars language. If hyperbole in the mundane sense conveys a picture that is "larger than life," the psalmists, in the quoted passages, speak as though they find mundane language "smaller than life"[6]—smaller than reality-as-experienced-in-joy-and-praise.[7] When the presence of the everlasting God is experienced in joy and praise, everlastingness can, as it were, take on a two-way reference, characterizing not only the God celebrated in joy and praise but also the joy and praise itself. In Isa 35:10 and 51:11, "the ransomed of the LORD shall return, and come to Zion with singing; everlasting joy [śimḥat 'olam] shall be upon their heads." Insofar as the joy and the praise are not simply something that the worshipper *does*, but more deeply, on occasions of utter exaltation, characterize the resonant quality of the worshipper's own *existence* in that act, the sense of everlastingness envelops the worshipper as well.

What, then, is "everlasting joy"? Does the adjective (in Hebrew the noun 'olam) indicate its temporal duration? Or does it serve to express its qualitative character? In pondering this and related questions I find myself drawn to Hans Loewald's comments, from a psychodynamic point of view (but no doubt informed also by his philosophical studies under Martin Heidegger), in his essay on "The Experience of Time." Loewald's essay is so rich in implication that it deserves more extended treatment than this

---

6. I am playing here on the distinguishable but inseparable meanings of the word "mundane": "1. Of, relating to, or typical of this world; secular; 2. relating to, characteristic of, or concerned with commonplaces; ordinary." See www.thefreedictionary.com/mundane.

7. Compare Paul's language in 2 Cor 4:17–18: "This slight momentary affliction is preparing for us an eternal weight of glory *beyond all comparison*, because we look not to the things that are seen but to the things that are unseen." The italicized words translate the piled-up expression, *kath' hyperbolen eis hyperbolen*. Each phrase in itself means, "in excess, extravagantly." The first, in the form, *kath' hyperbolen*, is the Greek grammatical name for an adjective in the superlative degree. For Paul, the "weight" of glory that lies ahead (compare 2 Cor 3:18) is such that a single superlative degree of description is inadequate, and he has to add the phrase *eis hyperbolen*, as though (as when driving through the Smoky Mountains) one glorious vista comes into view only to be surpassed, when cresting that ridge, by another vista arising beyond it, in "one degree of glory after another." Here, so-called exact language is in fact inexact: only hyperbole will, with the least inadequacy, report the sense that one seeks to convey.

paper has room for. But at least the following may be drawn from it, in bare summary.[8]

Experiencing Time, Mundane and Otherwise

Normal waking consciousness displays, in whatever shifting display of focal vividness and ambient vagueness, at one and the same time an awareness of past, present and future. (As I revise this section of the paper, for example, I keep one eye on the clock so as not to miss a late-afternoon appointment; and my sense of dissatisfaction over my inadequacies in dealing with the present topic is ameliorated by a residual sense of the pleasure brought by a noon-hour telephone conversation with one of my brothers.) But under certain circumstances this three-dimensional structure of consciousness is interrupted by experiences of such a qualitative character as dramatically to re-configure consciousness in suspending the awareness of temporality altogether. The sense of duration, the sense of a past and a future circumstancing the present, gives way to a complete absorption in "now"—except that the word "now" is misleading insofar as it implies a "then," so that what one needs is a word that does not connote such a contrast. As Loewald says,

> scholastic philosophers speak of the *nunc stans*, the abiding instant, where there is no division of past, present, and future, no remembering, no wish, no anticipation, merely the complete absorption in being, or in that which is.

He goes on to say,

> In conditions of extreme joy or sadness, sometimes during sexual intercourse and related orgastic [sic] experiences, at the height of manic and the depth of depressive conditions, in the depth of bliss or despair, the temporal attributes of experience fall away and only the now, as something outside of time, remains.[9]

---

8. Loewald, "The Experience of Time," reprinted in *The Essential Loewald*, 138–147. If Peter Manchester and Robert Neville are right, many modern attempts to deal adequately, whether exegetically or hermeneutically, with the biblical texts and themes discussed in the present paper suffer from a loss of appreciation for the integral relation, in experience and reflection, between time and eternity. See Manchester, "The Religious Experience of Time and Eternity," 384–407; and Neville, *Eternity and Time's Flow*.

9. Loewald, "Experience of Time," 141 and 142.

PART TWO: Forays into a Biblical World

How does one talk of such experiences? They arise at a certain point in time; and when they have passed they leave one once again at an identifiable point in time. Because the experience, "while it lasts" (a misleading expression!), displays within itself no features of temporal successiveness, but rather all presences within the experience display themselves as ingredient in one unitive reality, in reflecting back on the experience once it is over one is led to draw on language pertaining to unchanging features of one's ordinary perception, like mountains and stars, and to speak of this experience, as one might speak of such features, in terms of "everlastingness." But, says Loewald, "the experience of eternity does not include everlastingness." It does not include it because eternity is devoid of the sense of the passage of time while everlastingness implies the passage of time. Yet the language of everlastingness is employed in the attempt to distinguish this experience from the ordinary consciousness of time as duration, time as marked by succession and differentiated into past, present, and future.

Here I want to pause and propose that, *pace* Loewald, it is not sufficient simply to say that "the experience of eternity does not include everlastingness." For insofar as the term "everlastingness" refers to that which is unending, it has as its polar opposite the notion of "limited duration." But in an experience "devoid of the sense of the passage of time," precisely because it is devoid of such a temporal sense it is also devoid of the sense that the experience will end. In that sense part of the quality of the experience is its endlessness. However inadequate or misleading the term "everlasting" may be in other respects, it does serve to attest to the experience under discussion as being devoid—while it occurs—of the sense that sooner or later it will end.

Loewald writes of this experience as marking both "conditions of extreme joy or sadness . . . the height of manic and the depth of depressive conditions . . . the depth of bliss or despair." For some time now I have been persuaded, on syntactic and exegetical grounds, that in passages like Psalm 13 we should translate utterances like *'ad-anah yhwh tiškaḥeni neṣaḥ* in one continuous sentence: "How long, O LORD, will you forget me forever?" Such a proposal is not unprecedented, but when addressed in the literature it is rejected as nonsensical. But it is nonsensical only at the level of abstract reasoning in the comfort of a study chair, to those who have never found themselves plunged into the sort of depressive despair that Loewald writes about, or who have never found themselves sitting alongside such a person for whom any talk of "it will be better in the morning" is cruelly meaningless, since "the morning" is not within the reach of such a consciousness. In such a state, the darkness, while it lasts, is "everlasting" in the sense that it displays itself as an unchanging, immovable "now," and obliterates all sense of a temporal horizon beyond which it might lift. In such a state, the suspension of a sense of time consists in part in being suspended over an abyss of utter meaninglessness. Conversely, the joy of which the psalms speak may be characterized as śimḥat 'olam, a *quality* of experience—in Mays' terms, a sense of life—articulable only by the use of the term 'olam. Insofar as that experience of joy and praise is qualitatively marked by an unawareness of its needing to come to an end, it participates in the reality which devotees attribute to God when they affirm God to be everlasting.

Where joy issues in, or arises within, the praise of God who is *me'olam we'ad-'olam*, "from everlasting to everlasting" (Pss 41:13; 90:2; 103:17; 106:48), that joy, that praise, opens one up to a presence Who qualifies the joy supremely. To put it this way, the joyous praise of God in the house of God can involve a sense of both spatial and temporal dis-location and re-location which is strictly ineffable but for which terms like those used in the above-quoted psalms passages are the least inadequate. Such terms can give expression to the sense that, by the grace of God, one is drawn into a participation in the life of God. As Psalm 16 will put it, God is experienced, in the present "moment," as one's inheritance and one's lot.[10]

But what of the phrase *le'orek yamim*, "for length of days"? Does this phrase, in the very plurality of its second term, not to speak of the durational connotation of its first term, suggest a quite ordinary sense of duration in time? The uses of this idiom elsewhere suggest that its tenor in Psalm 23 may not be confined to such a "plain sense."

Connotations of the Phrase, "Length of Days," and Its Cognates

The phrase *le'orek yamim* occurs eight times in the Hebrew Bible. According to Job 12:12, "Wisdom is with the aged, and understanding in length of days ['orek yamim]." In Prov 3:2 the hearer is promised that observance of the sage's teaching will win "length of days ['orek yamim] and years of life and abundant welfare [šalom]." Likewise, Prov 3:16 promises that "long life ['orek yamim] is in [wisdom's] right hand; in her left hand are riches and honor"; and Ps 91:16 reads, "With long life ['orek yamim] will I satisfy him, and show him my salvation." In these four passages, nothing invites

10. See further Chapter 17.

PART TWO: Forays into a Biblical World

a construal of the phrase as referring to anything more than a long life within the normal constraints of human mortality.

In Ps 93:5, the phrase takes on a different connotation. RSV renders the verse, "Thy decrees are very sure; holiness befits thy house, O LORD, forevermore [leʾorek yamim]," and NRSV follows suit with "forevermore." In light of references elsewhere to the sanctuary in Zion as everlasting (Ps 78:69; 132:14), it is difficult to suppose that the phrase *leʾorek yamim* here connotes a limited duration, however extended. So RSV and NRSV are appropriate if interpretive. We may recall Towner's comment on the ending in Psalm 23: "Had the psalmist wanted to close the poem with the word 'forever,' a perfectly good Hebrew word, *leʿolam*, was available. Instead, the choice was made to use the phrase *leʾorek yamim*, 'for length of days.'" But the usage here in Ps 93:5 shows that this is an insufficient argument. For here likewise the psalmist could have used *leʿolam*, or *laʿad*. Yet, while clearly intending such a meaning, the psalmist chose *leʾorek yamim*, as though this phrase, standing by itself, could carry such a connotation. To be sure, such a connotation need not be inherent in the phrase itself, but probably derived from its use with reference to the sanctuary, which, as Psalms 78 and 132 attest, would be understood as everlasting. But this instance shows how the phrase can be used in contexts where its connotations exceed its plain sense.

The occurrence in Lam 5:20 depends on no such connotation-by-association: "Why dost thou forget us forever [laneṣaḥ], why dost thou so long [leʾorek yamim] forsake us?" The order of the two phrases is striking. One would expect that the sequence would be reversed (see below, on Ps 21:4), with the more extreme expression coming last for rhetorical emphasis. While the parallelism certainly does not establish intrinsic semantic equivalence, it is noteworthy that once again *ʾorek yamim* is associated with foreverness, but this time not by association with anything that is intrinsically everlasting.

In Ps 21:3, the people celebrate the king as God's crowned servant. Verses 4 and 6 are relevant to the present topic:

> He asked life of thee; thou gavest it to him,
> length of days [ʾorek yamim] forever and ever [ʿolam waʿed]. (Ps 21:4)
>
> . . .
>
> Yea, thou dost make him most blessed forever [berakot laʿad]; thou dost make him glad with the joy of thy presence [beśimḥah ʾet-paneka]. (Ps 21:6)

The king is not literally to live forever. Yet the language of foreverness is employed. In v. 6, I suggest that the *berakot laʿad* refers not simply to the bestowal of kingly rule, but to the sense that the rule thus bestowed embodies God's everlasting kingship and in some fashion participates in it. Likewise, the gladness evoked by that bestowal is qualified as the joy of God's presence. Here we verge on the thematics discussed in connection with the psalms passages cited in the previous section, and may invoke that part of the discussion revolving around Hans Loewald's reflections on the experience of the sense of the suspension of time. How, then, are we to understand the usage in v, 4? As hyperbole for the request for a long reign? As an exuberant "long live the king"? Possibly, but it is interesting to note how *ʾorek yamim* is glossed with *ʿolam waʿed*. Taken together, Ps 93:5, Lam 5:20, and Ps 21:4 alert the reader that *ʾorek yamim* can carry connotations beyond its plain sense.

There is one more occurrence of this precise idiom in Deut 30:20; and this occurrence connects with another form of the idiom, the usage of which will bear in a distinctive way on the issue under discussion. The verse reads as follows:

> loving the LORD your God, obeying his voice, and cleaving to him; for that means *life* to you and *length of days* [*ʾorek yameka*], that you may dwell [*lašebet*] in the land which the LORD swore to your fathers, to Abraham, to Isaac, and to Jacob, to give them.

These lines come within the passage in which Moses sets before the people "life and good, death and evil . . . blessing and curse" (30:15, 19). The "death and evil . . . and curse" consist in the prospect that "you shall not live long [*loʾ taʾarikun yamim*, literally "lengthen days"] in the land which you are going over the Jordan to enter and possess," (30:18). If we compare this prospect with the words in v. 20, the most natural conclusion is that the two verses refer, on the one hand, to Israel's extended habitation in the land, in accordance with the oath sworn to the ancestors, and on the other, to Israel's expulsion from it. How long is that promised habitation envisaged to endure?

In Deut 4:26, as in Deut 30:19, Moses calls heaven and earth to witness to what he is setting before the people. The two passages are thus closely related thematically. In 4:26, as in 30:18, Moses warns the people that if they fall into idolatry after having long been in the land, "you will soon utterly perish from the land which you are going over the Jordan to possess; you will not live long upon it [*loʾ-taʾarikum yamim ʿaleha*, "you will not lengthen days upon it"], but will be utterly destroyed. And the LORD will scatter you among the peoples." Then, after assuring them that, despite such a dire judgment consisting in exile from the land, when in exile they turn back to God, God will mercifully take them back mindful of the covenant sworn with the ancestors (4:31). Then, in 4:49, Moses exhorts the people to "keep his statutes and his commandments, which I command you this day, that it may go well with you, and with your children after you, and that you may prolong your days in the land [*ta'arik yamim 'al-ha'adamah*] which the LORD your God gives you forever [*kol-hayyamim*, NRSV "for all time"]." Here, *kol-hayyamim* (literally, "all the days") clearly is intended to connote a duration without a temporal limit; and the implication is that covenant obedience would result in Israel "prolonging its days" in the land "days without end," that is, "forever."

The point to be noted is that, while in some instances the verbal form of this idiom may refer to individual longevity (Josh 24:31; Prov 28:16; Eccl 8:13), in numerous other places (as shown in Chapter 10) it refers to the question of Israel's residence as a people in the land in accordance with God's promise to the ancestors, a land which God intends them to have "forever/for all time" (*kol-hayyamim*). Though individual Israelites will perish, Israel as a people is promised an endless occupation in the land. So stated, the promise is analogous to the phrase in the Priestly tradition where the land is said to be given as an everlasting possession (*'aḥuzzat 'olam*; e.g., Gen 48:4). In these passages, the connotations for Israel as a collective entity of the verbal form of the idiom are similar to the connotations of the nominal form of the idiom for the psalmist as an individual in Ps 93:5, Lam 5:20, and Ps 21:4.[11] How might this bear on the construal of the last line of Psalm 23?[12]

---

11. I shall comment further, in the addendum to this essay, on the relation between the community and the individual in respect to "everlasting dwelling," which now finds support in Levenson's *Resurrection and the Restoration of Israel*.

12. It should be noted here that the verbal form in v. 6b, *wešabti*, is unusual if it is a form of *yašab*, "sit, remain, dwell"; and some take it to be a form of the verb, *šub*, "return" (or, perhaps, another verb meaning, "dwell"). Insofar as the verbal idiom occurs in Deuteronomy almost a dozen times, always with reference to prolonging days in the land, and the nominal idiom occurs in Deut 30:20 with the verb, *yašab* ("that means life to you and length of days [*'orek yameka*], that you may dwell [*lašebet*] in the land"), we may take the last line in Psalm 23 to (a) convey the notion of *dwelling* somewhere for "length of days," and (b) to be taking the stock Deuteronomic usage and employing it in an elevated sense, on analogy with the way in which Psalm 16 takes the theme of land allotment in Joshua and gives it a theological connotation. In Psalm 23, then, the "promised land" (so to speak) becomes the house of God.

On Some Axes of Affirmation Converging in Psalm 23:6b

I want to suggest that the connotations of Ps 23:6b arise from thematics converging on it along several inter-textual axes. One is the way in which, as Mays traces in his commentary, the psalm may be read as playing off of well-known elements in the story of the exodus, the wilderness trek, and the entry into the land. In that connection, the epic thematics of journey (with attendant dangers) and arrival (with connotations of safety in the land)—of motion and rest—provide a backdrop for the themes in Psalm 23 of journey and arrival, motion and rest, danger and safety; and thereby they provide a backdrop specifically for the reprise of the transition in v. 6, from journey in v. 6a to dwelling in v. 6b.

But this opens a connection to another theme that we see appropriated in a transformed way in Psalm 16. While Israelites generally were allotted plots of land after their entry into it, Levites did not receive such plots, but were dependent for their support on offerings to the sanctuary where they served. Their "portion" or "inheritance" lay in the vocation to serve the sanctuary and its precincts. Theologically speaking, then, just as other families dwelt in the land they received, the sanctuary is where they could be said to "dwell." Such a theology is pressed further in Psalm 16 where, in vv. 5–6, the psalmist confesses,

> The LORD is my chosen portion [or "inheritance," *ḥeleq*]
>   and my cup;
> thou holdest my lot.
> The lines have fallen for me in pleasant places [*neʿimim*];
> yea, I have a goodly heritage [*naḥalah*].

Here, the psalmist "pushes the envelope" of the connotations of the language of land-holding whose "plain sense" is clearly demarcated in the last half of the book of Joshua. The "portion" or "inheritance" by which this psalmist's body "dwells secure" [*yiškon labeṭaḥ*] is identified not simply as the material offerings received in connection with temple service, but as the LORD. What does such a bold extension of the plain meaning of "portion/inheritance" intimate? What does it mean to "inherit" God, or to "dwell within the boundaries" of the divine life?

It is widely recognized that the four lines quoted above are a transformative appropriation of the theme of material grant of land to individual Israelite households. The expression, "dwells secure," may reflect in another way the thematics of Israel's dwelling in the land. For while the verb *škn* adverbially modified by *beṭaḥ* can refer simply to individual security in a general way (Prov 1:33), it frequently refers to Israel or one of

PART TWO: Forays into a Biblical World

its tribes dwelling securely in the land allotted to it (Deut 33:12; 33:28; Jer 23:6//33:16; cf. Judg 8:11). In view of the psalmist's appropriation of land allotment imagery earlier in the psalm, this phrase too may be taken as an existential appropriation of the related motif in Deut 33:28 and elsewhere:

> Therefore my heart is glad, and my soul rejoices;
> my body also dwells secure [*yiškon labeṭaḥ*].
> For thou dost not give me up to Sheol,
> or let thy godly one see the Pit.
> Thou dost show me the path of *life*;
> in thy presence there is fullness of *joy*,
> in thy right hand are pleasures [*neʿimot*] forevermore [*neṣaḥ*].[13]

In reference to a mundane inheritance, the connotations attaching to a portion chosen by "lot" as falling to an Israelite "in pleasant places [*neʿimim*]" would be those of a fertile portion of land affording economic plenty. These connotations are carried over in Psalm 16 to the speaker's inheritance of God, with connotations that come to a hyperbolic climax in the last verse, where the psalmist says, "in thy presence there is fullness of joy, / in thy right hand are pleasures [*neʿimot*] forevermore." Thus the themes of dwelling in God's presence, life, joy and everlastingness come to particularly thick expression. What does the psalmist understand by these words? We need not attribute to him an articulated understanding, to appreciate that for him the experience that goes with his "inheritance" fills him with a sense of vibrant joy and life in the presence of God that is felt throughout his whole body, a sense of a communion with God that the threat of Sheol and the Pit cannot diminish.[14] It is such a sense, I think, that also informs the psalm passages canvassed in the first section above. It makes good sense, then, as others have done, to attribute Psalm 16 to a Levite, or one speaking with a Levitical persona. In this connection we may note the words of Hans-Joachim Kraus where he comments on Ps 23:6:

> The expression of trust and confession [in v. 5] then goes over to the concluding statement, "I will remain in Yahweh's house—as long as I live." This is not "pious illusion, born of exuberance" (thus L. Köhler). Verse 6b could originally have been a Levitical

---

13. As in Ps 23:6, this psalm ends with the motifs of motion ("the path of life") and rest ("thy presence," "thy right hand"). On the intimate relation between Psalms 16 and 23 within Psalms 15–24 as a carefully edited group, see further below.

14. In post-biblical Jewish usage, one way of referring to God is with the phrase, *'En Sof*, "The Infinite, Endless, Boundless" (see Chapter 4). What would it mean, to have the *'En Sof* as one's "unboundaried portion" or "allotment," one's *Lebensraum*?

confession (cf. the comment on Ps 16:6f). This Levitical confession would then have been spiritualized and recoined as a "sublime cultic mysticism" (G. von Rad).[15]

I take von Rad's phrase "sublime cult-mysticism," as quoted by Kraus, to be apt, insofar as the term "sublime" brings into linguistic view those liminal aspects of human experience that are not susceptible of clear assessment and confident evaluation by the rational and empirical means fitted to ordinary mundane experience.[16] From a psychodynamic point of view, what is at issue may be explored with the help of Hans Loewald's reflections in his short monograph, *Sublimation*, as well as the concluding reflections in his monograph *Psychoanalysis and the History of the Individual*. Loewald's discussions help to indicate a way of coming at the subjective dimensions of religious experience which does not simply play them off against, or dismiss them in favor of, objective reality, but rather takes them seriously, albeit cautiously and diffidently, as holding possibilities for a deeper awareness of and participation in the nature of things than ordinary waking consciousness is habituated to and than the forms of speech and epistemological criteria tailored to the needs of mundane consciousness and its concerns enable us to do justice to.

Apropos of Kraus's proposal that Ps 23:6b be read within the same "Levitical" frame of reference as aspects of Psalm 16, I would draw on another line of reflection opened up by recent developments in Psalms studies. I have in mind the interest in the shape of the Psalter as a whole and of smaller groups within it. Sustained work of my own on Psalms 93–100 (see chapter 8), together with careful study of the commentary on these psalms by Hossfeld and Zenger[17] and in particular of the monograph by David M. Howard, Jr.,[18] has persuaded me that the editorial process by which this group achieved its present form involved much more than mere collection and incidental editing. The inter-relations between the psalms in this group, especially as Howard traces them, give clear evidence of such religious and theological reflection and editing that the group as

15. Kraus, *Psalms 1–59*, 309.

16. One may recall Whitehead's words toward the end of his essay in cosmology, *Process and Reality*, V.II.II where, following a chapter in which he addresses issues of "permanence and flux, time and eternity," and arriving at the question of God, he writes, "[a]ny cogency of argument entirely depends on elucidation of somewhat exceptional elements in our conscious experience—those elements which may roughly be classed together as religious and moral intuitions." See also ibid., V.I.II.

17. Hossfeld and Zenger, *Psalms 2*, 6–7, 446–98.

18. Howard, *The Structure of Psalms 93–100*.

a whole stands before us as a theological testimony greater than the sum of its parts and certainly not reducible to the sum of the respective theological testimonies of the individual psalms in prior stages of self-standing existence.

How does this bear on Psalm 23:6b? Patrick Miller's essay on Psalms 15–24 as a coherent group,[19] while not as reticulated as Howard's monograph on Psalms 93–100, is as rich in its implication. Drawing on the work of Pierre Auffret, and of Hossfeld and Zenger, Miller explores the structural and thematic inter-relations of these psalms to show how their individual theological testimonies are not exploited fully until those testimonies are heard as voices in one symphonic choir moving through several phases. While he emphasizes the ring pattern in which they are organized, with 15 and 24 as the outer ring, 16 and 23 (!) as the next ring inwards, and so on, one may also discern a progression from 15 through 24.

I shall comment on the latter progression in a moment. First, I note the way in which the editing of this group positions Psalms 16 and 23 in the same ring. This positioning suggests that the editors (the redactors whom Martin Buber, following Franz Rosenzweig, proposed that we should acknowledge as Rabbenu, "our Teacher")[20] discerned in these two psalms themes advanced in common albeit each in its own way. Miller observes that, for example, only in these two psalms do we find anywhere in the Hebrew Bible the precise phrase, *kosi*, "my cup" (Pss 16:5; 23:5). By itself, this might appear incidental. But Miller's observations on the numerous connections between these two psalms lead him to the conclusion that "[i]n the first song of trust in this collection (Psalm 16) and in the song of trust that matches it near the end (Psalm 23), the psalmist uses the same unusual image to speak of God's beneficence."[21] This is to say that distinctive word-choice ("my cup"), genre (song of trust), sanctuary-inspired mystic vision (Kraus) and editorial positioning (same ring), all lead to the conclusion that these two psalms exegete one another. This reinforces the intimation of unbounded aspiration in Ps 23:6 as resonating with the intimation of unbounded expectation in Ps 16:11.

But we may note also, in respect to a progressive reading from 15 to 24, the possible implications of Psalm 23's position between Psalm 22 and Psalm 24. First, a comment on Psalm 22. Ellen F. Davis has shown in a brilliant analysis of its concluding verses, how this psalm, by its "extravagance

---

19. Miller, "Kingship, Torah Obedience, and Prayer," 279–97.
20. See Avnon, *Martin Buber*, 50 and n. 6.
21. Miller, "Kingship, Torah Obedience, and Prayer," 289.

of language" and "exuberance of . . . poetic vision," "explodes the limits of Israel's traditional understandings." "The psalmist," she writes, "lays a poet's claim to a previously unimaginable future, insisting that not even death can defeat the basic power of Israel's life, the power of faithful praise."[22] In respect to the present paper, I would simply say, we must read Ps 23:6b not only in relation to Psalm 16, but also in relation to Psalm 22 as Davis has so deftly read it, where the psalmist's call to praise God extends in widening circles until it draws even the dead—those "silent" ones elsewhere said to be unable to praise God—into its choir.

Next, let the comments of Alan Cooper on Psalm 24 and its "everlasting doors"—doors shown to be the gates of Sheol that give way before the victorious divine King—reflect back on Psalm 23. Cooper writes, in part,

> [I]t would be simplistic to claim that early Christian exegesis preserves or revives the original meaning of the text . . . Nevertheless, it might not be too far-fetched to suggest that, in their adaptation of Ps 24:7–10, the early Christian commentators realized, even actualized, a potential which is inherent in the imagery of the psalm. In other words, the mythological structure of Christian interpretation is fundamentally compatible with the content of the psalm.[23]

And consider, then, how this may bear on our construal of the phrase *ge' ṣalmawet* in Ps 23:4. Is this merely "the darkest valley," as NRSV has it? Or—at the very least, by a folk etymology or a play on words—are we invited to hear a reference to the valley of the shadow of death?[24] We may recall that in the Hebrew Bible the word *ṣel*, "shade, shadow" refers not only to what a standing object casts on the ground on a sunny day, but also—as in Ps 91:1—to a powerful figure under whose aegis or umbrella (little shadow) one may find protection from dangers and enemies. But as the latter psalm shows, God's protection is imaged repeatedly in terms of refuge in God's sanctuary. Now, if the sphere of power exercised by Death (a divine royal figure in Canaanite mythology, like Pluto in Greek lore) may be imaged as "the valley of the shadow of death," Death being, so to

---

22. Davis, "Exploding the Limits," 103–104. I myself would express the point a bit more reservedly, though in the same direction, by using the current expression (as I have earlier in this essay), "pushing the envelope." From the point of view of the writer of Luke-Acts who quotes Ps 16:8–11 in Acts 2:25–28, what "explodes the limits" comes later when "God raised [Jesus] up, having loosed the pangs of death, because it was not possible for him to be held by it" (Acts 2:24).

23. Cooper, "Ps 24:7–10," 59–60..

24. See Michel, "*ṣlmwt*, 'Deep Darkness' or 'Shadow of Death'?" 513.

speak, the warlord entrenched in the steep declivities in the hills around Jerusalem, Psalms 22, 23 and 24 each in their own way bring to a climax the progression of the themes of kingship, torah, and prayer that run through Psalms 15–24.

Drawing Matters to a Conclusion

What, then, of translations like "forever" in RSV and NIV, rooted in the KJV and and, before that, Coverdale's rendering of the Psalter? (Since Coverdale did not read Hebrew, we may assume that his rendering reflected translations that he worked from.) Do these represent what Towner several times, in reference to English paraphrases and hymn compositions, refers to as "Christianizing" of the text—a reading back into the text of understandings derived from the testimony of the New Testament writers? The issue is complex. Such a "retrospective" reading of the Old Testament in the light of the testimony to the resurrection of Jesus is explicit already in passages such as Luke 24:32, 45 where the risen Christ is portrayed as "opening" the disciples' eyes to a new understanding of the Old Testament and also "opening" those scriptures in a new way to them. But the rise of convictions as to immortality or resurrection of the dead within certain strands of Second Temple Judaism implies that some Old Testament texts were taken on other grounds to be patient of such a construal.

Among those other grounds, many have pointed to the pressure of the persistence of evil and the apparently un-redressed sufferings of the innocent and righteous faithful, as driving theological reflection to arrive at these convictions. However that may be (and we may suppose that the grounds for the rise of such convictions were of many sorts), in this essay I am proposing that one entirely positive ground is to be sought in the experience of the act of praise and its attendant joy, a praise and joy on occasion issuing in a suspension of all sense of time, a dance of sheer presence before the everlasting God in which the dance itself, while it lasts, seems everlasting.

The momentary transformation of one's ordinary sense of time into a sense of everlastingness has an analogy in—or perhaps is another dimension of—one's ordinary sense of existence in space. Normal waking consciousness includes the sense of existing in a finite space surrounded by other forces and agents; and such consciousness can take the form of awareness that one is surrounded by great danger and animosity (e.g., Pss 3:1–2; 22:12–21). In such circumstances, the divine sanctuary takes on the character of a fortified "tower" (*migdal*, Judg 9:46).²⁵ Though the claims to inviolability intrinsic to any material sanctuary present no ultimate security against military might (see, e.g., 2 Kgs 10:18-27), the psalms are full of affirmations of unqualified trust and security "in God's house" and in the presence of God that house represents (e.g., Pss 3:3-4; 4:8; 91). Psalm 61:2-4 combines these themes of everlastingness and absolute security:

> Lead thou me to the rock that is higher than I;
> for thou art my refuge,
> a strong tower [*migdal 'oz*] against the enemy.
> Let me dwell in thy tent forever ['*olamim*]!
> Oh to be safe under the shelter [*beseter*] of thy wings!

The thrust of this paper raises epistemological issues concerning the force of testimony grounded in privileged experiences—what Whitehead refers to as "somewhat exceptional elements in our conscious experience . . . religious and moral intuitions." These issues are exemplified in Psalm 73, which begins with the speaker's perplexity over the prosperity of the wicked and their apparent immunity from the troubles that strike the faithful (v. 5). This perplexity leads to weariness in the attempt at understanding (v. 16), and (as with Naomi and Job) to bitterness of soul (Ruth 1:20-21; Job 3:20 and often). Then the psalmist enters the sanctuary of God, and sees matters in a completely different light (Ps 73:17): Whereas the apparent prosperity of the wicked is as ephemeral as a dream (vv. 18-20), the psalmist's sense of the presence of God (v. 23) leads to an affirmation that in v. 26 culminates in an echo of Psalm 16: "My flesh and my heart may fail, but God is the strength of my heart and my inheritance [*ḥelqi*] forever [*le'olam*]." Here, I suggest, the affirmation arises not as the conclusion to a theological reflection wrestling with a conundrum, but as an articulation of the psalmist's sense of existence in being caught up into the presence of God. I would go so far as to suggest that the key to the shift from the acknowledgment in v. 26a to the affirmation in v. 26b lies in the shift from the theme of motion to rest, journey to arrival, in v. 24: "Thou dost guide me with thy counsel, / And afterward thou wilt receive me to glory [or: "in honor"]." This is the same shift that I see in Pss 16:11 and 23:6.²⁶

---

25. And prayer itself can "take place" *beseter*—within God's "shelter"—amid a social context which deprives one (as in Psalm 69) of any mundane, or other existential, "foothold" (see Chapter 9).

26. It is as though the historical themes of exodus, journey in the wilderness, and entry into the land where Israel experiences rest from its enemies, have become

PART TWO: Forays into a Biblical World

Are such realizations exclusive to the sanctuary, peculiar to what von Rad dubs "cult mysticism"? Erich Zenger's comment on associations between Psalm 73 and the book of Job opens the way to a wider possibility. He writes, "Like Psalm 73, the book of Job also presents, as a way out of the existential crisis, a 'vision of God' (cf. Job 38:1; 42:5; 19:26)."[27] The comparison is apt. But there is this difference: While the psalmist comes to his new awareness as a result of his sense of the presence of God within the sanctuary, Job comes to his awareness in the context of a natural phenomenon—the arrival of a fall sirocco (the "whirlwind" of traditional interpretations) from off the south-eastern desert that presages, and then shortly gives way to, the first fall rains that refresh the parched earth and bring all creation to renewed and vibrant life.[28] Tellingly, the first section of what Job hears God say to him through these phenomena ends with the words, "when the morning stars sang together, / and all the children of God shouted for joy" (Job 38:7). It is this lyrical note, reversing the piping lament through which Job has voiced his present situation (30:31), that establishes the utterly positive tone of the divine speeches and that leads Job, finally, to the resolution of his plight. So then, while Zenger is right in drawing a parallel between the psalmist and Job, the difference I have noted attests that the sort of experience possible within the sanctuary is not necessarily confined to the sanctuary, but—especially for a soul *in*formed by "cult mysticism" (von Rad)?—can arise within creation itself. That this should not surprise us is perhaps an implication of Jon

---

internalized to the extent that they have become part of the matrix of the religious consciousness of those who produced psalms like 16, 23 and 73. Such a matrixal consciousness may also underlie Psalm 95 which, although Israel has already entered and long inhabited the promised land, portrays God as addressing the members of the covenant community with the warning that, like the wilderness generation, they too may yet fail to enter God's "rest." Whatever this rest may consist in, it no longer consists simply in dwelling in a physical locale safe from surrounding political foes. The "plain sense" of the earlier meaning of "rest," such as we find in, for example, Deut 12:9; Josh 1:13; and 2 Sam 7:1, has taken on connotations the character and scope of which are open for further exploration and discovery.

27. Zenger in Hossfeld and Zenger, *Psalms 2*, 225. Zenger's commentary as a whole, on this psalm, and on its significance for one's reading of the whole "Asaph" psalm-group which it introduces (Psalms 73–83), is rich in implication for the thesis of the present paper. Note, for example, his concluding sentence in the section titled "Context, Reception, and Significance": "Thus the 'counter-world' proposed in Psalm 73 acquires a fundamental hermeneutical significance for the entire collection" (ibid., 237).

28. For my analysis of the divine speeches in Job, see Janzen, "The Lord of the East Wind," 247; and Janzen, *At the Scent of Water*.

D. Levenson's exposition of the significance of the sanctuary as a microcosm of the cosmos,[29] which suggests that being "at home" in the house of God is a cultic pedagogy for being at home in God's creation, and that the "counter-world" of which Zenger sees Psalm 73 speaking may in fact be a New Heaven and a New Earth. In this respect the "house of God," and the "green pastures" and "springs of water," in Psalm 23, may be taken to exegete one another.

Addendum

The above essay, and in particular my heuristic exploration of the relation between the individual and the community in respect to everlasting dwelling in the land or elsewhere, was written before Levenson's *Resurrection and the Restoration of Israel* appeared. This addendum serves two purposes: (1) First, and most importantly, I wish to draw attention to Levenson's book which, combining consummate historical-critical scholarship, literary subtlety, and theological sensitivity, shows how

> [w]hen the belief in resurrection finally makes an unambiguous appearance in Judaism [in Dan 12:2], it is . . . both an *innovation* and a *restatement* of a tension that had pervaded the religion of Israel from the beginning, the tension between the LORD's promise of life, on the one hand, and the reality of death, on the other. In the case of resurrection, the last word once again lies not with death—undeniably grievous though it is—but with life. Given the reality and potency ascribed to death throughout the Hebrew Bible, what overcomes it is nothing short of the most astonishing miracle, the Divine Warrior's eschatological victory . . . Our exploration into the rabbinic doctrine of resurrection has traced its ultimate origin to the transformation that nature undergoes as a result of the Divine Warrior's astonishing victory. That transformation replaces sterility with fertility, childlessness with new descendants (and the return of lost descendants), hopelessness with a radiant future—death with life. Within the religion of Israel from the earliest time that we can identify it, the hoped-for transformation could not have been thought complete, or even real, without a restoration of the people Israel itself, and this perforce entailed a recovery from humiliation and defeat, a reconstitution of the broken nation, and its rededication to its redeemer and restorer. The affirmation that such a

---

29. Levenson, *Creation and the Persistence of Evil*, 78–99.

restoration could even bring back the dead was both *innovative* and *deeply conservative*.[30]

(2) While Levenson's analysis of the biblical tradition establishes a complex thematic matrix within which I would now nestle my analysis in the present paper, I draw attention to one theme on which our analyses run hand-in-hand. Following Mays' and Freedman's identification in Psalm 23 of the epic themes of exodus, wilderness wandering, and entry into the land, and noting how Psalm 16 has drawn on the theme of the allotment of land and transposed it into a new theological register where God is the psalmist's allotted "portion," I have proposed to read the last line in Psalm 23 as similarly transposing into a new theological register the Deuteronomic theme of Israel's "prolonging its days" in the land of promise. And I have suggested that, insofar as the latter idiom in Deuteronomy implies the possibility of dwelling in the land for an unlimited duration, this gives additional support to the implication at the end of Psalm 23 that dwelling in the house of God is endlesss.

One of the many strengths of Levenson's analysis is that he does not approach the biblical text with our contemporary notions of individual resurrection or immortality and ask whether they are to be found in the Hebrew Bible. Rather, he starts where the Bible starts, with communitarian understandings of human life, and with the emphasis on possibilities for survival and flourishing of family, clan, tribe, and people as a whole. At such a time, the fundamental themes of restoration, and the language of resurrection (as in Ezekiel 37) arise in relation to Israel as a whole. With the emergence of issues of individual standing in the community and before God (I would point to the prominence of the psalms of individual complaint and praise, and, as in Chapter 12, to prophets of embattled solitariness and solidarity like Jeremiah), these fundamental themes are available for transposition to the individual in a way that eventually becomes explicit in Dan 12:2. I hope to have shown that the testimony of Psalm 23 marks one moment in that development, and that, as Alan Cooper has said of Psalm 24, we may say of Psalm 23 that the subsequent "structure of [Jewish and] Christian interpretation is fundamentally compatible with the content of the psalm," as drawing upon "a potential which is [already] inherent in the imagery of the psalm [itself]."[31]

---

30. Levenson, *Resurrection and the Restoration of Israel*, 216–17 (italics added).
31. Cooper, "Ps 24:7–10," 59–60.

part three

The Standpoint of Two Prophets

12

Solidarity and Solitariness in Ancient Israel

The Case of Jeremiah

PARTICULAR TEXTS, LIKE PARTICULAR experiences, arise and have their meaning partly as the point of convergence of diverse vectors of energy flowing from diverse contexts. But such particulars are not merely the sum of such vector forces. In some measure, such vector forces combine to produce a new force, which transcends the sum of its parts, and which in its turn may then become a vector in subsequent occasions. In this essay I will seek to identify contextual vectors, ancient and modern, within which the particular meaning of Jer 20:7–18 may be explored by the reader. There are no less than seven.

In the *first* place, Jeremiah draws upon the general form of Israel's psalms in which the worshiper confesses loyalty to God and expresses bewilderment as to why God has dealt otherwise than the worshiper had been led to expect. This general form is elsewhere nuanced in two somewhat different contexts: among the prophets, where the complaint arises over the prophet's fate while pursuing the prophetic vocation (e.g., Habakkuk),[1] and in Job, who becomes the archetypal confessor/plaintiff. In the *second* place, Jer 20:7–18 comes as the climax, or rather the nadir, of a series of such confessions in Jeremiah, the previous instances of which are 11:18—12:6; 15:10–21; 17:12–18; 18:18–23. *Thirdly*, these five poetic complaints give an inkling of Jeremiah's experience as narratively portrayed later in the book, where Jeremiah's arrest is followed by increasingly grave stages of confinement, from house arrest to prison to pit, then brief respite at the hands of the Babylonian occupiers, to final descent into

---

1. See Chapter 13.

PART THREE: The Standpoint of Two Prophets

Egypt. Writing of this narrative *via dolorosa*, Gerhard von Rad has suggested that "Jeremiah's death apparently formed no part of this account."[2] To the contrary, given the connotations of Egypt in the Bible generally, it is no coincidence that the confessions which "shade off into darkness" and show him walking "a road which led ultimately to abandonment,"[3] find their nadir in Jer 20:14–18, an exact confessional analogue to the narrative ending in Egypt. One might even hazard a thematic if not chronological correlation between Jeremiah's brief liberation by the Babylonians and the brief surge of praise in 20:13 before the final descent.

A *fourth* context is provided by the call account in 1:4–18, itself a particular reflex of the wider context of prophetic call accounts. Among other items in which Jeremiah's call resembles that of Moses are his reluctance to go and speak and God's repeated assurance that, having sent him (1:7), God will be with him (1:8, 19). This polar tension between sending and being with (cf. Exod 3:10, 12) will be seen to lie at the heart of Jeremiah's dilemma as voiced in his confessions. Meanwhile, the analogy between Moses and Jeremiah heightens the irony of his biographical ending in Egypt, after the assurances in his call, and helps us to appreciate all the more the reasons for his darkness in 20:14–18.

A *fifth* and wider context must now be brought into view. According to Thorkild Jacobsen three millennia of Babylonian religion may be traced in terms of three fundamental metaphors for the gods.[4] In the fourth millennium the gods were powers immanent in the phenomena of nature, powers willing to come to specific form as the phenomena. In the third millennium the gods transcended nature and society as royal figures who had created nature and society as artefacts and slaves to serve them. In the second millennium, among some Babylonians, the gods became also personal deities, divine parents of their human children, responsible for their birth, nurture, protection, and guidance. Within this metaphor the god was said to be with the human devotee as the power within the individual for success. Jacobsen maintains that, whereas in Mesopotamia itself this "personal religion" gave way in the first millennium before the resurgence of older modes of perception of the divine activity, through Israel's ancestors (Abraham/Sarah, et al.) it gave Israelite religion much of the latter's definitive character; though in the latter development Yahweh became the

---

2. Von Rad, *Old Testament Theology*, 2:207.
3. Ibid., 2:203 and 206, respectively.
4. Jacobsen, *Treasures of Darkness*.

divine father not of the individual but of the community and nation as a whole (e.g., Exod 4:21–23; Hos 11:1; Isa 1:2–3).

The eccentric but seminal thesis of Julian Jaynes contributes to this picture.[5] In his view, the period 1500–800 marks the rise of individual consciousness in the ancient Near East. At the beginning of this period the gods still speak loudly and clearly as vocal presences within the head of the barely-individuated members of the community. By the end of this period the divine voices no longer speak clearly or univocally, and people become distracted by the conflict in what they hear or by the silence which increasingly displaces the immanent voices. What emerges is a growing solitariness within the individual and with it a growing self-consciousness vis-à-vis the gods and the community. At this point we may return to consider a *sixth* context for Jer 20:7–18.

The pericope comes immediately after Pashhur's challenge to Jeremiah. Just as Zedekiah had struck Micaiah for his solitary opposition to the 400 prophets (1 Kings 22), so Pashhur beats Jeremiah who stands in solitary opposition to the prophets of national well-being. Who speaks, any more, for Yahweh? At times it is not clear even to Jeremiah (20:7). Having publicly excoriated Baal-worship as cistern-building (a social, structural-functional form of religion) which forsakes Yahweh the fountain of living waters (2:13), he privately accuses Yahweh (15:18) of being a disappointing/deceitful brook whose waters fail (*lo' ne'emanu*, "are not reliable/faithful"). In response to Yahweh's questionable reliability, both the confessions and the biography show Jeremiah enacting a Joban steadfastness in which doubt and patience define one another and in which even the momentary wish for non-existence is but the dark coloration of the light of faith and unquenchable vocation. Von Rad's words are apt at this point:

> It is still Jeremiah's secret how, in the face of growing scepticism about his own office, he was yet able to give an almost superhuman obedience to God, and, bearing the immense strains of his calling, was yet able to follow a road which led ultimately to abandonment . . . Again, if God brought the life of the most faithful of his ambassadors into so terrible and utterly uncomprehended a night and there to all appearances allowed him to come to utter grief, this remains God's secret.[6]

---

5. Jaynes, *Origin of Consciousness.*
6. Von Rad, *Old Testament Theology*, 2:206.

PART THREE: The Standpoint of Two Prophets

The secret has, perhaps, to do with that aspect of any experience (or text) in which it in some measure transcends its context and achieves the standpoint from which to transform its inheritance into a novel legacy.

But these words lead us to consider a *seventh* context for Jeremiah's confessional complaints: the divine milieu. I have suggested that his dilemma lies in the tension between "I am with you" and "I send you." Who is it that is with Jeremiah? It is not only Yahweh the dread warrior (20:11; cf. Exod 15:3); it is also Yahweh the Solitary One (*'eḥad*; Deut 6:4; Zech 14:9; Job 23:13; and 31:15). This Yahweh, who has called the people into the most intimate (Exod 24:9–11) and comprehensive (Deut 6:4) of covenants, has been abandoned by the covenant people (Jer 2:1–6, 13), is no longer known by them (2:8), and is left to a divine grief which no one can share (8:18—9:3).

If Heschel is right in interpreting the prophetic experience as a participation in the divine pathos,[7] are Jeremiah's complaints evidence of such a participation? But if the divine pathos consists not only in an ontological solitariness of Yahweh as God, but also in an existential solitariness as deserted covenant partner, how does Jeremiah share in that pathos? Is it possible that only in the blackness of his own sense of isolation from both people and Yahweh can Jeremiah enter most deeply into an incomprehensible fellowship with God in suffering for the people? And is not the depth of such suffering and the refusal to heal it lightly (15:18a; cf. 6:14) the measure of Jeremiah's loyalty both to the people and to Yahweh? Conventionally, "witness" connotes presence and "sending" absence. Is it possible that the two in this instance are one? That the sense of absence communicates, strangely, a presence?

Whitehead gives three pithy characterizations of religion where it has emerged beyond what he calls herd-psychology.[8] (The first is often

---

7. See Heschel, *The Prophets*.

8. The following sentiments from Whitehead, *Religion in the Making*, 60, are important for understanding the relation between individual solitariness and universal community in Whitehead's three characterizations of religion, and underlie one of the basic intuitions informing the present collection of essays: "Religion is the art and the theory of the internal life of man, so far as it depends on the man himself and on what is permanent in the nature of things. This doctrine is the direct negation of the theory that religion is primarily a social fact. Social facts are of great importance to religion, because there is no such thing as absolutely independent existence. You cannot abstract society from man; most psychology is herd-psychology. But all collective emotions leave untouched the awful ultimate fact, which is the human being, consciously alone with itself, for its own sake . . . Religion is solitariness; and if you are never solitary, you are never religious."

quoted—in ignorance or in despite of the other two—usually in order to criticize its excessive individualism.) (1) "Religion is what the individual does with his own solitariness." (2) "Religion is world-loyalty." (3) "The world is a scene of solitariness in community . . . The topic of religion is individuality in community." If Israelite religion emerged in the first instance as a communal version of the personal religion of second-millennium Mesopotamia, the individual dimension nevertheless also developed *pari passu* with the communal. This was reflected in the second personal singular form of the Decalogue, for example, and in the fact that the single archetypal prohibition in Genesis 2 is given, not to the community but to the individual; yet, that community has its function in the mutual help to be given in fulfilling the earthly vocation. This individual dimension is portrayed not only in the great figures of Abraham, Sarah, Jacob, Rachel, Moses, Miriam, Samuel, Elijah, Jeremiah, Job, the Servant of Yahweh, and Lady Wisdom, but also in the large numbers of the psalms of individual complaint and confession, psalms which, in their character as "common prayer," bespeak the existence in Israel of numbers of individuals who, anonymous to the public and to history, are known only to Yahweh as loyal participants, wittingly or not, in the divine pathos.

The dimensions of the contemporary context for Jer 20:7–18 are manifold; only brief suggestions in some directions can be offered here. One two-headed issue is perhaps germane. In some circles still, religion is a strictly individual affair, issuing in a purely transcendent salvation of the soul; in other circles, religion is chiefly a communal affair, bent on an immanent salvation of society. My characterizations are of course gross and oversimplified, but the reader will recognize the issues to which I refer. On the one hand, there is an overlooking of the concomitant feature of the individual dimension of biblical religion, in the burden of solitariness and suffering as a participation in the life of the community and of God. On the other hand, there is observable in some quarters a distinct tendency to reduce Yahweh to functionalist categories under the aegis of sociological

---

In support of Whitehead's assertion, I would point simply to the issues analyzed earlier in my discussion of Psalm 69 (Chapter 9); also, the remarkable fact that the largest single genre in the Psalter consists in psalms of individual complaint/lament. These psalms are themselves a bulwark and a window on transcendence vis-à-vis the sheer weight of social consciousness and its "herd-psychology" in any culture. It may be noted that Psalm 69, which at the beginning articulates so graphically the "awefulness' of solitary consciousness, in its concluding verses, articulates what Whitehead calls "world loyalty." Is it perhaps the very transcendence-as-isolation, signaled in the initial solitariness vis-à-vis the community, that both indicates the condition of, and finds the foothold or standpoint for, such world loyalty?

analysis. Yahwism becomes a "servomechanism" for the society (Norman Gottwald)[9] or the humanly constructed hero-figure in a story which Israelite society tells in constructing a social world of meaning as a way of overcoming the chaos of ultimate non-meaning. The story, however, does not report on something beyond the manifold social context, does not make an ultimate reference, but is a humanly constructed web of meaning in "an essentially meaningless environment."[10] Such currently-emerging views of religion accurately analyze what was earlier described in Jer 2:13a. They cannot account, finally, for the Israelite experience attested in Jer 2:13b, an experience of a power and a presence energizing social renewal and cohesiveness (Exodus and Sinai) and at the same time enabling the emergence of a solitary consciousness whose extreme form occurs where all sense of social support and cohesion, human or divine, either is negatively experienced as against one or is experienced as absent altogether, yet where the harassed/abandoned solitary individual persists in solidarity with the divine and human covenant community, toward renewal and transformation (e.g., Jer 31:31–34). Such a solitary consciousness serves a Presence which, in its covenant loyalty to the community, at times must announce the end of all orders and structures of meaning and life (Jer 4:23–26), a Yahweh who not only leads Israel out of the chaos of the wilderness into the orders of Canaan,[11] but who as often leads or drives Israel out of corrupt orders and back into the wilderness. It is time for biblically-oriented communities and leaders to take us beyond the false dichotomy inherent in inherited metaphysics of matter/spirit bifurcations, beyond the split between social-material functionalism and individual-spiritual transcendentalism, in a covenanting path which incorporates the genuine concerns of both and wherein Yahweh is both high God of people and world and personal God of the individual.

It is, finally, poignant that the final redactor of Jeremiah—this prophet to the nations (1:10) who displayed such an intensely solitary suffering consciousness as the interior substratum of his socially active life—should have appended to the final scene of the biographical narrative a personal vignette involving just Jeremiah and his biographer. (Cf. 1:10 with 45:4; and 45:1–5 with 44:26–30.) In response to Baruch's own complaint, which but for this passage would have remained one of the many anonymous ones, Jeremiah offers a pastoral word which for all its stringency may be

---

9. See Gottwald, *The Tribes of Yahweh*, 646–48 and esp. 786 n. 569.
10. Thompson, "The Jordan Crossing," 343–58.
11. Ibid.

supposed to have offered the deepest comfort. This is but a firstfruits of how Jeremiah's transcendent solitariness issued as a social vector of transforming energy, available for the wider world of subsequent occasions (cf. 2 Cor 1:3–4).

13

Eschatological Symbol and Existence in Habakkuk

IN AN EARLIER STUDY, I attempted to "bring to the determination of the meaning of Hab 2:4b in its own context a degree of precision made possible by recent proposals as to the meaning of other parts of the passage 2:2–4."[1] That study further presented "an analysis of the rhetorical form and content of 2:2–4 as a whole" and made preliminary proposals concerning the connections of this passage—understood in the manner which I have urged—with other prophetic traditions. But if my reading of Hab 2:2–4 and my proposals concerning its connections with other prophetic traditions are in any significant degree valid, this has important implications for the way in which we should read the Book of Habakkuk as a whole. In the present chapter, accordingly, I shall offer a reading of Habakkuk which I believe to be consonant with the results of my analysis of 2:2–4. Admittedly, the reading will arise through a consideration of only select features, since the scope of the chapter precludes a close reading at every point. Even so, the features selected for attention here are pivotal or representative and provide primary or representative clues to that overall understanding of the book within which all the other features will be seen to yield their meanings most felicitously.

The present sketch will presuppose the broad lines and the specific details of my previous study. However, a few of the crucial results will be recapitulated, by way of reconvening the discussion, so to speak: (1) In 2:3a I read, not "for *still* the vision awaits its time" (RSV), but "for the vision *is a witness* to the end." (2) In 2:3b I read, not "it *hastens* to the end" (RSV), but "it *is a testifier* to the end" (so also Ehrlich, Loewenstamm; compare KJV). (3) In 2:4b I read, not "the righteous shall live *by his faithfulness*"

---

1. Janzen, "Habakkuk 2:2–4 in the Light of Recent Philological Advances," 53.

Eschatological Symbol and Existence in Habakkuk

(*be'emunato*), but "the righteous through *its* reliability shall live," i.e., the third masculine singular pronoun suffix on *'emuna* refers not to *ṣaddiq* earlier in the line, but to an antecedent earlier in the passage. Depending on how one reads 2:3c-d, it is possible grammatically to identify God, or the appointed time/end, or the vision itself as that antecedent (see the discussion below). But if the six proverbial parallels are taken as an interpretive guide,[2] the antecedent must be the noun *ḥazon*, "vision." For it is the witness and the testifier which, in three of the proverbs, are described as giving reliable evidence in contrast to lying and false witnesses. Habakkuk 2:2-4 (and indeed, in my view, the whole of Habakkuk) arises in connection with problems concerning the question of the truth or falseness—and hence the reliability—of a prophetic word or vision. Precisely in Habakkuk's time, this problem is seen to arise, not only between a self-proclaimed prophet and that person's hearers, but within the prophetic consciousness itself. (4) In 2:4a I read, not "Behold, he whose soul is not upright in him shall fail" (RSV), but "as for the sluggard, he does not go straight on in it." This line is generally considered to be corrupt, and numerous restorations have been proposed. I have adopted one of the proposals for the emendation of the word *'uppelah*, construed the following predication in a manner which is more consistent with usage of the terms in question, and tried to show the integral connection of my restoration and construal with the rest of the rhetoric of 2:2-4. The proposal is tentative and heuristic and does not constitute a basis for the rest of my interpretation. Indeed, in the following discussion I shall explore one or two slight modifications of it.

On the basis of these textual, philological, and syntactical proposals, 2:2-4 is to be read within a context that includes the following components: (5) the general existential issue at stake in all instances of witness-bearing, that is, the question as to whether the testimony is true or false, life-sustaining or life-destroying; (6) the quite specific dilemma internal to the prophetic consciousness about the reliability of the received vision and of the God who has granted that vision (the visionary scene painted by Micaiah in I Kgs 22:19-22 served that prophet well in that context, but such a scene can cut both ways, as Jeremiah's dilemma shows); (8) the prophetic tradition initiated by Isaiah (cf. Isa 8:16-20a), and transmitted by successive generations of disciples (reaching as far down as Second Isaiah), concerning a prophetic word characterized as *torah* and written down on a *tablet* as a faithful *testimony, forever* (Isa 30:8) to the promise

---

2. The texts are Prov 6:19; 12:17; 14:5, 25; 19:5, 9. See the discussion in ibid., 57-62.

PART THREE: The Standpoint of Two Prophets

of God's action to set things right at some future time. With this summary re-statement, I shall turn now to sketch a reading of Habakkuk as a whole.[3]

From Despairing Complaint to Affirmation in Hope

1:1–4

The superscription in v. 1, however it is to be explained editorially, aptly introduces the book because the relative clause with its verb "saw" (*ḥazah*) identifies 2:2–4 with its repeated noun "vision" (*ḥazon*; cf. *ṣippeh* and *ra'ah* in 2:1) as the rhetorical and hermeneutical center of the book.

It is customary to interpret the opening questions of Habakkuk in 1:2–4 on the basis of an understanding of v. 4 as referring to a breakdown in the administration of *torah* and *mišpaṭ* in Judah. In this view, Habakkuk may be read against the background of the so-called deuteronomic reforms and may be taken as a complaint to God that those who are responsible for seeing that covenant *mišpaṭ* and *torah* are done in Judah are falling down on the job. The result of this breakdown is evident in widespread social wrongdoing—wrongdoing in which "the wicked surround the righteous" and which Habakkuk is forced to witness to his great pain and indignation. But I believe that 1:4 is to be taken in a quite different way, and that 1:2–4, accordingly, is to be taken as belonging to a specific genre which, as it happens, is eminently exemplified by Habakkuk's contemporary Jeremiah.

---

3. There is no space here to defend in detail my approach to the book as a whole. I should simply register my view that analyses which relegate substantial portions of the book to secondary phases in its growth, proceed from a misconception of the issues to which the materials are oriented, and from attempts to interpret specific parts of the text in terms of this or that specific historical referent. It seems to me that a successful attempt to read the book as an integral whole constitutes, already, one sort of argument for integrity. Given the nature of such a task, there is no space, either, to enter into consideration of other scholarly analyses in general or on specific points (for extensive presentation of the history of scholarship on Habakkuk, see Jöcken, *Das Buch Habakuk*).

On one critical point, however, a brief comment is necessary, concerning chap. 3, which (like the Prayer of Jonah) often has been held to have been added to the book at a later date. A number of more recent studies have shown how the prayer of Jonah is woven integrally into the larger context. I would argue the same for Habakkuk. Whether this poem was appropriated as it stood, or adapted to Habakkuk's needs, or written with older models in mind, I am not able to decide. What is clear to me is that chap. 3 continues the fundamental rhetorical features of the preceding two chapters and that it resolves the issues posed at the beginning of the book.

To begin with, the noun *mišpaṭ* need not refer to covenant law in its administration, but bears a variety of specific connotations depending upon the context in which it is used. In Hos 6:5, for example, it refers to God's words of judgment as spoken through God's prophets:

> Therefore I have hewn them by the prophets,
> I have slain them by the words of my mouth,
> and my *mišpaṭ* goes forth (*yeṣe'*) as the light.

In other words, *mišpaṭ* can be used to designate a specific type of prophetic utterance.

But what of the parallel word *torah*? Does this word not anchor *mišpaṭ* securely in the covenantal-judicial realm of discourse? Not necessarily. Like *mišpaṭ*, *torah* also occurs in a variety of contexts—covenantal/legal, priestly, sapiential—within which it takes on different specific connotations. Joseph Jensen has argued that in eighth-century Isaiah the term is employed in a way which trades specifically on its sapiential connotations.[4] In Jensen's view, Isaiah in his debate with the royal court and its counselors, claims that their so-called wisdom is bogus and that his own message, his own *debar-Yhwh*, is the true wisdom-teaching or *torah*. Jensen's argument is highly plausible. Moreover, I think the wisdom elements in Hab 2:2–4, together with the general connection between 1:2–4 and 2:2–4 along with the latter's connection with Isaiah, help to corroborate Jensen's conclusions. But even if his identification of *wisdom* connotations in Isaiah's use of the word *torah* should be thought untenable, the fact remains that in Isaiah the prophet's word several times is characterized as *torah*.[5] Just as the word *mišpaṭ* can mean, in relevant contexts, a word or verdict from God as judge, as announced by a prophet, so, on the basis of Isaianic usage, in relevant contexts *torah* can designate a word or teaching from God through a prophet. Thus, in Isa 42:1–4 the words *mišpaṭ* and *torah* appear to refer to the utterance of God's servant-prophet to the nations; while in Isa 51:4–5 the *torah* // *mišpaṭ* that goes forth for a light to the peoples takes the form of that deliverance // salvation (*ṣedeq* // *yeša'*) by which God will rule (*šaphaṭ*) the peoples. The way in which the ending of Isa 51:5 ("the coastlands wait for me, / and for my arm they hope [*yeyaḥelun*]") parallels the last line in Isa 42:4 ("and the coastlands wait

---

4. Jensen, *The Use of tôrâ by Isaiah*.

5. Ibid., 65–121. According to Jensen, the word *torah* acquires what I have called a "speaker's meaning" (Janzen, "Habakkuk 2:2–4," 64 and n. 9), that is, a new connotation by virtue of the use of this old "wisdom" word in new contexts. Jensen argues that this meaning remained peculiar to Isaiah, whereas I hold that it was continued in Habakkuk, as also, in my view, in Deutero-Isaiah.

PART THREE: The Standpoint of Two Prophets

[*yeyaḥelu*] for his *torah*") reinforces the connotations of *mišpaṭ* and *torah* as referring in these contexts to the prophetic/visionary announcement of God's actions to set matters right in the world. It is my contention that Hab 1:2–4, read in conjunction with 2:2–4, constitutes precisely such a context.

Habakkuk addresses God in a form and in words similar to those used in those psalms where we hear a cry of complaint and an appeal for help. While it is possible to read 1:2–3 as a complaint concerning the general social situation and an intercessory plea (or even a representative petitionary plea) on behalf of those who are suffering at the hands of others, it seems to me that Habakkuk here may most naturally be taken as crying out on his own behalf. Certainly v. 2, taken by itself, looks like an instance of a typical petitionary cry in the psalms. The verbs *tar'eni* and *tabbiṭ*, then, are to be taken as meaning that Habakkuk is suffering violence and destruction, in the form of strife and contention, at the hands of those who oppose him and the message he is delivering. This does not rule out a concomitant reference to the wider scene of social wrongdoing (cf. Jer 11:18—12:6). The force of the causative *tar'eni*, "why have you caused me to suffer wrongs?"[6] lies in the fact that Habakkuk's troubles can be traced to the word which Yahweh has given him to utter. Why have you done this to me, God? Why do you countenance (*tabbiṭ*) and tolerate the trouble which this word is bringing on me? Habakkuk's questions lead him to a conclusion which he places before God as a double-edged accusation: God does not listen, God does not save, *mišpaṭ* never goes forth; and this can only mean that the *torah* God has given him to utter has no effective power but has "gone slack,"[7] and that God's word as *mišpaṭ* will not prevail.[8]

---

6. The verb *ra'ah* sometimes means "to experience, to be exposed experientially to." Notice, for example, how closely Hab 1:3, "why dost thou cause me to see wrongs and look upon trouble (*'amal*)?" parallels Jer 20:18, "why did I come forth from the womb to see trouble (*'amal*) and sorrow and spend my days in shame?" This parallel, so far as it goes, supports my contention that Hab 1:2–4 is to be taken as a prophetic complaint in the manner of the so-called confessions of Jeremiah. On this connotation of "seeing," cf. also Job 3:10 (again, with *'amal*).

7. The verb in Hab 1:4, "So the *torah* becomes slack [*tapug*]," along with its derived nouns, elsewhere always functions intransitively, characterizing something about its grammatical subject: either that that subject is faint (Gen 45:26), spent (Ps 38:8), helpless (Ps 88:15); or, in respect to determined or unrelieved effort, without wearying (Ps 77:2), without rest (Lam 2:18), without respite (Lam 3:49). In Hab 1:4 the implication is that God's *torah* has no staying power to accomplish its ends. (Compare the claim in Isa 40:8; 55:10–11.)

8. Compare Jeremiah's charge against God: "Truly, you are to me like a deceitful brook, like waters that fail" (Jer 15:18; NRSV).

Eschatological Symbol and Existence in Habakkuk

When we read 1:4a–b as an accusation that the divine word uttered through the prophet does not possess the vitality to endure and to prevail, 1:4d comes into a new focus. It may be taken as the charge that, as a word that goes forth twisted (*me'uqqal*), it does not reach its intended goal (cf. again Isa 55:10–11), and as such is deceitful and deceiving. It is this accusation which is answered in 2:3 with the reassurance that the visionary word *lo' yekazzeb*, "does not lie/deceive."

1:5–11

At first glance, the opening plural address (*re'u . . . wehabbitu*) appears to contradict my reading of vv. 2–4 and to support a reading which takes Habakkuk to be complaining about the general state of injustice in Judah. For these plurals direct the divine word not to Habakkuk but to the people, as though attesting the divine intent to rectify the situation about which Habakkuk complains. But let us examine the opening lines more closely.

To begin with, as is characteristic of the dialectic of this book, key words of Habakkuk's complaint (*ra'ah, hibbit*) are picked up and re-used in a slightly different sense. But what is the force of this divine call to the people to "look and attend to"? The tone is one of sarcasm tinged with irony. The people are to look, yet they will not see. For they will be bewildered, stupefied. For such is the work of God that, when told it, they will not believe it. The language and the rhetoric are reminiscent of vintage Isaiah: The verbs *ra'ah* and *hibbit* occur together with the noun *po'al* or its synonym *ma'aseh* in Isa 5:12 and 22:11, describing how the people disregard what God has done or is doing. The verb *tmh* "to become astounded, bewildered" occurs in Isa 29:9 along with references to blindness, describing the people's inability to hear the prophetic message. The latter passage, of course, continues the theme of unperceptive seeing announced in Isa 6:9–10. The expression *lo ta'aminu* picks up the negative note of Isa 7:9 (with which contrast the use of the verb in Isa 28:16). As in the time of Isaiah, so now, the people being addressed will not, indeed cannot, respond positively to the message of Hab 1:6–11. That message, for its part, portrays the Chaldean as the scourge of Yahweh who for fierceness, rapacious greed, and self-worship presents a portrait to match the depiction of Assyria in Isa 10:5–19.

What is the point of these rhetorical allusions by which the divine word now to Judah is compared with the divine word long ago to the same people through Isaiah? Are these allusions for the benefit of the Judean hearers of this message, perhaps to jolt them out of their complacency before they too become stupefied? That may be. But I suggest that the message is also for the benefit of the prophet himself. If Habakkuk is filled with doubt as to the efficacy and the truth of the divine word, it is because that word has had no effect on its hearers, except to rouse them to contentious animosity against him. The opening lines of God's address, by the multiple allusions to the Isaianic theme of contentious resistance through unperceptiveness brought on by a stupefaction which ultimately has its roots in the mystery of the divine work, serve to inform Habakkuk indirectly that the people's response in the form of strife and contention is not a sign of the word's weakness and crookedness, but just the opposite: that response shows that the divine word is achieving its intended effect! This, indeed, is part of the point of the second plural address to the people, rather than a response directly to Habakkuk, as was the case in Jer 15:19–21. In the latter passage, as in Hab 1:2–4, the prophet's complaint comes to the point of accusing God of the weakness and crookedness (Habakkuk) or the deceitfulness and ephemerality (Jeremiah) of God vis-à-vis the prophetic word. But whereas in Jer 15:19–21 God's response takes the form of words addressed directly to the prophet, in Hab 1:5–11 God's response takes the form of words addressed to the people. The overall effect of v. 5 as addressed to the people is to bring about what it describes. At the same time, this verse serves to answer Habakkuk's complaint in 1:2–4, in this sense: As directed also to Habakkuk, v. 5 in effect poses to him the question, whether he, by his negative attitude to this matter, wishes to count himself as one of those who do not believe even if told. Now, if v. 5 asserts the divine efficacy through this "strange work," vv. 6–11 assert the divine efficacy in another way, through the rhetorical vigor with which God describes the terrible energy of the Chaldean, an energy so great as to become idolatrous in its own eyes.

1:12–17

Habakkuk's response to God's address is not without irony. The opening couplet seems to offer a hasty, perhaps strategic, "touché" to Yahweh's vigorous rebuttal, as Habakkuk affirms the unflagging character of the divine vitality:

> Art thou not from everlasting, Yahweh?
> My God, my Holy one, thou shalt not die![9]

---

9. The common translation in 1:12, "we shall not die," is based upon the MT which is the result of one of the *tiqqune soperim*. I adopt what surely was the original reading.

Eschatological Symbol and Existence in Habakkuk

The twofold temporal affirmation *'atta miqqedem, lo' namut* seems to retract Habakkuk's twofold assertion in 1:4a–b as those lines are here being interpreted. This retraction seems to be elaborated in the last half of this verse: "Yahweh, you have ordained him as *mišpaṭ*; O Rock, for *hokiaḥ* have you established him." In v. 7 (*mimmennu mišpaṭo . . . yeṣe'*), God had picked up the rhetoric of 1:4b,d, and by applying it to the fierce power of the Chaldean had contradicted the earlier charge. Now in v. 12c–d, *mišpaṭ* and *hokiaḥ* continue that word-play, matching the earlier *torah-mišpaṭ* pair. The two words in v. 12 refer, of course, in the first instance to God's actions of judgment upon Judah. But these actions have not yet been accomplished; they have only been announced. In particular, we should recall that, in wisdom usage at least, *hokiaḥ* refers not only to the act of judgment but also to its significance—it is a chastening, a form of teaching or instruction. In this sense, *hokiaḥ* describes a specific form of *torah*. Again then, as in 1:4, so in 1:12cd we detect a reference not only to the divine act but also to the message announcing that act. At one level, Habakkuk seems to be saying, "Yes, God, I see that you have given me this powerful message, this *mišpaṭ* and *hokiaḥ*, you who are a Rock."

But with what follows, the seeming acquiescence may be observed to prepare itself for an ironic shift to what Habakkuk *really* feels about the divine message. He picks up the pair of verbs that Yahweh had taken from his mouth; with an undertone which belies their surface meaning, he says, "You who are of eyes too pure to look on idly while wrong is perpetrated and cannot countenance trouble . . ." If v. 12 seems to retract v. 4, v. 13ab seems to retract vv. 2–3. But then the ironic reversal comes with the "why?" of v. 13c, which resumes the "why?" of v. 3a with all of its force. This message concerning the Chaldeans—how will it help the *ṣaddiq*? If within Judah the wicked surrounds the righteous, the coming of the Chaldean will mean only the exchange of one wicked power for another, who will not only surround but will swallow the righteous. God seems to have made humankind like so many fish of the sea, like so much "fair game" for this fisherman with an expensive palate and an insatiable appetite. Because he acknowledges no power but that of his own implements (v. 16; cf. vv. 7b, 11b; and Isa 10:13–14), he recognizes no limits to his reach and his grasp. If now the "why?" of v. 13c resumes the "why?" of v. 3, the "how long?" of v. 2 is resumed in v. 17, which thereby functions

---

Habakkuk's ironic affirmation, after all, is not bolder than Jeremiah's charge (not a question!) to Yahweh that "you have become to me like a deceitful brook, like waters that fail," the implications of which are clear from a comparison with Jer 17:13d and 2:13.

225

PART THREE: The Standpoint of Two Prophets

doubly as an *inclusio*. It concludes the second accusation begun in v. 13c; and vv. 13-17 return the dialogue to the point on which it had begun.

Let us summarize to this point: In 1:2-4 Habakkuk complains that the message he has been given is weak and ineffectual, lacking in enduring vitality. This he infers from the fact that it has made no difference among the people, but has brought only misery upon his own head. In 1:5-11 God answers that that message is vigorous enough, both in the stupefying effect it has on the faithless hearers and in the portrayal it conveys of the fierce energy of the Chaldean as God's arm of *mišpaṭ*. One indirect question hidden in v. 5 has already been alluded to: If v. 5 really is directed to Habakkuk, one detects the undertone of a question, whether he, by his negative attitude toward this message, is to be counted as one of those who do not believe even if told. In vv. 12-17, Habakkuk opens with what appears to be a hasty acquiescence in the reality of God's word of promised action. But this seeming acquiescence is subverted by vv. 13-17, which imply that God's agent of *mišpaṭ* is indeed so strong that God is left to look on (v. 13ab) in helplessness or indifference while the Chaldean continues without end. The original charge stands: indeed, if anything it has been intensified.

2:1-4

If the rhetorical elements of vv. 13c-17 signal an end to the first phase of the dialectic between prophet and God, the next few lines signal a beginning to the second phase. Habakkuk's first complaint was followed immediately, and without introductory rubric of any kind, by the divine response. Likewise, Habakkuk's second complaint followed this response without introductory rubric. In contrast to this, the dialectic at this point is interrupted in two ways: (1) by the brief soliloquy serving as an interlude in 2:1, whereby not only Habakkuk but the reader is set for what is to come; and (2) by the rubric "and Yahweh answered me" at the beginning of v. 2. Having received one response which was deemed unsatisfactory, Habakkuk forewarns himself, and whoever cares to know, that he will "answer" Yahweh's second response one way or another, in accordance with its adequacy. By this means, our anticipation of the second divine response is further intensified.

The answer given in 2:2-4 has been analyzed fully in my previous study, in such a way as to show its relevance to Habakkuk's reiterated concerns. I offer a summary re-statement of that analysis here and add one or two incidental comments.

Eschatological Symbol and Existence in Habakkuk

ketob ḥazon
uba'er 'al-halluḥot
lema'an yaruṣ qore' bo
ki 'od ḥazon lammo'ed
weyapeaḥ laqqeṣ welo' yekazzeb
'im yitmahmah ḥakkeh lo'
ki bo' yabo' lo' ye'aḥer
hinneh 'aṣel lo' yašerah napšo bo
weṣaddiq be'emunato yihyeh

Write the vision,
make it plain upon tablets,
so that he may run who reads it.
For the vision is a witness to a rendezvous,
a testifier to the end—it does not lie:
"If he tarries, wait for him;
he will surely come, he will not delay!"
As for the sluggard, he does not go straight on in it;
but the righteous through its reliability shall live.

Like Isaiah before him, Habakkuk is to write God's message on tablets as a witness (Isaiah 30). That vision, so written, is to serve as a witness and a testifier to the fact that God has set an appointed time, which will have the character of a rendezvous, and which will bring to an end the period of judgment announced in 1:5–11, and concerning which Habakkuk had complained in vv. 13c–17. In 1:4d Habakkuk had impugned the divine message as *me'uqqal*, crooked and deceitful, disappointing its bearer and its hearers. Here God asserts, in a manner reminiscent of several proverbs, that this witnessing and testifying vision does not lie and will not disappoint. The two lines beginning, "If he tarries . . . ," contain the content of the vision. Grammatically, of course, the antecedent to the third person pronoun elements in the couplet is unclear: it may be the vision, it may be the *mo'ed / qeṣ*, or it may be God. In any case, if it is the end, that end will be marked by God's coming (compare 3:3!); if it is God's coming, that coming will bring about the end; and if it is the vision, that vision concerns an end marked by God's coming. I do not rest a great deal on my preference, but I am inclined toward it for reasons given in the previous study, and because it dovetails nicely with 3:3 and may be interpretively reflected in Isa 40:10 (in which case Isa 40:10 and 40:31 jointly carry forward the language of these two lines in Habakkuk). Now this rendezvous lies sufficiently far in the future that the divine and human parties to it are called to comport themselves toward it in two ways. The divine partner will *come*,

PART THREE: The Standpoint of Two Prophets

without procrastination or reluctance. But there is a factor—presumably some aspect of the human situation—which would make an immediate coming untimely. (For Deutero-Isaiah that factor no longer obtains [Isa 40:2], and accordingly God's coming can be announced as imminent, in 40:10-11, and as not tarrying, in 46:13.) Hence, if there is no procrastination, there is nevertheless a tarrying until the *right* time (compare Isa 28:23-29 and 30:18; and contrast Isa 5:18-19). This twofold modality of action and patience is to be met by a corresponding human existence in waiting ("wait for him," with which compare Isa 8:17) and in running.

But the response of the hearers of the message of Isaiah, and of the hearers as presupposed in Hab 1:2-4 and as anticipated in 1:5, should prepare the hearer of this vision for two kinds of response. There will be the response of unbelief. (2:4a, as I read it, picks up the theme sounded in 1:5.) That response, however, is now characterized in a fresh way, by the use of a term and its associations which are widely familiar from the proverbial tradition—precisely that tradition in which we find the closest parallels to the inter-relationship of our key terms *'ed, yapeaḥ, kizzeb,* and the noun forms from the root *'mn*. The sluggard is one who through fear or laziness, or both, cannot or will not venture forth to enterprises called for at the present time, even though those enterprises are the necessary condition for participating in an eventually fruitful outcome. By this device, the figure of the sluggard is transposed from the realm of the quotidian to the realm of eschatological concern. (Though it may be suggested that already, by inclusion in proverbial—that is to say sapiential—contexts, this figure takes on more than quotidian significance, takes on a typical significance in the realm of moral and religious existence, a significance which the use in Habakkuk only nuances with eschatological dimensionality.) But conversely, the vocabulary and the psychology of eschatological existence are enriched by the introduction of this figure.[10]

---

10. My tentative construal of 2:4a turns upon an emendation of the Hebrew text *'uppelah* to *'aṣel*, involving a change of the medial *p* to medial *ṣ* and elimination of the final *h* which, then, presumably arose secondarily by attraction to the gender of the noun *nepeš*. In conversation, Rabbis Jonathan Stein and Elliott Rosenstock suggest an emendation in which the final *h* is simply attached to the following negative, to give the text: *hinneh 'aṣel helo' yašera napšo bo*, "Behold the sluggard (that well-known malingerer)—will not (even) *his* soul go straight on in it? And the righteous (all the more) by its faithfulness will live." The rhetorical point in such a reconstruction would be that the vision is so powerfully attractive that it will stir even the sluggard to hope and action: how much more, then, will the righteous continue to live through it? I find this construal very attractive, and I am tempted to adopt it. But proverbial literature is full of two-line contrasts between positive and negative types (as in many of the

Eschatological Symbol and Existence in Habakkuk

Now if the term *'aṣel* opens a fresh perspective on the interiority of those who refuse or are unable to respond to the vision, by its rhetorical and semantic contrast to the *ṣaddiq* in the following line, it opens up a fresh perspective on what is involved in being a *ṣaddiq*. The latter is one who "runs" in accordance with the divinely given vision for the future, and who "waits" for God to come. The *ṣaddiq* thereby shows that he or she meets the vision with a belief in its reliability.[11] It is through such a response, and at the same time—though perhaps one may say at a deeper level of efficacy—it is through such a reliability[12] that the *ṣaddiq*, in spite of the twin threats portrayed in 1:4c and 1:13c–e, shall *live*. One detects in the two lines of 2:4, again, an indirect challenge to Habakkuk himself. The prophet styled himself a *ṣaddiq* in the act of accusing God of giving a weak and crooked word, and in the act of challenging God's first response. God now defines the *ṣaddiq* in such a way, vis-à-vis the vision, as to leave Habakkuk with a choice which transforms his complaining consciousness or leaves him in the company of the sluggard! One is reminded of the less indirect thrust of Jer 15:19–21, coming as that divine response does after the charge in 15:18 that Yahweh is become to Jeremiah like a deceitful (*'akzab*) brook, and like waters that fail (*lo' ne'emanu*), and that this is the cause of the fact that Jeremiah's pain is unceasing (*neṣaḥ*). As I have argued, this last word in 2:4 forms a semantic *inclusio* with the words *mo'ed / qeṣ*. For the latter words announce a future time toward which each party is to move, in modes of action and patience. The rhetoric symbolizes the message, by the way in which nouns, pronouns, and verbs referring to these parties set up two rhetorical vectors which move through the passage until they arrive in the last line, are named one last time—*ṣaddiq* and *be'emunato*—and then converge on the fulfillment described in the verb *yihyeh*.

---

six proverbs cited in n. 2 above). I am inclined to see in v. 4a a resumption of the theme sounded in 1:5. Still, such a striking reversal of a proverbial theme would fall well within the capability of the writer of this book and would further enhance the portrayal of the vigor of God's vision (in face of Habakkuk's charge in 1:4). Given the notorious difficulty of this half-verse, and the variety of solutions which have been proposed, this possibility certainly merits consideration.

11. One could construe the pronoun suffix on *be'emunato* as referring to God, especially if one takes God as subject of the verb *bo'*. In any case, since the vision is from God, God's reliability is bound up with that of the vision. But in view of the six proverbs cited in n. 2 above. it seems clear the the suffix in 2:4b refers back to the vision as a *reliable witness*.

12. See Janzen, "Habakkuk 2:2–4," 71 n. 14.

PART THREE: The Standpoint of Two Prophets

2:5-20

In the previous section, which constitutes the center-piece of the book and its message, God assures Habakkuk and those who heed his visionary word of the coming of that appointed time which will vindicate both God and the *ṣaddiq*, a coming which is God's own coming. Now the divine word turns to address the other side of the issue raised by Habakkuk's questions in 2:13c-17. The images in v. 5ab are not at all obscure (*pace* the RSV textual note) but pick up where the portrayals in 1:5-11 and 1:15-16 left off. In 2:5a, *we'ap ki hayyayin boged* effectively characterizes the Chaldean as one who is drunk with his own power—a characterization which lies just under the surface of the earlier descriptions. Moreover, whereas that intoxication had turned him into a treacherous dealer (1:13 *bogedim*),[13] by a fine proverbial irony that very intoxication will turn out to have betrayed and misled the arrogant Chaldean (in contrast to the vision which does not deceive but is reliable). Though he seems securely domiciled within the luxuries achieved through his excesses, his dwelling place, in fact, is unstable (cf. Job 18:15). The *boged* of 1:13c-16 may indeed be a swallower whose greedy appetite is as insatiable as Sheol and Death. But the very nations and peoples whom he gathers for himself and collects as his own will turn and take up a *mašal* against him, a *meliṣa ḥidot*, by which his own fate will be set forth. Interestingly enough, the root *luṣ* occurs most often in Proverbs and in other contexts displaying wisdom features. (In Prov 20:1 wine is described as a scorner; and the root occurs in Isa 28:14; 29:8; 29:20.) These aspects of its usage give the term *meliṣa* in Hab 2:6 a peculiarly apt rhetorical force following the wine imagery of v. 5, and vv. 2-4 with their wisdom rhetoric and Isaianic background.

Whereas, then, the imagery in 2:5c-f recognizes the force of Habakkuk's description in 1:14-16, it re-interprets that description by the effective use of a number of conventional wisdom motifs, to show that the very power and greed of the Chaldean will be his undoing. The scorning taunt of the peoples then follows in the form of five woes. (The use of the *woe* form is nicely ironic, following as it does upon the images of Sheol and Death as descriptive of the Chaldean's insatiable appetite.) As so often throughout this book, wherein each speaker in turn takes up words and themes of the preceding speaker and gives them a new rhetorical twist, the first woe takes up the twice-uttered question of Habakkuk (1:2, 17)

---

13. In the time of Josiah, Judah and Chaldea had in common at least the enmity of Assyria. Is it possible that the word *bogedim* in 1:13 describes a Chaldean power that has treacherously forgotten such a common cause?

Eschatological Symbol and Existence in Habakkuk

and, in the mouth of the nations, turns it mockingly upon the head of the Chaldean—"for how long?" The placement of this mocking question upon the lips of the erstwhile victims effectively gives the lie to the pretended everlastingness of arrogant power. The imagery of the vulture's nest in 2:9 resumes the vulture imagery in 1:8, as the house imagery throughout vv. 9–11 resumes the themes of the instability of the dwelling place in 2:5b (*yinweh*). The Chaldean thought his power was divine. 2:13 establishes Yawheh as God, in relation to whom the labor of the Chaldean (and all the nations) is seen to lead only to the fire of judgment and to futility (*riq*) and weariness (*ya'ap*). These last two terms answer Habakkuk's earlier charges concerning the weakness of God's *torah/mišpaṭ*.

Looking back over the text surveyed to this point, and looking forward to features still to be noted in the text which follows, one is led to take the book as a whole as a debate concerning the true locus and character of ultimate vitality versus that energy whose final fate gives the lie to its appearances. As such, and given the rhetorical prominence of the words for seeing (1:3, 5, 13; 2:1, 2 [*hazon*] with which compare 1:1; 2:15; 3:6, 7), the debate may be taken as turning (as in, say, Job and Genesis 2–3) on the question of reality and appearance.

The fourth oracle again picks up the imagery of strong drink and brings it to a climax. The Chaldean's intoxicating sense of power will turn into the wrath of God which the Chaldean himself will be forced to consume. The appetite which could not satisfy itself on the death of others (2:5d *welo'yisba'*) finally will be sated (2:16 *śaba'ta*) by the experience of its own demise. It is as though the power of Death, self-aggrandized to the point of ultimacy, can be satisfied by the taste of a death no smaller than its own. The final woe finely satirizes the idolatrous (see 1:7b, 1 lb, 16ab) character of the Chaldean power, in terms which anticipate Deutero-Isaiah.[14] By way of contrast with the undeceiving vision given by Yahweh in 2:2–4, the idol is portrayed as a teacher of lies (*moreh šaqer*), a *dumb* object of wood to which the Chaldean vainly appeals, "Awake! Arise!" as if it could give revelation (*yoreh*). The twofold use here of these cognates of *torah*, in a divinatory context which itself stands in deceiving contrast to the reliable vision of 2:2–4, tends to corroborate the construal of *torah* in 2:4 as connoting a prophetic message. With this quick survey of the rhetoric of 2:5–20, let us now look to see (in accordance with 2:1d) how Habakkuk answers Yahweh's response to his second complaint.

---

14. On the idol passages in Deutero-Isaiah, see now the study by Clifford, "The Function of Idol Passages in Second Isaiah," 450–64.

PART THREE: The Standpoint of Two Prophets

3:1–19

The transition from Habakkuk's second complaint to Yahweh's second response had been prepared by a first-person intentional comment on the part of the prophet, a comment which served also to intensify the reader's interest in 2:2–4 and 2:5–20. Now the transition from Yahweh's response to Habakkuk's final word is made in a different way, by the use of the sort of psalm title which one finds so often in the Psalter. In the Psalter, of course, such titles are often secondary to the original compositions. It is by no means clear that this is the case in the poem of chap. 3 or that the poem itself is a secondary addition to the Habakkuk materials. As I shall show, at point after point the language and the thrust of the poem, while no doubt arising out of earlier models and conventional usage, relates most aptly to what precedes it. Is it not possible that by the use of a specific formal feature of Israelite psalmistry, the title, Habakkuk signals the character of his final word? His first word of complaint, while it issued from the mouth of a prophet, was couched in the rhetoric of a specific type of psalm. His final word is similarly couched, but this time rendered specific by superscription and postscript. The effect of this, I suggest, is twofold. It brings the whole of the preceding dialectic to a conclusion by bringing it, as it were, into the setting of formal worship. But by this means Habakkuk's own response gains more than merely personal relevance, in becoming available for wider use by anyone who wishes to respond to the vision in the way Habakkuk has done. Here again, then, we see a rhetorical anticipation of Second Isaiah who, as Westermann has shown, uses the forms of the Psalter as a means of drawing hearers and readers of his visionary message into the sort of response that he himself has already made.

How then does Habakkuk respond? The opening two lines, in 3:2ab, lead us to ask what "work" is here referred to. It is unlikely that this is the same work as was set forth in 1:5–11, for Habakkuk has already responded to that divine word in 1:12–17. Moreover, the references back to this work of 3:2ab in the following three lines establish the character of this work as one of *raḥem*, "mercy." Rather, the "work" of which Habakkuk has heard a "report" (*šemaʿ*) is the work set forth in what has immediately preceded, in 2:2–20.

Now the term "report" seems generally to describe, not the experience of a first-hand witness, but one's reception of the testimony given *by* such a witness. In chap. 2, Habakkuk does receive the vision directly from God. Nevertheless, the terms *ʿed/yapeaḥ*, by the way they describe the vision in 2:2–4, suggest the appropriateness of Habakkuk's use of the

Eschatological Symbol and Existence in Habakkuk

term to describe his own reception of the word of the vision. If in the first instance Habakkuk is the *mediator* of this testimony, in the second instance—and especially after he has written it down and is left with the question as to what kind of response he should make to it (2:1)—Habakkuk may be said to be the first *recipient* of, or respondent to, this message. Will he "believe this report" (cf. Isa 53:1)? As we shall see, Habakkuk is not only the first but the exemplary respondent. If 1:4 characterizes him as a ṣaddiq, then the psalm as a whole further portrays him as the exemplary ṣaddiq of whom God (challengingly) speaks in 2:4b. He begins by saying:

> Yahweh, I have heard the report about you,
> and your work, Yahweh, do I fear. (3:2a)

The report of Yahweh's work of judgment in 1:5 will be met among the populace by stupefied bewilderment and unbelief. Even Habakkuk at first has protested its obvious implications. But now his response to the report of God's saving work is one of positive piety. With one eye on the intervening work of 1:5–11 (what Isaiah called God's "strange work," Isa 28:21), Habakkuk prays that the saving work may come soon. He both petitions and intercedes:

> in the midst of the years revive it.
> in the midst of the years make it known,
> in wrath remember mercy. (3:2b)

Habakkuk accepts in principle the necessity of a time of trouble, he even acknowledges the prospect of a years-long interval, but he prays that this time may be shortened, that Yahweh's saving work may be revived. We see here both a connection with and a shift from his earlier rhetoric. It is not that Yahweh's word is weak and ineffectual, but rather that Yahweh's word and work observe times and seasons, in which judgment and salvation succeed one another. By this verb *ḥiyyah*, "revive" (cf. 1:12; 2:4b), Habakkuk asks that the interval of judgment be shortened by a recurrence ("remember!") of Yahweh's action in the mode of mercy. By the use of the verb *ḥayah* in the Piel, "revive," a connection already is made between the envisioned future "work" of Yahweh and the old saving acts which brought the nation into existence. Now Yahweh is called upon to "revive" that work by "remembering" it.[15] But then Habakkuk himself goes on to do the very

---

15. In this context we find presupposed a realistic doctrine of memory rather than one in which memory is the mere throwing up on the screen of consciousness the empty, flitting, and essentially lifeless shadows and spectral shapes of the past. To remember, rather, is to recall, to call forward into the present, the efficacies which are a

PART THREE: The Standpoint of Two Prophets

thing which he calls upon Yahweh to do: he *remembers*; he recalls into his own present consciousness the old saving work which is to be revived. He does this in a specific way, by reciting a version of the old hymnic celebration of those saving acts.

The first line of this recitation includes the verb *bo'* which, as I have proposed, constitutes the core of the vision in 2:3cd. Of course, its occurrence in 3:3 does not arise merely by way of response to 2:3, for the verb occurs also, for instance, in the opening line of the old poem in Deuteronomy 33. But since it may well be that the verb in 2:3 itself is to be taken against the background of such materials as Deuteronomy 33, what we may well have is a tight complex relation among all these materials, by virtue of the rhetoric of the Book of Habakkuk as a whole. The new merciful work of God reported by the vision, a work in which God will "come" against the Chaldean enemy, is a remembering and a reviving of the old work in which God came from the wilderness abode to deliver Israel from the enemies of old.

But in the act of calling upon God to remember this work of old and to actualize what has been promised, Habakkuk himself performs this act within himself in 3:3–15. This material itself is clearly an archaic hymn or an archaizing imitation of genuinely archaic models. But the existential function of the rhetorical device of using archaizing material at this point is highly strategic—it is an archaizing with a view to contemporizing. For the archaizing features are a way of rhetorically reaching back in a realistic act of calling up and calling forward, from the depths of the past where they lie in the recesses of one's heart and mind, one's unconscious memory as it were, that depth at which one participates in the life of the nation throughout the span of its experience. There is here a calling forward of those efficacious energies of initiative and response, divine and human, in forgetfulness of which the present would be only bleak with crisis and uncertainty or the bleaker certainty of doom. But this remembering has the effect of revivifying that old work already in Habakkuk's own imagination and from there in his own life. In the midst of the years Habakkuk himself has participated in the reviving of Yahweh's work. Moreover, by virtue of his part in the process by which the whole book has come to us, Habakkuk himself has participated in the act of making that work known. In the midst of wrath Habakkuk too has remembered mercy. What we see here

---

constituent part of the self through one's past experience, and which are woven irrefragably into those events and those other parties being recalled. Such, for example, I take to be the import of Exod 2:24, "And God remembered his covenant with Abraham, with Isaac, and with Jacob."

Eschatological Symbol and Existence in Habakkuk

already is Habakkuk's own exemplification of one aspect—the energetically active "running" aspect—of the human vector which runs through Hab 2:2–4. The redemptive act to which the vision points may lie in the future. Yet not merely so. For that future act will come as a realization of divine energies of intent which already have come to expression in the vision which points to that future; and that vision is itself a present fact, a rhetorical and symbolical energy present and efficacious in Habakkuk. In the person of Habakkuk, then, divine and human faithfulness converge in a manner which already, as a sort of first-fruits, attests the trustworthiness of the concluding lines of the vision. We have arrived at a position, then, from which we can identify the progress of the central theme in this book: the theme of *vitality*, which is introduced negatively through *tapug/lo' laneṣaḥ* in 1:4, is affirmed through the rebuttal of 1:5–11, is acknowledged in v. 12 only to be challenged by the counter-vitality of the Chaldean in 1:13c–17, is promised in 2:4b and contrasted with the sudden end to the power of the Chaldean in 2:5–20 and, having been besought anew in 3:2, is now preliminarily exemplified by Habakkuk himself. The transformation of history begins in the transformation of the prophetic consciousness.

Following his remembering of the old saving work, through the hymnic forms in 3:3–15, Habakkuk goes on in v. 16 to speak of his response—"I hear" (as in 3:2). The first four lines of v. 16 convey the sense of numinous dread or *tremendum* which was announced already by the verb "fear" in 3:2b. (For portrayals of such *tremendum*, not upon hearing of Yahweh's salvific end but in the context of such an end, see Hos 3:5 and 11:10.) The last two lines of v. 16 convey the other side of that "fear"—the *fascinosum* in which Habakkuk is drawn into an attitude of acceptance and commitment. All his old structures of awareness, concern, and understanding are melted down and transformed into a structure of quiet confidence by virtue of which he is able to "wait" for the promised day of deliverance, to wait through the intervening time of trouble upon his own people. That interval of wrath he portrays (v. 17) in agricultural and pastoral figures common to judgment passages in the prophets and in Deuteronomy 28. Through this interval he will wait (cf. 2:3; Isa 40:31). In 1:2–3 he had cried out "how long?" and had accused God's *torah/mišpaṭ* of weakness. Now he testifies to the strength which he has derived from the vision in 2:2–4 and its sequel in 2:5–20. This strength makes his feet like hinds' feet enabling him to run (again, cf. 2:3; Isa 40:31), actively to traverse the unknown interval. In full recognition of the hard character of that interval, he is able to affirm in it the strange work of God, and to hope in the saving work, in that eschatological joy which knows that salvation is not yet, and yet somehow already is.

On Existing Eschatologically within and for the Present Time

In this concluding section, I should like to offer the outline of a hermeneutical and theological meditation on the book of Habakkuk as here interpreted.

Eschatological existence is existence in time, toward an appointed time. It is such for God as well as for the people of God. Faithfulness is the moral and spiritual character of such existence, and patience is its exercise. Patience exercises itself in two modes: in action and in passion.[16] The interrelation of these two modes—on the one hand, the polar relation of the two modes within the individual party; and on the other hand, the chiastic interrelation of active and passional with passional and active modes between covenanting partners in eschatological existence—arises from the character of reality as a radically temporal field of power originating in the aboriginal life of God ("I will be who I will be") and in which all are given a share. The power of action is shared; and the power of passion is shared. Each is called upon to act; and each is called upon to wait and to suffer the acts of others. Each does what only that one can do; and each waits upon the other for what only the other can do. Each is called to faithfulness exercised in action and passion; and each is called to trust in the faithfulness of the other.

Eschatological existence is too much for the sluggard, whose fatal flaw lies not finally in laziness or fear, but in the weakness and the attenuated reach of the temporal imagination. But the temporal imagination is the symbolic imagination. For it is to the symbolic imagination that the vision of salvation is addressed—that vision whose rhetorical form and dynamic itself abides in and partakes of the eschatological reality which it enunciates.[17] It is the symbolic imagination which—unlike the sluggard who bur-

---

16. I am using the term "passion" and its adjective "passional," not in the sense it would have in, say, a Harlequin romance, but in the sense it has in the religious and theological tradition, as for example in the phrase "the passion narratives" or "the passion of Christ." In spite of possible confusion, the term "passion" docs seem preferable to passive, which would leave open construals with overtones of inertness (cf. the sluggard!).

17. I am using the term "symbol" here specifically in the sense defined by Coleridge, "[A symbol] always partakes of the Reality which it renders intelligible; and

ies his hand in the dish but by that act is too wearied to bring it back to his mouth (Prov 26:15)—reaches toward the vision as eschatological symbol and, appropriating the symbol, appropriates the eschatological energy which it conveys. And it is the symbolic imagination which, under the impetus of this eschatological energy, extends itself backward, archaizes itself in the reach of memory, to unite memory and hope, past and future, history and eschatology. Thereby the one who exists eschatologically, the so-called eschatological visionary, does not escape from the realm of so-called "plain history"[18] into a purely imaginal heterocosm. Rather, thereby one is freed from bondage to the constraints of the present, constraints which, viewed in abstraction from salvific past and future, take on the character of mere coercion and necessity. One is freed for a recovery of the concrete dimensionality of temporal existence as rooting in a past shaped and indwelt by the presence of the God of Abraham and of Isaac and of

---

while it enunciates the whole, abides itself as a living part in that Unity, of which it is the representative" (*Lay Sermons*, 30). Indeed, as I have attempted to interpret Hab 2:2–4 and the figure of Habakkuk himself in this essay, they exemplify beautifully what Coleridge is talking about. It may be noted that Coleridge's definition occurs precisely in that work where he attempts to provide an adequate *philosophical* context for the proper interpretation of the Bible in its bearing upon the social and political life of contemporary society. My estimation of the efficacy—the saving efficacy—of literary symbol generally and of the vision of Habakkuk in particular, stands in the sharpest opposition to the estimate of J. A. Miles, Jr., in his essay "Gagging on Job, or The Comedy of Religious Exhaustion," 71–126. The general drift of Miles's argument, brilliantly and exquisitely executed, is that the Bible, and more broadly literature (in spite of the agenda announced by Matthew Arnold), "cannot save" (p. 120). But Miles's universe is one in which salvation as such is a chimera, unless it may be said to consist, somehow, in the courageous willingness to come to terms with a universe the lifeless and bleak character of which is announced through modern physics (see his key paragraph, 79–80). The point of this note is to draw attention to the fact that, as promising as rhetorical criticism of the Bible may be, the fruits it bears for theological reflection turn on the philosophical—I should say the metaphysical—views which such criticism explicitly or implicitly presupposes. In his day Coleridge saw this issue clearly, and in the work quoted above and elsewhere addresses himself to it. Interestingly enough. Miles draws heavily, in "Gagging on Job," on the work of I. A. Richards, a superb literary critic and a sensitive interpreter of Coleridge, but, for all that, philosophically a descendant of Coleridge's contemporary Jeremy Bentham. It is tempting, and perhaps a form of cheap grace, to suppose that, whether in the name of Liberation Theology, or in the name of Rhetorical Criticism, we can dispense with the metaphysical task.

18. The phrase "plain history" occurs in Hanson, *The Dawn of Apocalyptic*, where it functions throughout as a *leitmotif* by way of contrast with what Hanson regards as the increasingly "escapist" character of visionary apocalyptic eschatology. My remarks here indicate my sharp disagreement with this aspect of his generally suggestive analysis and interpretation. In my view, Hanson's "plain history" perpetuates what Whitehead terms the fallacy of misplaced concreteness.

PART THREE: The Standpoint of Two Prophets

Jacob, and vectoring toward a future which is grounded in the future of the One whose name is "I will be who I will be." Thereby, one is freed to undergo and to engage the necessities of the present in the salvific modes of patience and joy.

part four

An Interlude

14

Toward a Hermeneutics of Resonance
A Methodological Interlude between the Testaments

If Scripture is ever again to be a living source for theology, those who practice theology must become less occupied with the world that produced the Scripture and learn again to live in the world that Scripture produces. This will require imagination and, no doubt, the willingness to leap. It will require the willingness to leap precisely because embracing imagination means entering truly into a world other than modernity. The Enlightenment banished imagination to the epistemological attic. Championing descriptive forms of knowledge that can be empirically verified, modernity considered the imaginative capacities of the mind as a matter of private fancy. Only slowly are we regaining a sense of imagination as a constructor of worlds. The more we appreciate the constructive power of imagination, the more we can begin to explore the proposition that the truth of Scripture is to be found less in its accuracy in describing the world than in its capacity in imagining a world. —LUKE TIMOTHY JOHNSON[1]

The IMAGINATION ... I consider either as primary, or secondary. The primary IMAGINATION I hold to be the living Power and prime Agent of all human Perception, and as a repetition in the finite mind of the eternal act of creation in the infinite I AM. The secondary I consider as an echo of the former, co-existing with the conscious will, yet still as identical with the primary in the *kind* of its agency, and differing only in *degree*, and in the *mode* of its operation. It dissolves, diffuses, dissipates, in order to re-create; or where this process is rendered impossible, yet still at all events it struggles to idealize and to unify. It is essentially *vital*, even as all objects (as objects) are essentially fixed and dead. —SAMUEL TAYLOR COLERIDGE[2]

---

1. Johnson, "Imagining the World that Scripture Imagines," 119.
2. Coleridge, *Biographia Literaria*, 1:304.

PART FOUR: An Interlude

> What is theory, after all, but explaining what it is you do by instinct, by instinct and through love?
>
> —CATHERINE M. WALLACE[3]

BEYOND THE FEW IMPRESSIONISTIC remarks offered in the introductory chapter, the preceding forays into this or that locale in the biblical terrain have been offered without the support of a clearly articulated method or theory of interpretation. In the author's preface I have taken refuge in Catherine M. Wallace's words concerning what one does "by instinct, by instinct and through love." But before moving into some New Testament treatments of some of the themes running through these previous essays, it is high time that I attempt to identify and reflect on some methodological assumptions implicit in how I have been going about these forays, and attempt to ground these assumptions in a wider, supportive context.

Most simply, I take it that the meanings words convey to their hearers and speakers depend on meanings those words have already been recognized as carrying in other, earlier encountered contexts. When four-year-old Lucy, helping her grandmother fold up a deflated air mattress, takes hold of a corner and says, "let's make this a team effort," one can hear in her words the echo of an appeal to her, on an earlier, similar occasion, from her mother. As in this illustration, I have frequently used the term, "echo," in referring to the semantic interaction between two or more biblical texts; but I have also used the term, "resonance," without indicating the difference in these two sonant figures of speech. The present essay offers a preliminary reflection on resonance as a feature of intertextuality, and beyond that as intrinsic to human life understood as embedded in the wider world, this wider world in turn understood as called into being and called toward its future through the animating Word of God. A fully-fledged "hermeneutics of resonance," such as might be undertaken by a Paul Ricoeur, might explore the function of resonance as it operates in every field of human experience and inquiry. Here, for limitations of space and personal competence (I to Ricoeur as a lightning bug to lightning, to use the words of Mark Twain), I confine myself to six areas: (1) the practice of biblical interpretation as represented by Richard B. Hays and Patrick D. Miller; (2) the poetry, poetic theory, and metaphysics of Samuel Taylor Coleridge; (3) the cosmological reflections of Alfred North Whitehead; (4) the role of resonance in the natural world according to Rupert

---

3. Wallace, *Motherhood in the Balance*, 157.

Toward a Hermeneutics of Resonance

Sheldrake; (5) the psychological theory and meta-psychological speculations of Hans Loewald; and (6) the function of resonance, as broached by Steven D. Smith, in liberating contemporary secular discourse from its "iron cage." My reflections are offered as a heuristic sketch, and make no pretense to offering a comprehensive or unified theory; but they do indicate the perspectives within which I understand what it is that I am about in the present volume (and elsewhere in my work, since several of the essays in the present collection anteceded its publication).

I conclude this "interlude" by recurring to one of its central passages—Coleridge's extended characterization of the nature of the work of the "poet in ideal perfection"—and in light of it explore the statement in Col 1:17 that "in [Christ] *all things hold together*," as that statement resonates with Old Testament passages bearing on the same theme. This foray will provide a segue to the following three chapters which deal with the New Testament.

Richard B. Hays on Intertextual Resonance

Since the publication in 1981 of *Echoes of Scripture in the Letters of Paul*,[4] Richard B. Hays has continued to explore the interpretive program set forth in that book. [5] For my purposes I shall confine my attention to his theoretical discussion in the first chapter of the 1981 book, where, although the literary term "echo" predominates, he uses the word "resonance" some thirteen times, in a manner that while unelaborated is highly suggestive.

Hays' "working hypothesis" is that "certain approaches to intertextuality" developed by recent literary critics "prove illuminating when applied to Paul's letters." Among such critics he cites Julia Kristeva, Roland Barthes, and Jonathan Cullers." For such critics, intertextuality is "the study of the *semiotic matrix* within which a text's acts of signification occur."[6] In Culler's words, "Intertextuality . . . becomes less a name for a work's relation to prior texts than a designation of its participation in the *discursive space* of a given culture."[7] Noting that "this kind of criticism . . . very quickly shades over into a sociology or anthropology," as describing a given culture's "system of codes or conventions that the texts manifest," Hays confines himself to "the phenomenon of intertexuality in

---

4. Hays, *Echoes of Scripture in the Letters of Paul*.
5. See, e.g., Hays, Alkier, and Huizenga, eds., *Reading the Bible Intertextually*.
6. Hays, *Echoes of Scripture*, 15 (italics added).
7. Culler, *Pursuit of Signs*, 103; cited in Hays, *Echoes of Scripture*, 15 (italics added).

PART FOUR: An Interlude

Paul's letters in a more limited sense, focusing on his actual *citations* of and *allusions* to specific texts."[8] He observes that "[t]his approach to Paul is both possible and fruitful because Paul repeatedly situates his discourse within the *symbolic field* created by a single great textual precursor: Israel's Scripture."[9]

In Hays' taxonomy, "Quotation, allusion, and echo may be seen as points along a spectrum of intertextual reference, moving from the explicit to the subliminal."[10] It would appear that for him "echo" identifies one extreme on this spectrum, for he goes on to say "as we near the vanishing point of the echo, it inevitably becomes difficult to decide whether we are really hearing an echo at all, or whether we are only conjuring things out of the murmurings of our own imaginations."[11] That he does not locate "resonance" on this spectrum, but sees it operating on a different plane of intertextual interaction, is implied in his discussion of Paul's affirmation in Phil 1:19 that "this will turn out for my deliverance," vis-à-vis the same affirmation on Job's lips in Job 13:16. It is clear that he takes Paul's affirmation as echoing Job's (the word "echo" occurs 18 times in this discussion), and what he means by "resonance" is suggested in the following statements:

> [S]criptural echoes lend resonant overtones to Paul's prose ... The reader whose ear is able ... not only to discern the echo but also to locate the source of the original voice will discover a number of intriguing *resonances* ... When the source of the phrase is read in counterpoint with the new setting into which it has been transposed, a range of *resonant harmonics* becomes audible. None of the correspondences between Paul and Job, or between Paul's rivals and Job's interlocutors, is actually asserted; instead, they are intimated through the trope of metalepsis. The trope invites the reader to participate in *an imaginative act* necessary to comprehend the portrayal of Paul's condition offered here.[12]

Clearly, "resonance" does not lie on the same plane as the spectrum from quotation to echo, but operates as an "overtone," perhaps subliminally. Moreover, in the example given it refers not to the verbal "point of

8. Hays, *Echoes of Scripture*, 15 (italics added).
9. Ibid. (italics added).
10. Ibid., 23.
11. Ibid.
12. Ibid., 21–22, 23 (italics added).

Toward a Hermeneutics of Resonance

contact" between the two texts, but to the interaction of elements in the textual fields within which the respective points of contact are embedded.[13]

This would fit in nicely with an observation of Reuben Brower on the poetry of Alexander Pope that Hays had earlier introduced into the discussion. Referring to the pleasure a reader may gain in "hearing echoes of earlier poets in [Pope's] verse," Brower observes that "through allusion, often in combination with subdued metaphors and exquisite images, Pope gets his purchase on larger meanings and evokes the finer resonances by which poetry (in [Samuel] Johnson's phrase) 'penetrates the recesses of the mind.'"[14] Here, despite Hays' aim to limit himself to "the phenomenon of intertextuality," we are plunged into social psychology. For the resonances of which Brower speaks, as penetrating the recesses of the mind, are resonances not simply between one text and another but between the meanings of the words on the page and the meanings of those words as embedded in the recesses of the reader's mind. Those recesses are shaped through an induction into and internalization of the "semiotic *matrix*" and "discursive *space*" of a culture as mediated through its symbolic codes. The symbolic codes are, thus, outward and visible signs of inward and spiritual space, a space that those codes work to shape in the inner life of a culture's members through the traditioning process.[15] Hays quotes John Hollander who says, "The reader of texts, in order to overhear echoes, must have some kind of access to an earlier voice, and to its *cave of resonant signification*, analogous to that of the author of the later text."[16] I suggest that, taken as a community's canon, Scripture as a whole may be thought of as a gigantic "cave-system of resonant signification," with innumerable niches and smaller caves, each with its local resonances, and innumerable linking passageways along which resonances may travel in varying degrees of audibility or subliminal effect. As the initiate into this cultural canon indwells it, the effect is to reproduce within the initiate, in lesser or greater

---

13. Compare my comment in Janzen, "Creation and New Creation in Philippians 1:6," 37, on the recurrence in the Bible of certain verbs as "nodal points of a narrative pattern that both arises out of and goes to foster a distinctive religious ethos, at such a deep level that the linguistic terms of that ethos operate much more frequently in a formative and expressive way than conscious awareness and intention alone can account for." Note also my subsequent comment that "however much [Rom] 11:36a may have the form of a 'Stoic type formula,' the terms of that formula should be heard as *resonating* with biblical themes" (41; italics added).

14. Hays, *Echoes of Scripture*, 18.

15. This point will connect with my use in chapter 16 of Mark George's work *Israel's Tabernacle as Social Space*.

16. Hays, *Echoes of Scripture*, 31.

degree, an imperfectly corresponding or analogous "cave of resonant signification." And it is within this "cave," these "recesses of the mind," that such resonances are evoked as linking one Scriptural passage to another. Needless to say, to a reader who has not been in-formed in this manner, others' claims to hear resonances will fall, quite literally, on deaf ears.

I may put my point in terms of the acoustic capabilities of stringed instruments. It is well known that "any fine stringed instrument is built of precisely calibrated wooden parts, all of which must 'learn' to vibrate sympathetically," while "the more a stringed instrument is used, the more responsive and resonant it becomes."[17] Analogously, I suggest, the more deeply one is immersed in the cultural memory of a community through immersion in its Scriptures, the more sensitively one will hear the resonances between its texts and resonate with them. An additional—though nowadays controverted—dimension lies in the direction of the sacred, as when, for example, one person strengthens another through "the encouragement of the scriptures" (Rom 15:4), and the latter person experiences in that encouragement the strengthening activity not only of the other person but also of God.[18]

In discussing the echo of Job's words in Philippians, Hays states that this echo "invites the reader to participate in *an imaginative act*" through which "a range of resonant harmonics" between the two larger passages, and between the situations of the two figures, "becomes audible."[19] Earlier, in commenting on Paul's statement in 1 Cor 10:4 that "that rock was Christ," Hays asks, "what poetic linkages of sound or imagery make this sort of imaginative leap possible, what effects are produced in the argument by it, and what sort of response does it invite from the sympathetic reader's imagination?"[20] In his chapter's concluding section, titled "Beyond the Hedges," he observes that, despite the "seven tests" he has proposed[21] for discerning genuine echoes in the text and distinguishing them from "the murmurings of our own imaginations,"[22]

> there will be exceptional occasions when the tests fail to account for the spontaneous power of particular intertextual

17. Delaney, "Using an Electronic Device To Break In a New Violin."
18. See my discussion of the threefold dimensions of encouragement in Chapter 10.
19. Hays, *Echoes of Scripture*, 14.
20. Ibid., 14.
21. Ibid., 29–32.
22. Ibid., 23.

Toward a Hermeneutics of Resonance

connections. Despite all the careful hedges that we plant around texts, meaning has a way of leaping over, like sparks. Texts are not inert; they burn and throw fragments of flame on their rising heat. Often we succeed in containing the energy, but sometimes the sparks escape and kindle new blazes, reprises of the original fire.[23]

Only reprises? Or kindlings of a new fire out of the old? I recall a small birthday party in our home for a forty-year-old seminary student, when, after dinner, I set about carefully lighting all forty small candles on the cake. At first each candle took a small, individual flame from the long, lighted wooden taper in my hand. But suddenly, before I had quite finished, what I later supposed was a sudden bellows-effect created by the updraft of all those individual points of heat, a huge single flame, accompanied by a startling *whoosh!* sound, shot up toward the ceiling and then as suddenly died out, leaving only the cake, the totally melted candles, and the small company now in a daze of departed glory. When the "meaning has a way of leaping over," does it leap only over the hedges we have planted around it, to kindle blazes that are only *reprises* of the original fire? But what was that original fire? Was it not, itself, a leaping over the hedges of previous understandings? Did the attempt to image that original fire not itself, in Brower's phrase, get a "purchase on wider meanings"? What I am getting at, and what I see Hays moving toward in speaking of the call for an "imaginative leap," may be suggested by the following aspect of resonance in music.

The resonance between a vibrating string and its sound chamber (as in a cello) intensifies and enriches the quality of the note it produces; and the same holds true for the resonance of the concert hall's acoustics. But resonance in such a context operates also on another plane, where a second-order resonance may arise. I am told that when a symphony orchestra plays well in an acoustically adequate concert hall, the resultant cornucopia of sound can display fine subtleties of sonority that arise from no single instrument but rather from the interactions of the sonorities produced by the several instruments. The average listener may not be conscious of such second-level sonorities, but the finely tuned ear of the conductor or critic may catch them, if not as discrete sounds, then as "the difference that makes the difference" between a routine and a transporting performance. A musically trained acquaintance puts the matter as follows:

---

23. Ibid., 32–33.

PART FOUR: An Interlude

> Overtones definitely mix together when instruments play together and with acoustics in a hall, so new sounds will be created that don't come from any single instrument. Even on one instrument alone, the way the instrument resonates within its own sound chamber will create overtones that are audible to attuned ears and impact all ears even unknowingly. Our son Andrew's percussion teacher was tuning his snare drum last week . . . and my husband Willie and I were acutely aware of overtones coming from the drum and resonating around the room as he tightened and loosened the drum heads, which changes the color and basic pitch of the drum. Musical performers are taught to listen for the wonderful series of overtones that results from a nicely tuned chord; that's one of the ways for players to know that they've settled into the right intonation.[24]

The question is whether texts that carry their own "resonant significations" can, through the interaction of those resonances within a larger "cave of resonant significations," produce second-order resonances hearable only as one, as it were, positions oneself in the imaginative space "between the texts," the imaginative space opened up by the juxtaposition of the texts. An example of this may be seen in the opening chapter of the letter to the Hebrews, where, in setting forth what a later generation of theological reflection called a "high Christology," the writer simply strings together a series of Old Testament texts. It is easy to show that each of the texts, taken by itself in its original Scriptural context, did not carry the meaning the writer seeks to have us hear in them as so juxtaposed. But if one ponders the texts within the "matrix" or "cave" or "discursive space" of the letter's overall argument, and with an ear out for the second-order resonance generated by their juxtaposition, one can begin to see what the writer is driving at. Here is not simply a "reprise of the original fire," but a newly kindled fire arising from the fusion of all the individual flames.

Philip Wheelwright—if I read him aright—discusses this sort of "new meaning through juxtaposition" under a special category of metaphor which he calls "diaphor." In a standard understanding, metaphor achieves semantic movement when the literal meaning functions as a "vehicle" for its figurative "tenor." Wheelwright dubs this mode of metaphorical function "epiphor;" and he distinguishes epiphor and diaphor as follows: The

---

24. Personal communication from the Rev. Deborah Grohman, a clarinetist with music performance degrees from New England Conservatory and DePauw University, and her husband, Willie La Favor, a pianist with music performance and pedagogy degrees from Indiana University, New England Conservatory, and Jacksonville University.

first "stand[s] for the outreach and extension of meaning through comparison, the other for the creation of new meaning by juxtaposition and synthesis." Again he writes, "The role of epiphor is to hint significance, the role of diaphor is to create presence."[25] I suggest that what makes diaphor work, what produces its effect on the reader, is the second-order resonance arising from the juxtaposition of its elements, that this resonance is synthetic in effect, and that the synthesis effected is not simply re-creative (as a "reprise") but, with reference to the elements out of which it arises, creative. It is this creative effect of the second-order resonance arising in the space between juxtaposed texts, I suggest, that a more traditional hermeneutical discourse—and behind that, a tradition of reading experience—points to in speaking of the *sensus plenior* of a text, a meaning that a text has come to bear beyond its earlier "plain sense."[26]

By way of a segue to Patrick D. Miller's use of "resonance" in his own exegetical and theological work with Scripture, I want to raise another point. From Hays' discussion in his methodological chapter one might, rightly or wrongly, gain the impression that resonance is properly to be identified only as accompanying a quotation, allusion, or echo—that it is, so to speak, only ever the widening and ever more faintly thinning out comet-tail of the echo connecting two texts. Whether or not he would so argue, I will maintain that such is not the case—that resonance may felicitously be discerned between texts that display no specific "node" of contact such as lies along the spectrum from quotation to echo. And I would add one more point here. Whereas "echo" is a one-way phenomenon, a later text echoing an earlier, "resonance" works both ways, in such a way as to intensify, nuance, broaden, focus, or otherwise modify both members of a resonant relation. And in another dimension, the growth in complexity and richness of a community's symbolic matrix, of its "cave of resonant signification," can itself evoke dimensions of resonant signification not evidently hearable in earlier periods of that community's existence. This way of viewing the dynamics of living tradition anchored in a canonical Scripture would be one way of understanding what the Puritan pastor John Robinson meant when, in addressing the members of his flock who would embark in the Mayflower without him, he was said to express his confidence, on behalf of the pilgrims, that "the Lord had more truth and light yet to break forth out of his holy word."

---

25. Wheelwright, *Metaphor & Reality*, 72, 91.
26. I refer the reader here to my discussion in Chapter 11.

PART FOUR: An Interlude

Patrick D. Miller on Resonance

In an essay on "The Hermeneutics of Imprecation," Miller speaks of the poetic, evocative power of Psalm 137 to "resonate with a distant audience"[27] such as ourselves at some remove from the literal situation of the Babylonian exiles. It is clear that by resonance he means in the first instance a poignant felt identification with the psalmist vibrantly awakened in our own hearts in hearing "how shall we sing the LORD's song in a strange land?" The lyre or harp, with its vibrant strings ("On the willows there we hung up our lyres"), is itself a quintessential image of such a resonant identification; as when, in Isa 16:11, the prophet cries (in the literal rendering of the KJV), "my bowels shall sound like an harp for Moab, and mine inward parts for Kir-haresh." In the latter passage, the plangent plight of Moab evokes in the prophet an emotion that reverberates organically "like a harp." So, we may say, the language of Psalm 137 in the first instance "plucks at our heartstrings."

But the burden of Miller's essay focuses on the "repulsive violence"[28] imaged in the psalm's grisly conclusion, imagery that fills us with "revulsion."[29] Such texts, he suggests, do not simply "serve to perpetuate violence" as a surface reading might gather, but rather "contribute to the exposure and unmasking of violence." He writes (following work by Norbert Lohfink and René Girard), "the text mirrors the violence that is universal and exposes the violence that is a part of the society and of its God in the context of the announcement of a counter-society and a counter-God."[30] Anticipating my later discussion of the place of resonance in sublimation, I want to elaborate here on this function of mirroring and unmasking, as depending on such a reading of Psalm 137 within the context of the whole of Scripture where other ways of responding to great suffering are lifted up, for it is only within that canonical context that the "announcement" Miller speaks of comes into view over against the immediacy of the psalmist's concluding vengeful rage. The "violence that is universal" is a violence that lies dormant within all of us, however distant as readers, like the "sin" that is "couching at the door"[31] of Cain's heart like

27. Miller, "The Hermeneutics of Imprecation," 196.
28. Ibid., 196.
29. Ibid., 197.
30. Ibid.
31. The verb "couch"—perhaps quaint to our ears—captures the connotation of the Hebrew verb as no other English word does, resonating as it does with the noun, "couch," which is a place of rest or repose. The Hebrew verb, occurring some 30 times,

a wild beast, ready to spring forth upon Abel as opportunity presents itself (Gen 4:7). So the Psalm's "attraction and revulsion" for us is not simply its power to attract our sympathy for the oppressed and our revulsion over the violence in the image of vengeance. As Freud has helped us to appreciate, the very vehemence of our revulsion over the concluding image may mask from ourselves the deeper attraction the psalm has for us in providing a vehicle for our own latent appetite for violent revenge. Simply to deny this deeper (because hidden) attraction is to allow it to smolder and "couch at the door" until it finds its own opportune occasion and form of expression in our present situation. But, as therapists know, it is as we admit the presence within us of such powerful but masked (suppressed or repressed) feelings, and engage them in a "safe place" where there is no danger of discharging them on some unfortunate (or even "deserving"!) victim, that the possibility arises of "mastering them" (as God counsels Cain), by integrating and reconciling them with other, more peaceable but equally powerful feelings, and in this way transforming them. It is the very repulsiveness of the concluding image, then, as penetrating to primitive levels of vibrant emotion (in "the recesses of the mind," to use Samuel Johnson's phrase), that gives this scripture the power to do its necessary work toward the realization of a "counter-society" and its "counter-God." I shall return below to the function of resonance in sublimation toward human transformation.

Later in the same essay, in commenting on Isaiah 34 with its imagery of "the bloody sword of the Lord," Miller notes that "one of the texts that resonates most with Isa[iah] 34 is the Book of Revelation at the end of the New Testament."[32] Here, texts in the two testaments resonate with each other not as one quoting or echoing the other, but as both participating in the same field of sensibility and cultural symbolic tradition.[33]

---

never connotes "lurking" in the sense of an alert, active readiness to attack. Rather, it indicates the state of lying down in repose (e.g., Pss 23:2; 104:22). The subtle implication here, displaying acute psychological insight, is that "sin" is an ever *latent* potential for action, currently in repose but susceptible of being roused to action if disturbed or provoked.

32. Ibid., 198.

33. Whitehead expressed his wish that the New Testament had been edited to end, not with the Book of Revelation, but with Pericles' famous Funeral Oration (see Price, *Dialogues of Alfred North Whitehead*, 17–18). An armchair Freudian might respond with the observation that the Book of Revelation does at least show war in all its lurid grisle, while Pericles' oration cloaks its gore in patriotic sentiments celebrating heroic virtue. From that point of view, Revelation, like Psalm 137, may serve a more "revelatory" diagnostic function than Pericles.

PART FOUR: An Interlude

The importance of Miller's interpretive move lies in his application of the term "resonance" to a semiotic and existential relation between texts not evidently ligatured through quotation, allusion, or echo.

Miller explores the question of resonance between biblical books as wholes in his essay, "Deuteronomy and Psalms: Evoking a Biblical Conversation."[34] The importance of his reflections on the inter-relation between these two biblical books lies in the fact that they belong for the greater part to two entirely different textual genres, and arise in relation to two different social domains: Deuteronomy arises within the domain of law, with its motivation, foundation, rationale, and administration; while the Psalter arises within the domain of worship including praise and prayer, with their motivations, foundation, rationale, and orchestration. The comment of Mario Cuomo, one-time Governor of New York State, that politicians campaign in poetry and govern in prose, applies aptly to these two biblical books. If one were to seek interconnections between them, one might identify echoes of one in the other here and there (depending on one's identification and dating of strata of composition and redaction), and one might note how Deuteronomy ends (except for its epilogue) with two poems, while in the Psalter one might here and there detect a Deuteronomistic redactoral thumb print, but hardly an outright quotation. Yet by the time Miller has finished listening in on what he hears as a conversation between them (informed by large chunks of a scholarly lifetime inhabiting each book exegetically and theologically in and for itself), he succeeds, to my mind, in justifying his claim that these two books resonate with one another across a broad range of common human concerns.

Though Miller in this essay more often uses the terms "conversation" and "dialogue" to indicate his sense of the interaction between these two Old Testament books, he explores his topic under the question, "is there a larger enterprise of intertextuality that seeks to discern the resonances between the larger segments of scripture designated by the term 'book?'"[35] And at one point he implicitly answers that question for himself in stating that "the Psalter and Deuteronomy are the books of scripture that most explicitly set within the community of faith a *cultural memory* that funds its

---

34. Miller, "Deuteronomy and Psalms: Evoking a Biblical Conversation," originally his Presidential Address at the 1998 Annual Meeting of the Society of Biblical Literature.

35. Miller, "Deuteronomy and Psalms," 4.

Toward a Hermeneutics of Resonance

identity and guides its life."[36] I repeat my earlier comments about initiation into such a cultural memory as a prerequisite for hearing such an intertextual conversation, in view of Miller's own concluding remarks concerning his own embeddedness in the Reformed tradition with its emphasis on a positive function of the Old Testament legal tradition and the prominent place of the Psalms in that tradition's public worship and domestic piety.

I want, now, to propose that what Miller "hears" between these two apparently quite different Old Testament books may exist also between the two Testaments that make up the Christian Bible. Funding the motivation for quotation, allusion, and echo of the Old in the New, but deeper than all these forms of intertextuality, and energizing them, is a resonance that at times takes the form of dissonance but at greatest depth persists as a matrixal and generative chorus of harmony or at any rate consonant polyphony. To put the matter in an image, the resonances between the Testaments are like the Pacific Ocean bed whose deep trenches may be mapped only by sonar probes but whose submerged ridges and ranges here and there break the surface in atolls, islands and archipelagos. When Miller approaches the conclusion of his essay in asking "whether through the varied voices of scripture it is possible to hear chords that create a harmony, even if the chords are sometimes dissonant ones,"[37] I take him to suggest a way of thinking of the Bible as a complex, loose but somehow coherent *canon* within whose "cave of resonant signification" a community may find its identity and guidance for its life. It is such an undertone of matrixal resonance, for those who can hear it, embracing and sustaining with "negative capability" the complex stresses and strains of its sometimes dissonant textual resonances, that may, when adequately read, constitute the Bible an instrument in the at-one-ing and reconciling of fractured human relations.

As a specific dimension of such an instrumentality, taken in the mode of textual, ethical, and existential resonance, one may study Miler's most recent work, his monumental study of *The Ten Commandments*. Occurring twelve (!) times in the book,[38] the hermeneutical importance of resonance for Miller may be gauged in the following quotations.

> What is before you is . . . an exposition of the Commandments in depth, seeking to give not only a reading of each commandment in its context but also to lay out the trajectory of its

36. Ibid., 13 (his italics).
37. Ibid., 17.
38. Miller, *The Ten Commandments*, xi, xii, 9, 48, 64, 148, 195, 216, 257, 345, 362, 424. I thank Miller for kindly furnishing me with these references.

253

movement and place in Scripture. In the process, the interplay and resonance of the Commandments with many other texts is uncovered, and the outcome is a thick description of the Commandments resulting in various theological and moral issues coming to the fore and receiving some illumination.[39]

In introducing his exposition of the third commandment, on "Hallowing the Name of God," Miller writes,

> Here, as elsewhere, the meaning and force of the commandment is not fully or adequately uncovered apart from an exploration of *the trajectory of meaning and effects that flow from it, the specifics and illustrations, and possible implications* that may be discerned in story and law, proverb, Psalm, and prophecy.... The individuality and particular focus of each commandment should not obscure the many ways in which it shows itself to be part of a whole, reflecting an order participating in various internal groupings, and revealing [in]numerable associations and resonances to other commandments.[40]

I take Miller to imply in these passages, and in his other references to the resonances of the commandments in Scripture, that community cohesiveness in covenant fidelity to God arises and is sustained not simply in flat-footed observance of the commandments in their literal meaning, but also, and perhaps more deeply, in and through an imaginative moral vision informed by a deep sense of resonance between what the commandments enjoin and what the contemporary situation calls for as "new occasions teach new duties" and "time makes ancient good uncouth."[41] In this sense, such "resonance" might identify phenomenologically what is meant by the "spirit" of laws vis-à-vis their letter.

In practicing what I am calling a hermeneutics of resonance, Miller does not claim a method with tight controls, but rather a method that depends always on two questions: Do you, too, hear what I think to hear? and, am I hearing a resonance, or is that just the tinnitus in my own ears and brain? The issue is perhaps another way of talking about "testing the spirits" (1 John 4:1).

---

39. Ibid., xi.

40. Ibid., 64 (his italics). Miller informs me that "numerable" in the published text is a misprint for "innumerable."

41. With reference to contemporary discussion within the churches on issues of human sexuality, see Miller's essays, "The Sufficiency and Insufficiency of the Commandments," and "What the Scriptures Principally Teach."

Samuel Taylor Coleridge on Resonance in the Nature of Things

What is the status and significance of resonance—the resonance of poetic speech, of the sort that "penetrates the recesses of the mind"—in the nature of things generally? Is such resonance, as the life-blood and felt pulse of connective meaning in human experience, inherent in reality at all levels and in all modes, or is it peculiar to human existence? Is the resonance we experience in our immediate felt relations with one another, and in the oral and written language through which we conduct those relations, a fundamental clue to the nature of things? Does it take us to the heart of things? Or is it a peculiarly human "add-on"? Has it any "truth" value, as Freud would assess truth in his classic 1927 essay, *The Future of an Illusion*, or does it, along with religious sentiments, aspirations and speech, come under his category of "illusion"?

One sort of response to this question was provided in 1926 by I. A. Richards—who shared Freud's acceptance of the materialistic epistemology in the science of his day—in his book, *Poetry and Science*. "Poetry, or rather . . . the possibility of emotional experience instigated, if not wholly controlled, through ordered words," offers what Richards calls "pseudo-statements," in contrast to the "statements" proper to science. By pseudo-statements, he labors to emphasize, he does not mean false statements, but rather, statements that relate to human interests and concerns arising out of human desires and purposes. "A pseudo-statement is 'true' if it suits and serves some attitude or links together attitudes which on other grounds are desirable . . . A statement, on the other hand, is justified by its truth, i.e., its correspondence, in a highly technical sense, with the fact to which it points."[42] In this analysis he reflects the classic basis of modern science in a philosophic naturalism or materialism for which the natural world is to be analyzed purely in terms of efficient cause, devoid of subjectivity, purpose, or meaning, and describable in the purely objective terms of mass, motion, and measurement. In such a vision, human emotion, and the resonance that is the life-pulse of human desire, purpose and meaning has no status, no significance, in the nature of things. Such resonance is relevant only to the human project within a universe in which, for Richards, it finds no home.[43]

42. I quote from the book's reissue, under the title, *Poetries and Sciences*, 60.

43. But note that Richards states earlier that "science, as our most elaborate way of *pointing* to things systematically, tells us and can tell us nothing about the nature of things in any *ultimate* sense" (ibid., 54).

PART FOUR: An Interlude

But the materialism of modern science, with its mechanistic model of the natural world on which Freud and Richards based their respective assessments of human aspirations and concerns vis-à-vis that larger world, was already under fundamental critique at the time they wrote. One sustained critique may be found in the books Whitehead penned following his first career as a theoretical mathematician, beginning with *Science and the Modern World* in 1925 and coming to final expression in "Nature and Life," the concluding part of his last book, *Modes of Thought* (1938). I will draw Whitehead himself into the discussion later. At this point I note his discussion, in *Science and the Modern World*, of a yet earlier critique, in the form of what he there calls "the Romantic Reaction" to the mechanistic materialism of the seventeenth and eighteenth centuries and continuing to inform the "second scientific revolution"[44] in the six decades spanning the turn into the nineteenth century. For example, Whitehead writes that William Wordsworth

> in his whole being expresses a conscious reaction against the mentality of the eighteenth century. This mentality means nothing else than the acceptance of the scientific ideas at their full face value. Wordsworth was not bothered by any intellectual antagonism. What moved him was a moral repulsion. He felt that what had been left out comprised everything that was most important.[45]

With the exception of the sentence referring to Wordsworth's lack of "intellectual antagonism," these sentences could be applied equally well to Samuel Taylor Coleridge, because Coleridge was both morally and intellectually exercised over the materialistic science and the sense-empirical philosophy of his day, and, deeply read in Continental philosophy (including Kant as early as the late 1790's), spent a good deal of his later life developing a philosophical and theological response to it.

Coleridge from childhood was possessed of a vivid sense of the whole vast universe in its interconnectedness. With a poet's eye for the merest details and the finest discriminations in the infinite variety of nature he combined a deep feeling for the unity in which all individuality is embraced and sustained; and he was preoccupied philosophically with the question, in his phrase, of "multeity in unity." As Thomas McFarland has

---

44. Cf. Richard Holmes, *The Age of Wonder*, xvi: "The first person who referred to a 'second scientific revolution' was probably the poet Coleridge in his *Philosophical Lectures* of 1819."

45. Whitehead, *Science and the Modern World*, 77.

Toward a Hermeneutics of Resonance

shown, Coleridge identified himself philosophically as a trinitarian theist, situating himself between pantheism on the one hand and deism on the other.[46] But always, his base lay in an intuitive sense for the interiority of things—for the interior life of reflection to which he sought to call his readers, and for the interiority of things in the world around him. This sensibility comes to poetic expression most vividly in his use, along with Wordsworth, of the image of the wind harp.[47] In "The Eolian Harp," he writes,

> O the one life within us and abroad,
> Which meets all motion and becomes its soul,
> A light in sound, a sound-like power in light,
> Rhythm in all thought, and joyance everywhere—
> Methinks, it should have been impossible
> Not to love all things in a world so filled;
> Where the breeze warbles, and the mute still air,
> Is Music slumbering on her instrument.

And, further on,

> ... what if all of animated nature
> Be but organic harps diversely framed,
> That tremble into thought, as o'er them sweeps
> Plastic and vast, one intellectual breeze,
> At once the Soul of each, and God of all?

Sonant words and images pile up in such a way that one is justified in summarizing them prosaically in the word "resonance."

Similarly, in "Lines Written a Few Miles above Tintern Abbey," Wordsworth speaks of

> that blessed mood,
> In which the burthen of the mystery,
> In which the heavy and the weary weight
> Of all this unintelligible world

---

46. Thomas McFarland, *Coleridge and the Pantheist Tradition*.

47. For a discussion of the importance of the image of the wind harp at this time, see Owen Barfield's essay, "The Harp and the Camera." Barfield's essays in *The Rediscovery of Meaning*, are deeply indebted to Coleridge's thought, and may be taken as one sort of philosophical underpinning for a hermeneutics of resonance. On the wind harp see also Hollander, "Wordsworth and the Music of Sound"; and note Hartman's summary comments in *New Perspectives on Coleridge and Wordsworth*, viii: "Wordsworth is haunted by 'images of voice,' by natural sounds that move beyond echo or physical ricochet into resonance and myth."

PART FOUR: An Interlude

> Is lighten'd:—that serene and blessed mood,
> In which the affections gently lead us on,
> Until the breath of this corporeal frame,
> And even the motion of our human blood
> Almost suspended, we are laid asleep
> In body, and become a living soul:
> While with an eye made quiet by the power
> Of harmony, and the deep power of joy,
> We see into the life of things.

Here the key phrase is "the power of harmony," and its interior answering "deep power of joy," again evoking the notion of resonance. It is easy to take these poetic passages as expressive of an exuberant nature pantheism, or simply as poetic hyperbole; but, while Wordsworth, for all his great poetic output, never wrote the great "philosophic poem" that Coleridge hoped for from him, Coleridge's own subsequent metaphysical and theological reflections, theistic and Trinitarian, never moved him to repudiate such early poetic effusions. So they may be read as consonant with such a theistic vision.

A close reading of Coleridge's later "Comment" on "Moral and Spiritual Aphorism VI," in his *Aids to Reflection*, will show the metaphysical seriousness with which we are to read his lines on the Eolian harp.[48] What vivifies and interconnects all things is the resonant presence of the "plastic and vast" "intellectual breeze" which we may parse in terms of his discussion, in the above-mentioned "Comment," of the divine Logos and the divine Spirit. His term, "plastic" has none of the current connotations of a non-bio-degradable bottle littering the streets and parks after the water contained in it has been consumed. Rather he means by it a *shaping power*, as when he speaks (in his own coinage) of the "esemplastic imagination"—the power of the imagination to "shape into one" the elements with which it works. But for Coleridge imagination is not only a *creative* power, it is also, and indeed first, a *power of perception*. If, in a mechanistic philosophy of nature, the elements in nature are perceived (by the eye as a "passive onlooker") as essentially lifeless bits of matter pushed or pulled about by gravitation, magnetism, and other such forces, for Coleridge it is the imagination as *active* in attentiveness to the natural world that, in

---

48. Coleridge, *Aids to Reflection*, 73–88. In this long, pivotal entry Coleridge canvasses the infinite variety in nature as displaying a *harmony* which attests to an "antecedent unity," and analogizes this to "the System of intelligent and self-conscious Beings, to the moral and personal World" (77), the realm or dimension of spirit as implicit in the actions of human free will.

Wordsworth's phrase, "sees into the life of things." Coleridge put the matter to Thomas Sotheby in this way:

> Metaphisics [sic] is a word, that you, my dear Sir! are no great Friend to but yet you will agree, that a great Poet must be, implicitè if not explicitè, a profound Metaphysician. He may not have it in logical coherence, in his Brain and Tongue; but he must have it by *Tact*[:] for all sounds, & forms of human nature he must have the *ear* of a wild Arab listening in the silent Desert, the *eye* of a North American Indian tracing the footsteps of an Enemy upon the Leaves that strew the Forest—; the *Touch* of a Blind Man feeling the face of a darling Child.[49]

In one of his public lectures on literature, given several years later, Coleridge put the matter this way, in the process using a metaphysical distinction that he often returned to:

> If the Artist painfully *copies* nature [that is, *natura naturata*], what idle rivalry! If he proceeds from a Form, that answers to the notion of Beauty . . . what an emptiness, an unreality. . . . The *essence* must be mastered—the natura naturans, & this presupposes a *bond* between *Nature* in this higher sense and the soul of man.[50]

Here, *natura naturata* refers to natural objects as perceptible to the senses and *natura naturans* as the process by which these objects come into existence, a process which he saw as involving dynamic activity in one or another mode whether inorganic or organic. It is this common "essence," this "bond," which for Coleridge is the key to the "unity" he senses as pervading the "multeity" in creation. Shortly afterward he says,

> To that within the thing, active thro' Form and Figure as by symbols . . . *Natur-geist* must the Artist imitate, as we unconsciously imitate those we love—So only can he produce any work truly *natural,*in the Object, and truly *human* in the Effect.[51]

---

49. Griggs, ed., *Collected Letters*, 2:810 (to William Sotheby, 13 July 1802).

50. Coleridge, *Lectures*, 2:220–21 (italics original).

51. Ibid., 223 (italics original). On "unconscious imitation" in respect to the formative influence of Scripture on a writer's thought and diction, see again my remark in Janzen, "Creation and New Creation in Philippians 1:6," 37, that "the linguistic terms of [Israel's religious] ethos operate much more frequently in a formative and expressive way than conscious awareness and intention alone can account for." For an example of this on a contemporary, everyday level, see the section on "Job and My Parents," in Janzen, *At the Scent of Water*, esp. 130–32.

PART FOUR: An Interlude

Elsewhere he distinguishes between a copy and an imitation in a way that suggests a mechanical transfer of lifeless form, and a rapport—one may say a resonance—between the imitator and what is being imitated.[52] As Coleridge notes (and as anyone may note in observing the behavior of infants and young children), such imitation of a loved person is often "unconscious." I flag this term here, along with the Coleridge's distinction between *natura naturata* and *natura naturans* and the unifying bond that I am indentifying as resonance, for recall when I take up the late metapsychological reflection of Hans Loewald.

Meanwhile, it is telling that of the three sensory modes that Coleridge identifies in his letter to Sotheby, he gives first place to hearing. In his poetry it is the image of the "organic" wind-harp in which he *hears* the presence of the all-animating "intellectual breeze" which is the ground of what he is wont to call reality's "unity in multeity." In "This Lime-Tree Bower My Prison," Coleridge, confined in a neighbor's garden by an accident that prevents him from accompanying his wife and two friends on a walk into the countryside, at first laments his inability to share the "beauties and feelings" that they must be enjoying. Then, reflecting on how one of the friends, Charles Lamb, now for once escaped from the "great City" of London with its "evil and pain / And strange calamity," and enjoying the Nature for which he had "pined / And hungered . . . many a year," Coleridge is able to rejoice in Lamb's present enjoyment; and, he concludes, "'Tis well to be bereft of promis'd good, / That we may lift the soul, and contemplate / With lively joy the joys we cannot share." But, in a sweet irony, this self-transcending empathy with his friend results in a transformation of his own present situation. For, as he sees a "last rook / Beat its straight path along the dusky air / Homewards," he imagines it passing, or having passed, before Lamb's gaze "or, when all was still, / Flew creeking o'er thy head, and had a charm / For thee, my gentle-hearted Charles, to whom / *No sound is dissonant which tells of Life.*"[53] It is in the resonant rhythms of sound—as when one hears (with the ears of "a wild Arab") the wings of a rook in the stillness of an evening—that one may hear the pulse of life that runs through all things.[54]

---

52. In a footnote to one such passage, the editors comment, "In general, as he uses the term, a 'copy' tries to duplicate the original so closely that it can almost be mistaken for it . . . [A] true 'imitation' duplicates the process of nature itself by presenting, in itself, the organizing spirit (*natura naturans*)." See Coleridge, *Biographia Literaria*, 2:72 n. 4.

53. Coleridge, "This Lime-Tree Bower My Prison" (italics added).

54. For a profound analysis of and meditation on this poem, including a reflection

Toward a Hermeneutics of Resonance

Even Wordsworth's "seeing into the life of things" is possible when the eye is "made quiet" by a "power of *harmony*" and a "*deep* power of joy." It is in such terms that we may deepen and broaden our sense for the potential scope and status of a hermeneutics of resonance.

Coleridge's most famous—and frustratingly brief—discussion of the human imagination comes in the paragraph cited as a caption to this chapter. For convenience I repeat it here:

> The IMAGINATION . . . I consider either as primary, or secondary. The primary IMAGINATION I hold to be the living Power and prime Agent of all human Perception, and as a repetition in the finite mind of the eternal act of creation in the infinite I AM. The secondary I consider as an echo of the former, co-existing with the conscious will, yet still as identical with the primary in the *kind* of its agency, and differing only in *degree*, and in the *mode* of its operation. It dissolves, diffuses, dissipates, in order to re-create; or where this process is rendered impossible, yet still at all events it struggles to idealize and to unify. It is essentially *vital*, even as all objects (as objects) are essentially fixed and dead.[55]

This is not the place to enter into a conversation with the reams of interpretation that have been offered on this paragraph. I shall simply venture my own take, encouraged to construe it against a biblical background by Coleridge's own discussion, in the preceding chapter of *Biographia Literaria*, of human consciousness on analogy with the divine pronouncement in Exod 3:14.

In its primary mode as *perceptual*, how may we understand imagination to be "a repetition in the finite mind of the eternal act of creation in the infinite I AM"? In Genesis 1, the God who at the burning bush had said, *'ehyeh ašer 'ehyeh*—"I will be who I will be"—said, *yehi 'or*—"let there be light," and so on. Each divine creative act is followed by a divine act of seeing that the created result is "good." In the case of the human imagination, the sequence is reversed. Imagination first sees what God has created—sees it *as* God has created it—and then, as *imago dei*, takes what it sees, "dissolves, diffuses, dissipates, in order to re-create." Whereas (as Coleridge never tires of saying elsewhere) the human understanding, informed by the perceptions of the senses unaided by imagination, takes

---

on how the poem implicitly inducts one into the art of reading a poem (or, for that matter, a Psalm, or the Scriptures generally), see Wheeler, *The Creative Mind in Coleridge's Poetry*, 121–49.

55. Coleridge, *Biographia Literaria*, 1:304.

PART FOUR: An Interlude

the objects of such sensory perception as "essentially fixed and dead," the imagination sees all as "vital," and so its re-creative activity itself issues in creations that resonate with the life of their human creator.[56]

With such a view of the perceptual imagination as "a repetition in the finite mind" of God's own perception of the results of divine creative action, we may compare Abraham Heschel's attempt "to attain an understanding of the [Hebrew] prophet through an analysis and description of his *consciousness*."[57] Put most succinctly,

> the fundamental experience of the prophet is a fellowship with the feelings of God, *a sympathy with the divine pathos*, a communion with the divine consciousness which comes about through the prophet's reflection of, or participation in, the divine pathos. The typical prophetic state of mind is one of being taken up into the heart of the divine pathos. Sympathy is the prophet's answer to inspiration, the correlative to revelation ... The prophet hears God's voice and feels His heart. He tries to impart the pathos of the message together with its logos.[58]

On this analysis I want to make the following observations. First, by the divine pathos Heschel means God's pathos over the state or condition of Israel at any given point. That pathos is portrayed programmatically in Exod 2:23–25, where we read,

> in the course of those many days the king of Egypt died. And the people of Israel groaned under their bondage, and cried out for help, and their cry under bondage came up to God. And God heard their groaning, and God remembered his covenant with Abraham, with Isaac, and with Jacob. And God saw the people of Israel, and God knew.

The sequence—if we may put it that way—is hearing, remembering, seeing, and knowing—verbs that match, point for point, the sequence, "groaned, cried out, cry, groaning." The pathos consists in a divine sympathy with the pathos of the people, and the prophetic consciousness arises as an

---

56. Consider again the sentences quoted earlier from Coleridge's essay "On Poesy or Art."

57. Heschel, *Prophets*, xiii (italics original). Heschel's approach through a phenomenology of consciousness, is similar to that of Coleridge, who argued—above all, in *Aids to Reflection*—that the evidences for the experience of God lie primarily within, open to examination by sensitive, disciplined reflection upon the receptive passivities and creative activities of the human spirit.

58. Heschel, *Prophets*, 26 (italics original).

Toward a Hermeneutics of Resonance

awareness of the divine pathos over the people's plight. Secondly, the first divine perceptual act is one of hearing. Similarly, if Moses is first attracted by the *sight* of the burning bush, it is when he *hears* a voice speak out of its midst that he becomes aware (3:7–8) of what the narrator had reported in 2:23–25 as the divine pathos. This sequence of hearing and seeing occurs repeatedly in the reports of prophetic experience. It lies also at the heart of the encounter of Job with God in and through the phenomena of the whirlwind, the ensuing rains, and the vibrant responsive life of nature epitomized by the lightnings that say to God, *hinnenu*, "here we are!" In other words, Job's sense of God's response to his plight comes through his re-awakened sense of the natural world as alive with a life animated by God's creative call. (Is the whirlwind a Joban version of Coleridge's "intellectual breeze"?!) Indeed, Job's first awareness of God comes hand in hand with his hearing of a transcendent choir as "the morning stars sang together, and all the children of God shouted for joy" (Job 38:7). When one compares this choir, and the lightnings' "here we are!" with the choral praise of God by all the elements of creation, in Psalm 148, one at the least can identify some of the biblical precedents for Coleridge's Eolian Harp passage, as well as for what Coleridge writes to young Sotheby.

But does the attempt to take such a poetic sensibility with epistemological seriousness not commit the pathetic fallacy? From the point of view of a sensibility weaned on generations of materialistic epistemology, it does indeed. The question is whether such a sensibility is stunted, and needs to be awakened to the deep resonance in the nature of things. Before I turn to the testimony of Whitehead and Loewald on this issue, I want to attend to one more celebrated passage in which Coleridge characterizes what it is that the poet does through creative activity involving genuinely "esemplastic imagination." Here it is:

> The poet, described in *ideal* perfection, brings the whole soul of man into activity, with the subordination of its faculties to each other, according to their relative worth and dignity. He diffuses a tone, and spirit of unity, that blends and (as it were) *fuses* each into each, by that synthetic and magical power, to which we have exclusively appropriated the name of imagination. The power, first put into action by the will and understanding, and retained under their irremissive, though gentle and unnoticed, control (*laxis effertur habenis*)[59] reveals itself in the balance or

---

59. The Latin here means "carried on with slackened reins." Our equestrian daughter tells me that the aim of the rider is to guide the horse through its jumps and paces in such a fashion that the guidance escapes detection by the judge.

> reconciliation of opposite or discordant qualities: of sameness, with difference; of the general, with the concrete; the idea, with the image; the individual, with the representative; the sense of novelty and freshness, with old and familiar objects; a more than usual state of emotion, with more than usual order; judgment ever awake and steady self-possession, with enthusiasm and feeling profound or vehement; and while it blends and harmonizes the natural and the artificial, still subordinates art to nature; the manner to the matter; and our admiration of the poet to our sympathy with the poetry.[60]

The passage has been subjected to repeated analysis.[61] Here I wish to draw attention to the following words and phrases: "in ideal perfection . . . the whole soul of man . . . diffuses a tone and spirit of unity . . . blends and fuses . . . synthetic and magical power . . . of imagination . . . in the balance or reconciliation of opposite or discordant qualities . . . * . . . blends and harmonizes." At the asterisk (*) I would insert a list of the sorts of "opposite or discordant qualities" so blended and fused, so balanced or reconciled, so *harmonized*, by a *tone and spirit of unity* that gives expression to the balanced or reconciled, harmonized, "whole soul" of the "poet in *ideal* perfection"—what nowadays we call the "implied author." For, we may note, the "tone and spirit of unity" achieved in and through the poetry is of a piece with the activity within the poet's own soul, as the poet "brings the whole soul . . . into activity, with the subordination of its faculties to each other, according to their relative worth and dignity." The unifying of the "opposite or discordant qualities" in the poem is the objective correlative to an action interior to the poet where analogous "opposite or discordant qualities" are likewise unified. This action, I suggest, may be understood in terms of Loewald's analysis of sublimation as I discuss it below. By "implied author," in this setting, we may understand the poet's momentary arrival at a hitherto unattained level or degree of integration or individuation as a person, from which of course the actual author may fall back upon emergence from the relatively unusual psychic state within which truly creative perception arises and formative creation issues. The "implied reader," then, is the actual reader who not only takes the words on the page in their conventional, literal meanings, but through an answering act of readerly imagination catches their "tone and spirit," the interactive resonances through which they are experienced as "living

---

60. Coleridge, *Biographia Literaria*, 2:15–17 (italics original).
61. See comparatively briefly, e.g., Richards, *The Portable Coleridge*, 45–54.

members" of one poetic body. In this way, and to this degree, the poem, as the expression of a sublimating moment in the life of the poet, can become a means (in terms of sacramental theology, an efficacious sign) to an incremental—and at times utterly transformative—moment of sublimation in the reader's personal history.

When I was a child, I was given a cheap pocket watch. Fascinated by its moving parts, I took it apart. To my dismay, I was unable to put it back together—the balance wheel, and the spring that drove it, resisted all the efforts of my fingers. In graduate school I learned to take the Bible apart into innumerable pieces. The activity was fascinating. The individual parts took on a new luminescent meaning in and for themselves. I shall be forever grateful for that training. But it has been through long continuing immersion in these texts, and gradual attunement to the resonances that arise between them—sometimes harmonious, sometimes dissonant—that I have arrived at the kinds of studies represented in the preceding and following forays into specific aspects of the biblical world. I may link these last remarks back to Miller's discussion of resonance between not only individual passages in Scripture, but also whole books, such as Deuteronomy and the Psalms. And I suggest that we may take the long passage just quoted from Coleridge as providing a model for conceiving of the diverse biblical texts as displaying a "tone and spirit of unity" in and through their various "opposite or discordant qualities."

Resonance and Alfred North Whitehead's Di-Polar Cosmology

Starting with *Science and the Modern World*, published in 1925, Alfred North Whitehead's philosophical writing may be taken as, among other things, a persistent critique of the philosophical naturalism or materialism at the foundation of modern science and an attempt to re-establish that foundation on a broader base that re-habilitates final causation, or "aim," as inherent in the nature of things and subjective interiority as an intrinsic dimension in all that exists. In respect to the last-mentioned point, Whitehead could go so far as to say, "the reformed subjectivist principle must be repeated: that apart from the experiences of subjects there is nothing, nothing, nothing, bare nothingness."[62] Grammatically speaking, the "of" in this statement functions both subjectively and objectively, so that the phrase, "the experiences of subjects," carries a double meaning:

62. Whitehead, *Process and Reality*, 254.

PART FOUR: An Interlude

(1) the experiences that a subject has, and (2) that of which a subject has experiences—that is, other subjects. This is to say that, for Whitehead, the objects that form the content of one's subjective experience are never bare, "vacuous" objects, but are objects that, considered in their own right, have an interior, a subjectivity, of their own.[63] Whitehead's radically temporal version of the "great chain of being," expressed in what he called his "critique of pure feeling," can be stated as a "great chain of feeling": The objects of a given subjective experience are themselves subjects having had experience of other objects that were themselves subjects having had experience of . . . and so on and so on. Early in the chapter that ends with his "reformed subjectivist principle" as quoted above, he puts it that

> The primitive form of physical experience is emotional—blind emotion—received as felt elsewhere in another occasion and conformally appropriated as a subjective passion. In the language appropriate to the higher stages of experience, the primitive element is *sympathy*, that is, feeling the feeling *in* another and feeling conformally *with* another.[64]

A page later, he writes,

> the primitive experience is emotional feelings . . . In the phraseology of physics, this primitive experience is "vector feeling," that is to say, feeling from a beyond which is determinate and pointing to a beyond which is to be determined. But the feeling is subjectively rooted in the immediacy of the present occasion: it is what the occasion feels for itself, as derived from the past and as merging into the future.[65]

With these statements one may compare Heschel's analysis of the prophetic consciousness as marked by a sympathy with the divine pathos which in turn arises in response to the perception of Israel's situation. In respect to the narrative in Exod 2:23—3:9, Moses becomes sympathetically aware of the divine pathos for the plight of Israel and drawn into the divine intention to deliver Israel.

One way of characterizing Whitehead's metaphysics is to say that it is "di-polar." By this, he means that there is no such thing as merely

---

63. To my knowledge, Whitehead never employed the older distinction between *natura naturata* and *natura naturans*, but his various discussions of objects and subjects resonate (!) with that distinction, especially his final discussion in the last two chapters of *Modes of Thought*.

64. Whitehead, *Process and Reality*, 246.

65. Ibid., 247.

"vacuous actuality," actuality with no interior dimension. All actuality displays a mental as well as a physical pole. The mental pole may not be conscious; indeed, at the most elementary levels it can only be posited. In this connection, he makes an interesting comment on language as a means of expression and communication. As he says, we might have resorted primarily to gestures and other visible signals rather than to sounds; but the problem there is that while one is gesturing one's hands are unavailable for other activity, whereas one can talk (or whistle) while one works. "But," he writes,

> there is a deeper reason for the unconscious recourse to sound production. Hands and arms . . . do not excite the intimacies of bodily existence. Whereas in the production of sound, the lungs and throat are brought into play. So that in speech, while a superficial, manageable expression is diffused, yet the sense of the vague intimacies of organic existence is also excited. Thus voice-produced sound is a natural symbol for the deep experiences of organic existence.[66]

Can one say that the "deep experiences of organic existence" are exemplified in the fact that all human (and animal) life arises within the body of the mother, a body that itself pulsates with a variety of resonances: neurological, muscular, circulatory, emotional, and so on? That one's capacity for sympathetic awareness has its origins in part in one's pre-natal awareness of these resonances within this particular, primal organic "cave of resonant signification"?[67] The point here is that we humans know first hand of subjective interiority in at least one part of the cosmos—namely, in the instance of our own existence. And on that analogy it is not difficult to recognize subjective interiority within other "higher" species. The question is on what basis we assert the absence of such subjective interiority—however attenuated, however strange and imperceptible to us—on any level of existence.

---

66. Whitehead, *Modes of Thought*, 32. This passage recalls Brower's observation, quoted above as per Hays (*Echoes of Scripture*, 18) and repeated here, that "through allusion, often in combination with subdued metaphors and exquisite images, Pope gets his purchase on larger meanings and evokes the finer resonances by which poetry (in [Samuel] Johnson's phrase) 'penetrates the recesses of the mind.'"

67. Recall the imagery of Ps 90:1–2, where it is the God who "brought forth" the mountains and "formed" the earth and the world (both verbs relating, in literal contexts, to giving birth) who is "our dwelling place [the word elsewhere refers to a lion's den or cave] to all generations."

PART FOUR: An Interlude

In his last book, *Modes of Thought*, Whitehead recapitulated the argument of the first six chapters with a two-chapter discussion under the general heading, "Nature and Life." In a chapter with the title, "Nature Lifeless,"[68] he offered a final critique of scientific materialism; he then expounded what he took to be a more adequate account in his final chapter, entitled, "Nature Alive." At the end of the latter chapter, in a succinct restatement of his di-polar metaphysics, he passes from a reference to "patterns of activity studied by physicists and chemists," where "mentality is merely latent," to "inorganic nature" in which "any sporadic flashes are inoperative so far as our powers of discernment are concerned," to "the variety of grades of effectiveness of mentality" in "the social habits of animals," arriving "finally in the higher mammals" and "more particularly in mankind" where "we have clear evidence of mentality habitually effective." Vis-à-vis the energetic activity observable in the physical world, he writes, "it is not necessary to assume that conceptions introduce additional sources of measurable energy . . . [T]he operation of mentality is primarily to be conceived as a diversion of the flow of energy." He concludes with a note on the possibility of constructing a "systematic metaphysical cosmology" based on such a view of things, and writes, "the key notion from which such construction should start is that the *energetic activity* considered in physics is the *emotional intensity* entertained in life."[69] What he here calls "energetic activity" he refers to repeatedly in *Process and Reality* as "vibration" or "vibratory activity," at times also as "emotion" (as in one passage cited above). In such an account of the nature of things, resonance in texts, as "penetrating the recesses of the mind," is not simply an inter-human affair, but touches, however vaguely and faintly, yet hauntingly, the recesses of all that exists.

But Whitehead's "radical empiricism," as going deeper than the evidence of the five senses, extends along an imaginative axis that makes suggestive connections with Coleridge's doctrine of imagination as the human "organ" of perception and creativity in relation to God. He gives

68. In the chapter by this title, he writes, "Our first step must be to define the term *nature* as here used. Nature, in these chapters, means the world as interpreted by reliance on clear and distinct sensory experiences, visual, auditory, and tactile" (Whitehead, *Modes of Thought*, 128) (Shades of Coleridge's three sensory modes, in his letter to Sotheby!) Throughout his philosophical corpus, Whitehead critiques such sense-empiricism as a superficial basis for full understanding of nature, insofar as it abstracts from those deeper modes of experience that he variously refers to as "intuition," "visceral feeling," and so on. See especially Whitehead, *Symbolism*, chap. 2 (30–59) and esp. section 4 (43–49).

69. The quotations come from *Modes of Thought*, 167–68 (italics added).

notice of this already in *Science and the Modern World* where, in the chapter on "The Romantic Reaction," he quotes from Wordsworth's "Elegaic Stanzas." The poem centers on a painting of Sir George Beaumont, "Peel's Castle," in which, below a cliff crowned by a ruined castle, a ship labors under heavy seas and a dark sky. Recalling that actual place, visited in more tranquil times, Wordsworth writes, "Ah, THEN, if mine had been the Painter's hand, / To express when then I saw; and add the gleam, / the light that never was, on sea or land, / The consecration, and the Poet's dream."[70] Whitehead, who quotes the third line in *Science and the Modern World*,[71] writes of human consciousness in *Process and Reality* as "ris[ing] to the peak of free imagination, in which the conceptual novelties search through a universe in which they are not datively exemplified."[72] By this he means that the "novelties" entertained in imagination ("the light that never was, on sea or land") do not (yet) have their counterpart in the world as data perceivable through the five senses. In Whitehead's view, such "novelties" have their origin in God who is the ultimate source of all creaturely creativity. One can imagine an interesting conversation between him and the Coleridge who wrote chapter 13 of the *Biographia Literaria*.

Rupert Sheldrake on Morphic Resonance

At this point I want to interject a brief note on Rupert Sheldrake's remarkable monograph, *The Presence of the Past: Morphic Resonance and the Habits of Nature*. To my knowledge, this is the most exhaustive, many-dimensioned treatment of resonance as a fundamental aspect of the natural world, both differentiating and unitive, manifest at all levels from physical energy through chemical interactions and the formation of crystals, and on "up" through biology, psychology, cultural forms and norms, and the mysterious aims and processes of the human psyche. In a nutshell, Sheldrake works with the notion, increasingly at home in various fields of analysis, that in place of unchanging bits of matter that then hurry and scurry about in a four-dimensional space-time "box," we should think of "rhythmic patterns of activity." These patterns ("morphs") are constitutive of acoustic resonance in music, electro-magnetic resonance (as in radio transmission on a given frequency), electron-spin resonance in physics,

---

70. William Wordsworth, "Elegaic Stanzas Suggested by a Picture of Peel Castle in a Storm Painted by Sir George Beaumont."

71. Whitehead, *Science and the Modern World*, 87.

72. Whitehead, *Process and Reality*, 245.

and so on. For Sheldrake, "[u]nlike these kinds of resonance, morphic resonance does not involve a transfer of energy from one system to another, but rather a non-energetic transfer of information."[73] He goes on to say later that

> [a]ccording to the hypothesis of formative causation morphic resonance occurs between such rhythmic structures of activity on the basis of similarity, and through this resonance past patterns of activity influence the fields of subsequent similar systems. Morphic resonance involves a kind of action at a distance in both space and time. The hypothesis assumes that this influence does not decline with distance in space and time.[74]

The purpose of the present interjected note is simply to draw attention to Sheldrake's study as a further resource—along with its ample reference to current work in related fields[75]—for reflection on the nature, function, and range of resonance as a potential key to a hermeneutics of texts that may itself be grounded in a "universal hermeneutics." Such a perspective would help to deepen our appreciation of the significance of approaches to the biblical text informed by the methods of form criticism and tradition history. For example, the identification of form-critical and tradition-history connections between textual representations of the call of Moses, Samuel, Isaiah, Jeremiah, and others need not remain at the level of "textual constructs," but reflect actual existential connections in which earlier call-experiences lay down a meaning-template into which later persons are drawn through their call-experiences and which in turn is deepened, elaborated, or otherwise modified by those latter experiences. "Others have labored, and you have entered into their labor" (John 4:38).

Hans Loewald on Resonance in Nature and in Human Becoming

In a preface to his *Papers on Psychoanalysis*, first published in 1980, Loewald writes, "Philosophy has been my first love. I gladly affirm its influence on

73. Sheldrake, *Presence of the Past*, 108.
74. Ibid., 109.
75. See, for example, Sheldrake's discussion (ibid., 303–5) of his concept of morphic resonance in relation to the physicist David Bohm's proposal that the classical model of the physical world, as an "explicate order" in space-time, be itself grounded in a deeper "implicate order"; and Bohm's own comments on Sheldrake, quoted in ibid., 305–6.

my way of thinking while being wary of the peculiar excesses a philosophical bent tends to entail."[76] While his papers are devoted largely to an understanding of dynamics and structures within the individual psyche, as informed by his deep immersion in the Freudian canon and his own clinical practice, one has the sense that this sustained focus has always in peripheral vision the wider cultural scene and the nature of things generally. In two late monographs and a remarkable final short companion essay, this peripheral vision gradually emerges into fuller view. As I hope to show, the interconnection between this endopsychic and these exopsychic realms at its living core is a resonant one.

1. I shall begin with Loewald's final essay, "Psychoanalysis in Search of Nature: Thoughts on Metapsychology, 'Metaphysics,' Projection" (1988). Observing that "older philosophers distinguished *natura naturans* (nature as active process) from *natura naturata* (nature as the assemblage of created objective entities)," and that, with Descartes, modern science "distinguished *res cogitans* from *res extensa* (the germ [respectively] of psychic and material reality),"[77] Loewald identfies such a distinction, and the resulting model of nature as a mechanism devoid of an "interior" life as a hermeneutic construct,[78] a construct that enables humans to reduce nature to the status of "simply an object of observation and domination by a human conscious mind, a subject."[79] But the recognition of the reality of the unconscious, and of unconscious mentation, involving instinctual dynamics embedded in organic processes, calls for "a deeper understanding of nature" that "will widen the horizon of a science of nature."[80] Such a deepening and widening will involve the recovery of a sense of *natura naturans*, of a living nature, of which human nature and human subjective consciousness are only an expression, albeit—from a human perspective— an eminent expression. The gist of Loewald's compact essay, as it bears on my discussion, may be presented in the following quotation: "Cosmic life evolves and contains human life, and human life as a microcosm can, in the mirroring of *conscire*, of consciousness, grasp something of the dynamics—the psyche—of nature."[81]

76. Hans Loewald, *Papers on Psychoanalysis*, reprinted in *The Essential Loewald*, xlii.
77. Loewald, "Psychoanalysis in Search of Nature," 50.
78. Ibid., 49.
79. Ibid., 50.
80. Ibid., 51.
81. Ibid., 49.

PART FOUR: An Interlude

If humankind may be taken as an inherent aspect of the natural world, then we know that in at least one instance—our individual selves, and others understood by analogy with ourselves—nature displays an active, subjective interior. It is reasonable to ask whether we alone have such an interior, or whether it is a dimension of all that exists. On the interpersonal plane, Loewald, observes, the psychic action of "projection" is "based on an obscure knowledge ... of the correspondence, the *resonance*, between endopsychic unconscious processes and unconscious processes of other individuals."[82] The notion of non-human nature as purely objective, and devoid of dynamic, resonant interiority, may be a projection onto that realm of our own defensive reactions generated in the first instance to defend against unconscious processes at the hands of which we do not wish to be vulnerable. In other words, the desire to dominate and control for our own conscious purposes may lie at the basis of the modern model of nature as a lifeless machine. On the contrary,

> [n]ature is ... an all-embracing activity of which man, and the human mind in its unconscious and sometimes conscious aspects, is one element or configuration ... By virtue of the *unison* and *reverberation* with the rest of nature we gain what understanding, including our own, we possess ... *Unison* and *reverberation*, as regards other human beings, is called *empathy*. But it would be erroneous to assume that this *empathic resonance* stops at the frontier of human mentality. Our knowledge of organic and so-called [!] inorganic nature is likely to derive from similar *attunements*.[83]

I note here the rapport between Loewald's "empathic resonance" and Whitehead's analysis of imaginative feeling as connecting emergent entities with one another in a universe in which, unendingly, "the many become one, and are increased by one." I note also how his identification of "unison" and "reverberation/resonance" accords with Coleridge's understanding of the unifying effect of the poetic imagination working on the objects of its perceptive attention. I venture the further suggestion that Loewald's understanding of empathic resonance, as an activity operative primarily at levels below human consciousness, may illuminate Coleridge's comments to young Sotheby concerning the poetic requisite of a hearing more attuned than the ear, a seeing more focused than the eye, a

---

82. Ibid., 53 (italics added). We may note the affinity of this statement with Coleridge's observation that "we unconsciously imitate those whom we love."

83. Ibid., 50 (italics added).

touch more tactful than the hand. A "postmodern science" that recovers a sense of *natura naturans,*

> appears to convey a new insight: we understand something about nature and reality, know something about them, by being open to their workings in us and the rest of nature as unconscious life, the openness being what we call consciousness. This insight—also very ancient—transcends, subsumes into itself the antagonism of "ego" and "reality."[84]

Over against a natural science that—hermeneutically rendering nature in merely objective terms and focusing on human nature primarily in terms of "(individualistic) consciousness"—"appears to enhance man's dominion of the world,"

> a deeper understanding of nature will widen the horizon of a science of nature and increase, one may hope, its power of mastery, a mastery that involves yielding no less than dominion.[85]

In respect to biblical interpretation, I suggest that this calls for a restoration of the balance between the eye and the ear: not only the analytic eye that subjects the text to its canons of evidence and intelligibility, but the ear that hears Scripture—and the resonances between its texts—under the rubric of the Deuteronomic *Shema*, "*Hear*, O Israel!" As Whitehead has said, it is through hearing that the excitations of "the vague intimacies of organic existence" are communicated from one to another.

2. Loewald's papers collected in the 1980 volume spanned a publication period from 1951 to 1979. He drew the fruits of these individual studies into the small monograph, *Psychoanalysis and the History of the Individual* (1978). As he says, psychoanalysis gives special attention to the individual's historicity: "The dimension of time plays an ever-increasing part in man's attempts to organize, master, and understand reality . . . This trend is connected with a deep modern interest in the nature of reality as process—in contrast to a static, substantive view—and with a pervasive tendency to understand what appears permanent and definitively structured in terms of becoming, that is, to reconstruct structures."[86] In such

---

84. Ibid..
85. Ibid., 51.
86. Loewald, *Psychoanalysis and the History of the Individual*, 535. In Janzen, "Qohelet on Existence 'Under the Sun,'" I have explored the theme of futility expressed in Ecclesiastes in connection with the iron rule of human time under Persian dominion with its overarching icon of the sun-disk, and I have suggested that the

PART FOUR: An Interlude

a frame of reference, Loewald analyzes the relation between the id, ego and superego as a temporal relation:[87] The id, with its inorganic, organic, and psychic energies, instincts, and appetites, he associates with one's past; the superego, which begins as the awareness of an ego ideal in the form of a parental or other formative "other" and then through internalization becomes one's own ideal ego, he associates with the future; and the ego, as organizing the energies, instincts and appetites of the id under the unifying vision of the superego, he associates with the present—that presence which, phenomenologically, William James described as "a blooming, buzzing confusion," a confusion as a turbulent interaction of diverse energies, but also as a con-fusing, a fusing together into single rich, complex "work of art." So viewed, the activity of the ego is one of sublimation, which Loewald will later define simply as "passion transformed." I note a profound congeniality of conception here with Coleridge's characterization of "the poet described in *ideal* perfection," who "brings the whole soul of man into activity, with the subordination of its faculties to each other, according to their relative worth and dignity."[88] Such poetic activity begins in an act of perceptive imagination—Loewald's "openness" to nature that "involves yielding" to its resonant activities; and it ends in an act of imaginative reshaping—the poem arising as a symbolic vehicle of the transforming of passion toward the ideal ego (the implied author) of the poet. And I note Coleridge's characterization of the imagination as achieving its unifying work by drawing together disparate elements under the "gentle and unnoticed control" (as of a horse managed *laxis effertur habenis*, "with slackened reins") of the will and the understanding.

In such a vision of sublimation (to be distinguished, finely but critically, from repression) the ideals of the superego and the vital energies of

---

acknowledgement in that book of God's brief gifts of joy represents the fleeting irruption of relief from that oppressive dominion—joy that I will now call momentary experiences of God's eternity. In light of my last chapter of the present book, which studies Eph 5:16, I might call Qohelet's willingness to accept and embrace these fleeting joys despite his prevailing sense of time's futility, as a willingness, in those momentary experiences, to "dance in time to the music of eternity." Meanwhile, I draw attention to Loewald's remarks in "Psychoanalysis in Search of Nature," which I have just reviewed, where he proposes that a more adequate, insightful understanding of nature may issue in an exercise of "mastery" that "involves yielding no less than dominion." Such a "mastery" would involve becoming more sensitive, more resonantly "attuned" to the deep tempos of the natural world and, deeper still, to the "times and seasons" of God's creating, sustaining and redeeming activity in and through and for the world.

87. Loewald, *Psychoanalysis and the History of the Individual*, 546–47.
88. Coleridge, *Biographia Literaria*, 2:15–17 (italics original).

Toward a Hermeneutics of Resonance

the id interact through the symbol-forming activities of the imagination, activities expressed in all the arts but, for Loewald, supremely through language. In this symbolic interaction, each can become a symbol, or metaphor, of the other: The vital energies of the id, in serving as vehicles for the expression of the aspirations of the superego, thereby are drawn into the shaping and unifying vision of the latter's ego ideals, to contribute their tone and color and lively energy to those ideals as realizing; while, conversely, the forms in and through which the ideals of the superego are entertained in-form the raw, turbulent energies of the id to channel those energies. Not to repress them? Not even just to suppress them? Such misguided actions lead to a flattening of affect and a desiccation of spirit, of form without life. As Loewald insists, sublimation involves a *channeling* of the vital energies of the id in the service of the superego, and a corresponding reception by the superego of those vital energies into itself. As an example, Loewald refers to the way in which, in the Middle Ages, mystic vision is sometimes expressed through erotic imagery, while sexual activity is at times experienced and spoken of in the exalted terms of religion.[89]

I realize, at this point, that in sketching Loewald's discussion of the psychoanalytic history of the individual as it pertains to sublimation, I have expanded on its latent forms with the help of his fuller discussion in the 1988 monograph, *Sublimation*. I shall turn to that monograph in a moment, where he again speaks of the key phenomenon of resonance. But first I want to comment on the chapter, "Comments on Religious Experience," in Lowewald's 1978 monograph. Noting Freud's negative assessment of religion as "illusion," and as symptomatic of childish wishes for "all-powerful parents" against "the heartache and the thousand natural shocks that flesh is heir to," Loewald offers his comments as a "tentative and fragmentary ... step beyond what psychoanalysis has contributed so far to the understanding—and misunderstanding—of religion."[90] He takes as his point of departure the conversation between Freud and Romain Rol-

---

89. For a discussion of the relation between sex and the sacred informed by deep immersion in the romantic tradition as filtered through the clear-eyed vision and concern of a mother of teenage children, see Wallace, *For Fidelity*, 99–100, and esp. 160–61 n. 1, where, riffing on the term Greek Patristic Trinitarian term "perichoresis" to refer to the "dancing" relation within God, she writes, "The coinherence of intimacy, desire, and blessing within an appropriate sexual union is also a dance, both spiritually and in a vivid, literal, erotic, physical way." In the terms that I shall explore more fully in the final chapter of the present volume, such an "appropriate sexual union" is a particularly resonant form of the human possibility of "a dance in time to the music of eternity."

90. Loewald, *Psychoanalysis and the History of the Individual*, 566.

PART FOUR: An Interlude

land on what they called "oceanic feeling," which Freud, who could never discover this feeling in himself, nevertheless described as "a feeling of an indissoluble bond, of being at one with the external world as a whole."[91] In comparison with this feeling, which he associated with the "unity of the infant-mother psychic matrix," Freud described a more developed and clearly demarcated "ego-feeling" as "only a shrunken residue." With this we may compare these lines in Wordsworth's Ode, "Intimations of Immortality from Recollections of Early Childhood": "Heaven lies about us in our infancy! Shades of the prison-house begin to close / Upon the growing boy." until / "At length the Man perceives it [the vision splendid] die away, / And fades into the light of common day." The light of common day is for Freud, of course, the light of enlightenment rationality and modern science as confronting us with reality. But, for Loewald, this

> later development . . . limits and impoverishes . . . unless it is brought back into coordination and communication with those modes of experience that remain their living source and perhaps their ultimate destination. It is not a foregone conclusion that man's objectifying mentation is, or should be, an ultimate end rather than a component and intermediate phase of vital significance to us.[92]

With these remarks we are ready to take up the topic of resonance in sublimation, in the symbols, (especially linguistic) in and through which sublimation is enacted, as well as in religious sensibility.

3. The body is the immediate organic locus of the activity of the id or unconscious. As we have seen, Loewald in his final short meta-essay proposes to take the resonant activities of the human unconscious as participating in the larger realm of nature as *natura naturans* a dynamic field of activity itself in some degree or other as displaying unconscious mentation or at least some dim analogue thereto.[93]

91. Ibid., 568.

92. Ibid., 569. Readers of Owen Barfield will detect here an analog to his evolutionary and developmental movement from Original Participation through an intermediate phase of detachment and alienation to Final Participation. Compare also Paul Ricoeur's movement from first naiveté through critical consciousness to second naiveté.

93. The paper was originally announced as forthcoming in a footnote at the end of Loewald's monograph, *Sublimation*. For some unaccountable reason the paper was not included in *The Essential Loewald*. I take the paper to be "essential" to an appreciation of the *Tendenz* of Loewald's latest thinking along the lines of his first love, philosophy.

Toward a Hermeneutics of Resonance

For Freud, the energies, instincts, and appetites of the body, as subjectively experienced through the activities of the id, come to most vivid focus in the sexual drive but are expressed already in the oral, tactile sensory, and anal exploratory outreaches of earliest infancy. A delightful and, to my mind, persuasive commonsense version, for those dubious of Freud's theory on this point, is given in Catherine Wallace's book, *For Fidelity*, under the heading, "Wholistic Views of Sexuality: Sigmund Freud."[94] Dubious herself about Freud's argument until she had children of her own, she couldn't help noting her infants' sensual pleasure in interacting with the concrete textures of matter, "gleefully smoosh[ing] peas or curds of cottage cheese one by one by one." She describes her toddler son who "would gently stroke the new growth on the arborvitae at the park and enjoy its fragrance. I could almost hear the tree purring in response. For years we could not walk by that hedge without pausing for the solemn exchange of greetings between boy and tree." She concludes,

> It doesn't demand uncommon empathic gifts to intuit the as yet unconscious and undeveloped origins of such intense delight in all the beauties of creation. Freud's hostility to religion notwithstanding, the world is indeed holy, and we are born in tune with it.
>
> Life with babies illustrated for me how true it is that sexuality underlies all sensuality of any kind, and sensuality is merely the experience of being bodies . . . That's why thousands of people stood five abreast in line in the August sun for hour hours to get into the recent Monet exhibit at the Art Institute of Chicago . . . All such delight, however deeply matured (or "sublimated") from the baby's diffuse sexuality, nonetheless still *resonates* all the way down to our deepest biological and psychological drives. Sexual pleasure per se, sexual desire as such, is merely the focusing upon reproductive function of the central delight we have in being alive, the joy of bodies that are alive and well and loose in the cosmos.[95]

---

94. Earlier in this book, Wallace narrates how, taking a class of students through Descartes' arrival at his (in)famous and fateful doctrine of mind/matter dualism by discovering that he could doubt the existence of his body but not of his mind, she—so pregnant at the time that she was unable to hold Descartes' text unless she rested her arms on the sides of her distended belly—suddenly felt the intrauterine living presence give a mighty kick, and realized that Descartes' capacity to doubt that he had a body was because he'd never been pregnant! See Wallace, *For Fidelity*, 30–31.

95. Wallace, *For Fidelity*, 34–35 (italics added).

PART FOUR: An Interlude

In this passage, Wallace captures perfectly Loewald's concern for the two-way traffic, conducted through the symbolic (and supremely through the linguistic) activities of the psyche, between the basic organic and instinctual energies and the so-called higher ideals and aspirations, values, and norms that give human existence its meaning and worth. In a following discussion, under the heading, "The Corruption of Wholistic Views: Hedonism," Wallace gives a salutary critique of what some cultural critics in the pages of Arts sections of at least one national newspaper are fond of calling, and lauding as, "transgressive" sensibilities. Wallace's overall argument is that sexual fidelity is not simply a matter of conformity to (increasingly outmoded) social proprieties, let alone obedience to religious or biblical mandates, but also "a practice *intrinsic* to the happiness of a happy marriage."[96] She illustrates what she means by "intrinsic" by noting that it is quicker and easier to make a simulacrum of cinnamon rolls by using biscuit dough. The latter taste okay in their own way; they are even popular with her family. But they are not cinnamon rolls. "Real cinnamon rolls, made properly with yeast, have a slightly chewy texture because the yeast interacts with the warm milk and the wheat protein in complex ways. Baking powder doesn't do that." She concludes, "Marriage requires sexual fidelity in the same way. Like yeast, fidelity is a growing, living thing that interacts with and transforms and reorganizes all the other ingredients of the relationship."[97] This sounds like Wallace-the-Coleridge-scholar describing fidelity in a marriage in the spirit of Coleridge's poet-in-ideal-perfection bringing his or her whole soul into the activity of unifying the disparate elements that will go into the making up of a poem. It is a way of describing how the erotic energy and color and delight of sexual love interacts with the aspirational ideal of monogamy to work a genuinely sublimating (and not repressive or prudishly dampening) alchemy in both members of such a relationship. This apparent digression into Wallace sets the table for my summary of Loewald's comments on the presence and possibly religious implications of resonance in the more advanced stages of sublimation.

Loewald begins with the observation, familiar to all, of how any bodily appetite may sleep, then is aroused by a sense of the absence of what it desires, and the arousal grows in intensity until it moves from a phase of initial pleasure—a pleasant itch—to a phase in which it becomes so insistent that, unless satisfied, it becomes unpleasurable. The satisfaction

96. Ibid., 13.
97. Ibid., 16.

comes as an orgastic climax of sheer ecstasy (think, even, simply of the first morsel of a favorite food exquisitely prepared as one's mother used to make it—say, lemon pie under sculpted mountains of golden meringue) before satiation, or "satisfaction," sets in to dull the appetite and it dies away. Loewald examines this phenomenon with his characteristically unhurried attention to detail and nuance, in the process noting how Freud's "death instinct" is connected thematically to the prevalence, in the literature and poetics of erotic experience, to the figuration of the post-orgastic state as a "dying." But then Loewald writes,

> One must consider that "satisfaction" . . . may be achieved by means other than elimination (or reduction) of a state of tension or stimulation. If there is such a thing as a *life instinct*, its "aim" would be satisfaction through the attainment of higher, more differentiated *unities*, in which tension is not eliminated but "*bound*"—satisfaction of a different kind.

(Think, here, of Coleridge's poet who achieves a "unity" between the elements of the poem that is not lost with the completion of the poem, but that persists in the poem for the attuned reader or hearer however often the poem is fitly read. A personal example: However often I read Wordsworth's Immortality Ode, I dare not read it aloud in public for fear of breaking down on attempting to read certain lines.)

In exploring the nature of this "satisfaction of a different kind," Loewald draws on D. W. Winnicott's work in the realm of the mother-infant matrix. Noting Winnicott's identification, in that relational matrix, of "phenomena that have no climax," he speaks of it as "an organization of excitation [he will call it "binding"] . . . in which discharge is not an essential element in the attainment of pleasure."[98] He writes,

> The ultimate aim of such binding of tension is not satisfaction in the form of the removal of the state of stimulation by discharge on or through the object . . . Is it perpetuation of tension, of the excitation inherent in living substance? One may speculate (and recent theories of physical science point in that direction) that inorganic matter and its constituent molecules, atoms, and so on, manifest a relatively permanent phase or form of bound energy . . . [I]n individual instinctual life the aim of perpetuation of tension by transindividual and endopsychic binding would fit

---

98. Loewald, *Sublimation*, 469.

better with the human wish for immortality than does the single aim of death and a complete state of rest.[99]

It is here that Loewald finally introduces the notion of "resonance." By this term I take him to characterize the quality of the satisfaction involved in non-climactic, steady-state excitation. As he writes,

> "Stimulation" should not be conceived exclusively as increase of the quantity of excitation; in many instances it can be thought of as *resonance with the "wave length" of a neighboring system.* Stimulation can result from an openness to the latter's level of binding or organization of activity. The neighboring system ... is simultaneously open to, in tune with, the "receiving" system. Stimulation, in such resonance, would consist in qualitative change of some kind ... As Freud said of the qualitative characteristic in pleasure and unpleasure, so is resonance a phenomenon implicated with rhythm, with time.[100]

Here Loewald trenches on his discussions elsewhere regarding the difference between subjective states marked by a sense of the passage of time and those states marked by a sense of "eternity" or "sempiternity" (everlastingness). The present discussion suggests that the latter sense arises, and seems "timeless" (in the ordinary sense), to the degree that the occasion is marked by a steady-state resonance. (One might call it a self-prolonging *nunc stans*, or "a dance in time to the music of eternity.") Loewald ends this section with the conclusion that "sublimation belongs in the area of ego development and of internalization as distinguished from defense. I conceptualize its dynamic quality broadly as *reconciliation*."[101]

There is not space here to go into Loewald's rich chapter on "Symbolism." I note only his developed point that the symbolic "arrow" of

---

99. Ibid. In so characterizing "rest," Loewald here overlooks that two bodies traveling through space-time "in sync" with one another are "at rest" vis-à-vis one another. Thus the biblical image in which one may find oneself in a state of weariness from the burden of one's task and perhaps friction with one's working surroundings until one comes under yokefellowship with another, such as the Jesus of Matt 11:28–30, whose pace and rhythm and resonance of spirit conveys a spirit of "rest" even as the work now vibrantly continues. See also Frost's wedding poem, "The Master Speed."

100. Loewald, *Sublimation*, 470–71 (italics added). Compare his application of the term, "open[ness]," here, with the use in his late "metapsychological" paper: "we understand something about nature and reality, know something about them, by being open to their workings in us and the rest of nature as unconscious life, the openness being what we call consciousness" (Loewald, "Psychoanalysis in Search of Nature," 50).

101. Ibid., 472 (his italics).

reference and of psychic movement between the "lower" and "higher" levels of human life points in both directions. This resonates with Frost's saying that the

> [g]reatest of all attempts to say one thing in terms of another is the philosophical attempt to say matter in terms of spirit, or spirit in terms of matter, *to make the final unity*. That is the greatest attempt that ever failed. We stop just short there. But it is the height of poetry, the height of all thinking, the height of all poetic thinking, that attempt to say matter in terms of spirit and spirit in terms of matter.[102]

Biblically, the earth can mourn (Hosea 4) and cattle along with humans can don sackcloth and cry out to God (Jonah 3:6), while the soul can cry out to God like a parched land (Ps 143:6). As Loewald sums up, "word-symbols in their more advanced form of functioning differentiate and connect (and so integrate) items of experience."[103] Similarly, there is no space here to go into Loewald's similarly rich chapter on "Illusion." I note only his own brief version of the positive treatment of the topic in Winnicott's work; in William Meissner's book, *Faith and Life*; in Paul Pruyser's *The Play of Imagination*; and in other works, all of which provide fertile ground for reflection on the transformative potential of religious symbolism under the "gentle and unnoticed control" of the will and the understanding. And I note that what Freud termed "illusion" was so deemed within the canons of his scientific materialism—an epistemological model that Loewald is at pains to re-conceive more adequately along the perceptual axis of the sort of resonance that may emerge through openness to nature as *natura naturans*—through willingness, one might say, of a stunted sensibility to cry out, like blind Bartimaeus along the roadside, for the gift of a wild Arab's ear, an American Indian's eye, and a non-sensuous touch like that of a blind man. It is the work of art, Loewald intimates, through its "illusions," to re-educate our sensibilities to our original and final connectedness with all things. (To the degree that contemporary poetry and literature so often shies from affirmative statement, and more often contents itself with "bewailing [our] outcast state" in an ultimately meaningless universe, one may suppose either that it has lost its vocation, or that its vocation is to call us to join Bartimaeus in crying out for renewed sight.)

In ending his analysis of sublimation with a discussion of subjectivity, Loewald returns to his critique of "the mechanistic view of nature in

102. Frost, "Education by Poetry," 723–24.
103. Loewald, *Sublimation*, 495.

PART FOUR: An Interlude

scientific materialism," claiming for his own discipline that "psychoanalytic theory still struggles with this heritage but is in the forefront of efforts to break the hegemony of the modern scientific *natura naturata* interpretation of reality."[104] In his last paragraph he writes of "those works of the human imagination—the productions of science, art, philosophy, religious thought and ritual—that are sublimations par excellence." Loewald's last sentences may be taken as an introduction to the last section of the present foray into the possibilities of a hermeneutics of resonance:

> Those [works] we admire most hold us spellbound, fill us with awe, partake of magic. May we assume that this magic is connected with the achievement of a reconciliation—with the return, on a higher level of organization, to the early magic of thought, gesture, word, image, emotion, fantasy, as they become united again with what in ordinary nonmagical experience they only reflect, recollect, represent or symbolize? Could sublimation be both a mourning of lost original oneness and a celebration of oneness regained?[105]

Resonance and *The Disenchantment of Secular Discourse*

In this section I want to trace in bare outline Steven D. Smith's diagnosis of the malaise in modern public discourse, and exploit for purposes of the present essay his modest but suggestive proposal of *resonance* as a way forward. As he puts it, resorting often to a phrase of Max Weber, secular public discourse today takes place within an "iron cage" that excludes appeal to moral values grounded in transcendent religious commitments. This iron cage has been constructed in two stages. First, the modern scientific enterprise, though including among its earliest practitioners many who were themselves avowed theists, undertook to account for the phenomena of the natural world without recourse to a transcendent "mover" of things, and without imputing subjective meaning or purpose, inherent in those phenomena, as explanatory of their behavior. Viewed on the analogy of a machine, nature's "moving parts" were to be explained as working on one another purely through efficient causation—like one billiard ball

---

104. Ibid., 516.
105. Ibid., 517.

impinging on another.[106] Secondly, the philosophers of the Enlightenment—especially in eighteenth-century France—proposed to ground their reformist vision of a just society on moral and ethical principles identified by pure secular Reason (capitalized!) having no need of "faith" or appeal to the notion of God (that supreme fiction and aegis of an unjust *ancien regime*). These two projects have worked to banish religious terms and commitments from the public square, and (where not proscribed altogether) to "privatize" them.

Smith's account of the reversal in the meaning of the word, "secular," is of interest for the fundamental, spatial image of the present volume, because it has to do with *where* different types of concerns and activities may *take place*, and conversely, what types of concerns and activities are proper to a given sort of place. In earlier times the word "secular" was used in the West to identify those clergy who, in contrast to the cloistered "religious," worked and ministered among everyday folk; and then it was applied to princes and kings and their this-worldly realms, in contrast to prelates and the pope, who exercised their leadership and authority in the spiritual realm. While these two realms could be thought of as "separate" jurisdictions, the secular was conceived of (certainly by popes and prelates, and implicitly by princes and kings insofar as they sealed their treaties and oaths in the name of God) as contained "within" the religious, and ultimately subject to it.[107] In the modern sense of the word "secular," this situation has been exactly reversed. Religion, where not proscribed, now occupies, in the old sexist adage, "a woman's place in a man's world"—that is, the privacy of home and neighborhood, without voice or vote in the marketplace, courts, academy, or legislative assembly.

But, as Smith shows, attempts to justify moral values in the public sphere purely on scientific grounds (the attempt to derive a meaningful, moral "ought" from a scientifically identified "is" that is by definition devoid of purpose or moral value), or purely on a universal rational calculus, either fail, or they "succeed" (that is, persuade a majority), by "smuggling in" value terms that are leftovers, vestigial legacies, from the pre-modern world, and depending on those terms to do the work they did in earlier times. Without such smuggling, the constraints on modern secular

---

106. See Whitehead's characterization and critique of this materialist foundation of modern science, as summarized earlier in the present essay.

107. It is tempting to compare this medieval vision, of the secular realm as *within* the spiritual, on analogy with Isaac ben Luria's idea of ṣimṣum, discussed in chapter 4, in which God who is everywhere contracts and opens up a space in God for the world to appear vis-à-vis God, displaying its own power-of-action which at the same time is God's gift constantly given to sustain the world in its dependent freedom.

PART FOUR: An Interlude

discourse make it virtually impossible to engage or satisfy those deepest values and commitments that variously drive all of us, religious or secular, in our concern for life's meaning and practice. One of Smith's examples will provide the means of connecting his analysis with my proposal for a hermeneutics of resonance. It has to do with John Stuart Mill's essay, "On Liberty," in which Mill offers what Smith calls "one of the most popular—and most powerful" of the "various instruments in the toolbox of modern liberal thought." That instrument is (in Mill's words) "one very simple principle" that should "guide and limit human society in its dealings with the individual in the way of compulsion and control," a principle holding that "the sole end for which mankind are warranted, individually or collectively, in interfering with the liberty of action of any of their number, is . . . to prevent harm to others."[108] In short, society should constrain an individual's liberty of action only where such action would harm another.

As Smith shows, the attempt to make this principle, or "instrument," do the work for which it was designed continues to run into problems. If one defines "harm" in subjective terms, words and actions that one person or group may object to as harmful may be considered by others to be unobjectionable and deserving of protection by the law. If one defines "harm" in objective terms, and identifies in ever-finer detail what constitutes objective harm, the word winds up carrying a specialized meaning so constricted in application that it no longer refers to what ordinary disputants over liberty mean by it. Yet, as Smith points out, ordinary Americans who may never have heard of Mill's principle as formally stated "resonate" to its substance as evoked in the expressions, "It's none of your business," "Don't stick your nose in where it's not wanted," and "Why can't I? It's a free country, isn't it?" As Smith says, "[r]esonating powerfully with familiar intuitions and adages and political theories, the basic idea seems almost self-evidently sound."[109] Yet, he asks, "if everyone in fact accepts this elementary idea, what exactly is going on when one party of advocates picks up the principle and uses it to club another party?"[110] Here, if both parties in such a dispute resonate with the basic idea, how is it that they do not resonate with each other's claims as to the way it should apply in a given instance?

108. Smith, *Disenchantment of Secular Discourse*, 70, quoting John Stuart Mill, *On Liberty and Other Writings*, 13.

109. Smith, *Disenchantment of Secular Discourse*, 76. See his use of the idea of resonance also at ibid., 72, 74, 74, 99, and elsewhere as noted below.

110. Ibid., 77.

Toward a Hermeneutics of Resonance

In exploring this question, it will be helpful—in view of Smith's use of the idea of resonance, to shift "containing" metaphors, from Weber's "iron cage" to "cultural cave," and to recur to John Hollander's notion, introduced earlier, that language, oral and written, functions meaningfully within a "cave of resonant signification," the cave itself consisting of a given culture (or sub-group in a culture), its natural setting, and the texts that have arisen and continue to arise as expressions of its core meanings and values. The question is whether the resonance that language has in such cultural or sub-cultural caves is irreducibly culture- or group-bound, or whether there is a universal resonance, however diverse its polyphonic parts,[111] in which all may participate. Is there such a thing as Truth (capitalized), in relation to which our respective conceptions and convictions are partial, fragmentary, and, in varying degrees, asymptotically convergent lower-case representations?[112] In metaphorical terms, can this secular world in some sense be thought of as a single "cave of resonant signification," or is such a possibility only visionary, a transcendent cave constituted by God?[113] The issue may be illustrated from Martha Nussbaum's project to "supply a justification for 'human rights' that, unlike traditional or classical approaches, is 'independent of any particular metaphysical or religious view.'"[114] Nussbaum's argument, as Smith reports it, is "that human *capabilities* can provide a universal normative standard for evaluating political practices and arrangements."[115] But, he notes, "[h]er program is very much in the Western 'liberal' tradition, as she acknowledges, so natu-

---

111. The term for this might be *polypoikilos*, literally "many-colored," used in Eph 3:9–10 for the variegated wisdom of God "made known" to the principalities and powers—those lower deities presiding each over their own "cave of resonant signification" according to Deut 32:8–9—and consisting, in part, in the way Jews and Gentiles are to co-exist in Christ, the "middle wall of partition" between them now being taken down. For a choral depiction of such a diverse choir, see Rev 7:9–12; 14:1–3.

112. Compare Coleridge's aphorism, "He, who begins by loving Christianity better than Truth, will proceed by loving his own Sect or Church better than Christianity, and end by loving himself better than all." See Coleridge, "Moral and Spiritual Aphorism XXV," in *Aids to Reflection*, 107.

113. Again, one may compare the imagery in Ps 90:1–2, where the Hebrew noun, *ma'on*, "dwelling," which elsewhere often refers to a lion's lair or den, is used of God's own dwelling. Or one may explore the suggestive potential, in John 14:2, of a single divine "dwelling" with many "rooms."

114. Smith, *Disenchantment of Secular Discourse*, 165, quoting Nussbaum, *Women and Human Development*, 101 (the context is Nussbaum's discussion "In Defense of Universal Human Values").

115. Smith, *Disenchantment of Secular Discourse*, 166.

rally it resonates with Western 'liberal' values . . . [W]ould the proposal elicit the same support in non-Western nations?"[116] What of counter-resonances emanating from long-held convictions in those nations embodied in different long-entrenched social arrangements?[117]

This brings me to Smith's modest constructive proposal. He begins by taking up the question, "what do we believe, really?" and explores it with the help of a conception of belief—and "hence, of the function and limits of reflective reasoning"—recently put forth by Joseph Vining.[118] In Vining's view, as Smith reports it,

> A belief is not simply a readily observable propositional piece on our cognitive chessboard: it is something less on the surface and instead more rooted in the depths of our being. Discovering what we believe . . . involves . . . a more serious and searching investigation of . . . what we *think* we believe, . . . but also of how we live, what we desire, what we would and would not be willing to do.[119]

In this case, the point of discourse and debate should not be simply to test and perhaps expose flaws in the other person's reasoning, or inconsistencies in that person's general position, but to identify "the deep *resonance* of genuine belief with and within a person's most central commitments—with and within her life."[120] Because this investigation is as much a searching out of what we ourselves really believe as it is of what our dialogical partners and dialectical opponents really believe, it "must be a cooperative enterprise." And any change of understanding that may occur in either party, in such an enterprise

> will typically come not because we are coerced by argument or evidence into repudiating our previous convictions, but because

---

116. Ibid., 183.

117. See, e.g., the article by Appiah, "The Art of Social Exchange," 22, which observes how the successful effort, in the nineteenth century, to change such entrenched practices as binding women's feet had to begin with a general respect for the culture and an effort to establish connection with attitudes, concerns, and efforts already present within the culture for such change.

118. Smith, *Disenchantment of Secular Discourse*, 195, drawing on Joseph Vining, *The Song Sparrow and the Child*. He finds Vining's understanding of believing "somewhat unusual." Coming to it from thinkers like Coleridge, on the one hand, and Hans Loewald on the other, I don't find it unusual at all. Regardless, Smith's report of Vining's view is a convenient restatement for present purposes.

119. Smith, *Disenchantment of Secular Discourse*, 195.

120. Ibid., 196 (italics added).

Toward a Hermeneutics of Resonance

we become able to acknowledge beliefs that at some level we have held all along without being wholly conscious of them, or perhaps without being willing to own up to them.[121]

In this enterprise, Smith goes on to say, Vining

> asks us to contemplate what he calls "openings"—realms of experience through which, if we pay close attention, we can sense the reality of something beyond the reductionistic world of material systems and can look into the world of what Vining calls, with misgivings, "spirit."[122]

As examples of the sorts of situations or activities that may provide such openings, Smith points to music,[123] language, land (or nature generally), death, or the distinction between "authority" and "the authorities," in those cases where "'authority' is something that we understand not as coercing us, exactly, but as having an authentic claim on our attention and respect."[124] Finally, "on a more intimate level," another opening is the "presence of other people," in meeting whom, in Vining's words, a "sense of life springs within us."[125] Or, one might say, something within our own depths resonates with something in the depths of the other person. If, even in the midst of vigorous disagreement on specific issues of mutual concern, we are attentive to the other as a person (one who in biblical terms is, like us, created in God's image), "it is always possible that, however much we may disagree with another person's worldview, *something* in that view will *connect* with something in our own that will result in constructive engagement."[126] The connection, I believe Smith would say, is through a resonance that identifies an area of common concern and potentially common ground that lies deeper than articulated concepts and entrenched positions.[127]

---

121. Ibid., 197. In the phrasing of Jesus' parable of the Prodigal Son, it will involve a "coming [home] to oneself."

122. Ibid., 205.

123. Cf. Arthur Lubow, who, in an essay on the Estonian composer Arvo Pärt, has noted that this composer does not simply "aim to ravish the ear or to tickle (or boggle) the mind ... [h]e wants to touch something that he would call the soul" ("The Sound of Spirit," 34–39). Though many of Pärt's works "bear a whiff of church incense," yet "the compositions resonate profoundly for the unconverted as well as the faithful."

124. Smith, *Disenchantment of Secular Discourse*, 206.

125. Ibid.

126. Ibid., 223.

127. As a side bar to the present discussion, I would refer to Bertrand Russell's

PART FOUR: An Interlude

I would cite, as a case in point, the Jerusalem council in Acts 15, where a controversy over the grounds and conditions for Gentile inclusion in the nascent Christian movement is settled in a manner summed up in an expression that is interestingly elaborated as it is repeated. I shall quote from the RSV because of its greater closeness to the Greek:

> v. 22: "it seemed good [*edoxe*] to the apostles and the elders, with the whole church . . . "
>
> v. 25: "it has seemed good [*edoxen*] to us, having come to one accord [*genomenois homothymadon*] . . ."
>
> v. 28: "it has seemed good [*edoxen*] to the Holy Spirit and to us . . ."

James's final reference here, to "the Holy Spirit," would seem to connect with Peter's preceding (v. 8) recollection to this council how "God who knows the heart bore witness to them, giving them [the Gentiles] the Holy Spirit just as he did to us." As intimated in Peter's "God, who knows the heart, and bore witness," the issues here go deeper than the surface meanings, or even the long and deeply-held interpretations, of texts central to the participants' self- and God-understandings; and it would appear that it was at that depth that they arrived at a harmony—*homothymadon*—that they could only attribute to the Holy Spirit.[128]

To suggest that a hermeneutics of resonance will provide a quick and easy resolution of controverted issues concerning the meaning of important texts or the conduct of life according to our deeply held values and commitments, would be naive folly. It is, rather, to indicate, with Smith, the *direction* in which we should look, in clarifying the nature of the difficulties that stand in the way of such resolution. As a modest but practical proposal, I would suggest that one way to help in discovering such

---

credo, "A Free Man's Worship," in which I hear strange resonances with Psalms 69 and 27 among other biblical texts (see chapter 9). Insofar as Smith quotes the same passage in Russell as I do, and takes it as a classic example of the "iron cage" of secular discourse, I would take the resonances between Russell and these psalmists as an extreme case of an "opening" of the cage through resonance between ostensibly profound differences in viewpoint and conviction.

128. Once again I would draw attention to Coleridge's remarks, in *Aids to Reflection*, 77–79, concerning the work of the Spirit within the human heart "at an unknown distance" from human consciousness, and in that depth where the human will initiates its "first acts and movements." These remarks find their Scriptural rootage in Rom 8:16 (where the Spirit "bears witness with our spirit") and 8:26 (where the Spirit "helps us in our infirmities" and "makes intercession for us").

resonant meeting-points is to *sing* with one another; or perhaps to begin by attempting to enter imaginatively into one another's favorite songs. "How shall I sing the Lord's song in a foreign land?" the psalmist asked (Ps 137:4). Such singing is truly impossible if "others" simply *taunt* one to sing (Ps 137:3). But what if room—place—is respectfully *made* for the one so to sing? And what if, when the other sings, one listens with a sympathetic ear, to see if one can identify with anything in that song? What if all conceptual discourse is for a time suspended, and only such singing "takes place," each party singing in turn, with no commentary preceding or following? When the Chilean miners trapped for seventy days rose from the depths of their mine, each one greeted with rejoicing, and when at the end all joined in singing the Chilean national anthem, was there anyone on the face of the earth who would not have said *Amen* if someone had said, "Tonight, we are all Chileans"? Perhaps the depth of that mine, and the depth of those miners' plight, is a symbol of the depth to which all must be willing to descend, and there remain for a time, "groaning in travail and in pain together," until, in matters of life's most painful controversies, we are to identify and embrace, in ourselves as well as in others, our deep, true humanity, and in it the image of God—the God of Exod 2:23–25, who hears, remembers, sees, and knows.[129] Perhaps it is in that groaning together that true resonance begins.

Resonance between the Testaments in Proverbs 8 and Colossians 1

I conclude this prolegomenon to a hermeneutics of resonance with an exploration of two biblical passages, themselves surely resonating with one another, and perhaps also with a fresh, second-order resonance when read through the lens of Coleridge's passage on "the poet in ideal perfection." I begin with the concluding words of Wisdom in Prov 8:22–31, a passage in which, beginning with a self-description as "the first of the ways of God" (*rēʾšit darko*; LXX: *archēn hodon autou*), Wisdom goes on to distinguish herself from the cosmos as God's handiwork, by the contrast between verbs describing the latter ("shaped, made, established, drew a circle,

---

129. KJV ends Exod 2:25 with "had respect unto them"; RSV: "knew their condition"; NRSV: "took notice of them"; NIV: "was concerned about them." The Hebrew text says simply *wayyedaʿ ʾelohim*, "and God knew." The resonant force of the Hebrew may be compared with the mother who responds to the cries and complaints of a distressed child, "I know, I know."

made firm, marked out") and verbs describing the coming into being of Wisdom ("created" [*qanah*], "set up" [or "woven," *nissak*],[130] "brought forth" [*ḥolal*]), which connote "begetting." Then Wisdom goes on to say (8:30–31) that "when" God set about creating the various components of the cosmos, "I was [*wa-'ehyeh*] beside him . . . I was [*wa-'ehyeh*] his day-to-day delight." But the Hebrew verbs here may resonate with two other passages. Given the extraordinarily exalted status and role given to Wisdom in this passage, are we invited here to hear a resonance with the *'ehyeh* of Exod 3:14 and 3:15, and also with the *yehi*, "let there be," of Gen 1:3 and following? Is this passage on the hermeneutical trajectory that leads to the Targumic grounding of the Memra as agent of God's activity in Exod 3:14–15?[131] And in the *wa'ehyeh 'eṣlo*, "I was with him," are we to hear the primordial resonance between Wisdom and God that is instanced in the *'ehyeh 'immak*, "I will be with you," to Moses in Exod 3:12, and, in Exod 4:12, as the maker of Moses' recalcitrant mouth, *'ehyeh 'im-pika*, "I will be with your mouth"?[132] However that may be, Wisdom is "beside God" in a two-directional resonant behavior: "rejoicing before him always," and "rejoicing in his inhabited world and delighting in the children of earth." I have not attempted to translate the variously interpreted expression, "like an *'amon*." But I note that in the LXX it is rendered with the participle, *harmozousa*. Whether *'amon* here means "master workman," or "child," or some other figure, the LXX takes Wisdom here to be engaged in "harmonizing." (Whimsically, one might imagine Wisdom here is the *sotto voce* humming of God at work in fashioning the world.) It is Wisdom, according to this passage, that, transcendently speaking, informs all truly imaginative activity. This, of course, comes to classic expression in Exod 31:1–6, where God provides Bezalel, Oholiab, and other artisans with

---

130. RSV and NRSV (following BDB) construe the verb as a Niphal form of *nasak*, "to set up, install," after the Qal form in Ps 2:6. *HALOT* takes the verb in Ps 2:6 as from a root *nsk* meaning, "pour out," here in the sense of "consecrate," and the verb in Prov 8:23 from a root *nsk* meaning, "to weave." With the latter we may compare Ps 139:13 where the related root, *skk*, occurs in the sentence, "you wove me together in my mother's womb." A meaning, "poured out," would also fit Prov 8:23, in light of the imagery in Job 10:10.

131. Recall the discussion of the relation between Targumic Memra and Exod 3:14/Gen 1:3 in chapters 3–4.

132. Note how, already in Hos 1:9, *'anoki lo'-'ehyeh lakem*, we may have a play on the divine name *'ehyeh* in Exod 3:14. Anderson and Freedman (*Hosea*, 143, 198–99) take *'ehyeh* here as a noun, and read the construction as a nominal sentence, "I am not Ehyeh to you."

divine Spirit, wisdom, and the like, to enable them to design the tabernacle in accordance with its coherent "pattern" (Exod 25:9) in the mind of God.

The second passage is Col 1:17, where it is said of the "image of the invisible God" and "firstborn of all creation" that "he himself is before all things, and in him all things hold together." When I juxtapose the presentation in Proverbs 8 of Wisdom as the "tone and spirit of unity" with the sentence in Col 1:17, "He himself is before all things, and in him all things hold together [*synesteken*],"[133] I cannot imagine that the two passages do not at least resonate with one another, and perhaps even achieve a second-order resonance. One such possiblity is this: While Col 1:17 occurs in a context that goes beyond the manifest vision of Prov 8:22-31, in speaking of the work of a "reconciliation" of dire "discordances" through "the blood of his cross," the notes of play and delight in Prov 8:30-31 ever so slightly may suggest that deeper than the gravity that characteristically informs traditional meditations on the Passion of Christ lies a continuing "tone

---

133. On the question of a connection between these two passages, note Peter O'Brien's recognition that many have "drawn attention to the similarities between this statement and the language of Platonic and Stoic philosophy" in the use of *synistemi*, whereas he follows others in taking "the parallels from Hellenistic Judaism, especially the LXX" as "much closer." In particular, O'Brien cites Ralph Martin who in turn follows R. B. Y. Scott in seeing a connection with the figure of Wisdom in Prov 8:30. See O'Brien, *Colossians, Philemon*, 47-48. On the last clause in 1:17, James D. G. Dunn comments, "Once again the *theme* reflects Jewish reflection on Wisdom" (*Epistles to the Colossians and to Philemon*, 93; italics added), citing the same LXX texts as O'Brien, including Prov 8:22-31. Acknowledging that the language is that of Platonic-Stoic cosmology, Dunn takes it that "conceptuality from contemporary cosmology seems to be loaded in an undefined way on Christ" through "poetic imagination, precisely the medium where a quantum leap across disparate categories can be achieved by the use of unexpected metaphor, where the juxtaposition of two categories from otherwise unrelated fields [as in Wheelwright's diaphor!] can bring an unlooked for flash of insight." Intriguingly, Dunn goes on to speak of the way in which, in Platonic and Stoic cosmology, "a rationality (Logos) . . . pervades the universe and holds it together," and "both the order and regularity of natural processes and the human power of reasoning *resonates* with this rationality" (ibid., 93-94; italics added). One may note here the "resonance" between a human interior mental faculty, aspects of processes in nature, and an underlying reality in which both participate. N. T. Wright likewise reads Col 1:17 primarily against an Old Testament-Jewish background. Following and improving on C. F. Burney's analysis of aspects of Col 1:15-20 as deriving from Prov 8:22-31, Wright argues that "the poem . . . asks to be read in the light not merely of one particular branch of Jewish tradition (i.e. 'Wisdom'), but of the entire Jewish worldview of which the wisdom-tradition was simply one of many facets," a worldview that, in respect to God's "final redemption" he finds instanced, for example, in Isaiah 40-55. See Wright, "Poetry and Theology in Colossians 1:15-20," esp. 108-9.

PART FOUR: An Interlude

and spirit," running through that unfathomable scene, intimated in the renaissance song about Christ's "dancing day."[134]

But let us return to the specific verb used in Col 1:17. The precise connotation of *synesteken* in this context may be hard to pin down—if indeed it is not to be taken as multivalent. It is formed from the verb, *histemi*, which in classical (non-biblical) Greek means, intransitively, "to stand," or transitively, "to cause to stand." The compound verb, *synistemi*, carries the following range of meanings in classical Greek:

> A. I. set together, combine. II. associate, unite. III. put together, organize, frame; (of an author) compose. IV. bring together as friends, introduce or recommend to one another. B. I. in the second aorist active voice [the form in 1 Col 1:17], with a passive connotation, stand together. II. In a hostile sense, be joined in battle. III. Of friends, form a league or union, band together.[135]

In the LXX, *histemi* translates 37 different Hebrew verbs, but in the overwhelming preponderance of its 705 occurrences there it translates either *qum* or *'amad*. Against the background of the range of connotations of *synistemi* in classical Greek, and my earlier discussion of *qum* or *'amad* in connection with "standing" in Psalm 69 (see chapter 9), the following occurrences of *'amad* in Isaiah 40–66 provide a suggestive "cave of resonant signification" for interpreting Col 1:17. Note the following passages:

In Isa 50:8 (a passage echoed, by most accounts, in Rom 8:33), the beleaguered servant of the LORD says, in law-court terms, "he who vindicates me is near. Who will contend with me? Let us stand up together [*na'amdah yaḥad*; LXX: *antisteto moi hama*]. Who is my adversary? Let him come near to me."

In the midst of the first cosmic trial scene (itself a precursor to apocalyptic scenarios), God says of an idol maker (Isa 44:11), "Behold, all his fellows shall be put to shame, and the craftsmen are but men; let them all assemble, let them stand forth [*ya'amodu*; LXX: *stetosan hama*], they shall be terrified, they shall be put to shame together [*yebošu yaḥad*; LXX:

---

134. I have in mind "Tomorrow Will Be My Dancing Day," as part of Stravinsky's "Cantata" (1952). Admittedly this song is about the day of Christ's birth; but, then, the Colossian hymn speaks of Christ not only as "firstborn (*prototokos*) of all creation" (Col 1:15; echoing Prov 8:22-25) but also as "firstborn from the dead" (Col 1:18; cf. Rom 8:29). This image will be picked up later in Eastern Trinitarian thought where the relation between Father, Son, and Spirit is spoken of as a *perichoresis*, literally, "a dancing around." See also the Shaker hymn, "Simple Gifts," and the more recent hymn, "Lord of the Dance," by Sydney Carter.

135. See LSJ, 1718-19.

aischynthetosan hama]." Two things are noteworthy here. First, idols are religious, or what Edmund Leach would call cosmological, symbols. As such, they are central to the way in which the culture achieves its cohesion, its "togetherness," in relation to the gods in their cohesion as a "divine council." Secondly, in this cosmic trial scene the so-called cohesion achieved through the making of idols has no "standing" before Israel's God and becomes, ironically, a "cohesion in shame."

The irony is continued in God's address to Babylon in that city's idol-making self-idolatry (Isa 47:12–13):[136] "Stand fast [*'imdi*; LXX: *stethi*] in your enchantments and your many sorceries, with which you have labored from your youth; perhaps you may be able to succeed, perhaps you may inspire terror. You are wearied with your many counsels; let them stand forth [*ya'amodu*; LXX: *stetosan*] and save you, those who divide the heavens, who gaze at the stars, who at the new moons predict what shall befall you." The solidarity in "shame" of such idol-makers will consist in the questionableness of their "standing" in the nature of things. One may note, further, the emphasis on "labor," leading to "weariness," involved in idolatrous systems of symbolic trafficking with reality. This note resonates with elements in the conclusion of the first long divine speech celebrating God's unfathomable wisdom and power in creation and universal history, where God is said to be unfainting and unwearying in these activities and where those who wait on, or hope in, this God will "renew their strength" to rise up on eagles' wings, run and not be weary, walk and not faint (Isa 40:31). The first cosmic trial scene opens, ironically, with the call to the nations, their gods and their witnesses, to "renew their strength," concluding, "let us together [*yaḥdaw*; LXX: *hama*] draw near for judgment" (Isa 41:1). Here, the "drawing near together," semantically equivalent to the verb *synistemi*, carries the latter's connotation of conflictual confrontation.

In contrast to the above scenarios, God says, in Isa 48:13, "My hand laid the foundation of the earth, and my right hand spread out the heavens; when I call to them, they stand forth together [*ya'amdu yaḥdaw*; LXX: *stesontai hama*]."[137] And in Isa 66:22 God says, "as the new heavens and

---

136. That the idolatry is of the city itself is indicated in the way the city, in Isa 47:8, 10, applies to itself language that elsewhere in Isaiah 40–55 is uttered self-referentially by Israel's God.

137. See my discussion of the rain and the lightning's "here we are," in response to God in Job 38:34, 35, as resonating with the responses to God's call to Abraham (Genesis 22), Moses (Exodus 3), Samuel (1 Samuel 3), Isaiah (Isaiah 6) and Mary (Luke 1:38), in Janzen, *At the Scent of Water*, 119–20. For other fraught occurrences of Hebrew *yaḥdaw*, "together," as expressing solidarity, see Gen 22:6, 8, 18; Exod 19:18;

PART FOUR: An Interlude

the new earth which I will make shall remain [*'omedim*; LXX: *menei*] before me, says the LORD; so shall your descendants and your name remain [*ya'amod*; LXX: *stesetai*]." If the current "heavens and earth" or "world-order" (including the nations' understanding and symbolizing of it) is doomed to wear out and vanish (Isa 48:6, 8), Israel is called to "fear not their reproach, nor be dismayed at their revilings" (48:7), for they who "wait for the LORD" already enjoy a "standpoint of faith" as a proleptic participation in that new creation where they shall enduringly stand. It is within such a proleptic standpoint that the servant of the LORD, who in 49:4 momentarily confesses, "I have labored in vain, I have spent my strength for nothing and vanity," is able to affirm the security of his cause in God (49:4) and confess that "my God has become my strength" (49:5).

It is on this basis that this servant is able to "sustain with a word one that is weary," and to say to his opponents and revilers, "he who vindicates me is near. Who will contend with me? Let us stand up together [*na'amdah yahad*; LXX: *antisteto moi hama*]. Who is my adversary? Let him come near to me" (50:4).

I, for one, have read Col 1:17 many times, contenting myself with the usual translations, without any specific sense of any resonances it might have for a consciousness in-formed by Scripture as a "cave of resonant signification." Having now spent a morning exploring that cave as a possible generative matrix for the subliminal symbolic world out of which Col 1:17 came to expression, I can no longer read the verb, its sentence, and the larger passage, without hearing in it something with multiple resonances from the Old Testament passages such as those canvassed above.

In respect to the canonical function of Scripture, I want to explore a bit further a subsidiary implication concerning the Old Testament as the primary (though not by any means exclusive) semantic "cave" within whose resonant significations one should read the New Testament. What I have in mind is the notion that, whatever the meaning of a given Greek word in classical Greek, its primary connotations in the New Testament should first of all be sought in the connotations that word has taken on through its occurrence in the LXX where the LXX translates Hebrew words and is further colored by its biblical context. This I take, not as a hard-and-fast, stipulative principle, but as and where it can be shown. (I have in mind, for example, the words *doxa* and *splangchnia/splangchnizo*.) Further, then, on the connotations of *synesteken* in Col 1:17: Given

---

and, eschatologically, Isa 11:6, 7; 40:5; 41:19, 20; 52:8, 9. See also my discussion, in Chapter 15, on the preposition, *syn*, "with," in Romans 8.

Toward a Hermeneutics of Resonance

its range of connotative possibilities in classical Greek, just how may we envisage the process and character of the "togetherness" that "all things" have in the one who is "before all things"? Do the texts in Second Isaiah canvassed above provide one range of possibilities? Given that Col 1:17 apparently has to do with origins, should we read *synesteken* as resonating with a text like Isa 48:13, where, speaking of the creation of the earth and the heavens, God says, "when I call to them, they stand forth together [*stesontai hama*]"? Given that creation through the commanding or calling word is a common Old Testament trope, would such a nodal expression of that trope provide a resonant niche within the Old Testament taken as the canonical cave of resonant signification, within which to hear the specific resonant implication of *synesteken*? Or, given that Colossians says of the figure in 1:17, in whom "all the *fulness* [*pleroma*] of God was pleased to dwell" (1:19; 2:9), that "in him are hid all the treasures of *wisdom* and knowledge" (2:3)—a wisdom to characterize, then, all this figure's followers (1:9, 28; contrast, dialectically, 2:23); a wisdom taught and imparted in a context of corporate singing of "psalms and hymns and spiritual songs with thankfulness in your hearts to God" (3:16)—given all this, are we to hear in *synesteken* an entirely or overwhelmingly undialectical *harmonious* joining together in response to this figure's "all-animating voice," as perhaps in Prov 8:30–31 (LXX), Psalm 148, and Job 38:7? or to what degree are we to hear this harmony as embattled, as in the various "standings together" that I have traced in Second Isaiah? Or are we to hear the embattled harmony as taking place on the plane of history while embedded within the matrix of an undialectical harmony standing behind and at the origin of all things, and subsisting in and through all historical dialectical oppositions as the matrix in which, despite their centrifugal tendencies, they still hold together?

In this last sentence, I have used the word, "dialectical," in a manner that may call to mind the dialectic logic of Hegel. But I am not a Hegelian. I prefer the polar logic of Coleridge, as a way of understanding dynamic process and emergence in nature and history, and indeed, as a way of reflecting on the Christian doctrine of the Trinity as rooted in the formulation in Exod 3:14. In Coleridge's polar logic, which he sometimes refers to as a logical "tetractys" and sometimes as a "logical pentad," dynamic processes may be thought of as originating in a *prothesis*. This *prothesis* is never visible, but comes to concurrent visible expression in the polar opposites to which it gives concurrent rise, polar opposites that Coleridge calls *thesis* and *antithesis*. (In my own image, the prothesis is like the milk in a cow's udder; the thesis and antithesis are like the cream and the milk that are separated out by a centrifuge.) Viewed simply as polar opposites, they may be exemplified in the positive and negative poles of a magnet, and stand in tensive relation with one another. (In politics, they are manifest for Coleridge in a party of progression and a party of conservation.) But the originating *prothesis* continues as the hidden midpoint or *mesothesis* between the polar opposites—in a magnet, the point of "indifference"; in politics, the "centrists" who are concerned to reach out to the polar(ized) parties and draw them into dialogical relation and potentially into a harmony of opposites. The process, where successful, culminates in a *synthesis* which, not merely a return to or recovery of the original *process*, but in some sense and to some degree a *novum*, becomes the *prothesis* for the next impulse in what, with some later "process theologians," one might call "the creative advance." The view that I am advancing, here, is that the resonance discernible between texts, whether harmonious or dissonant, participates in and is the symbolic expression of a social and cultural resonance that, however mixed in its display of local and temporal harmonies and dissonances, are rooted in the harmony of a primal *prothesis*, and are "held together" (*synesteken*) through the course of their tumultuous history by a continuing *mesothesis* that is the promise of "everlasting joy" (Isa 35:10; 51:11; 61:7) in their eschatological *synthesis*. In a "hymnic/creedal" passage such as Col 1:17, where every word carries a great freight of meaning and connotation, I suggest that (with a respectful but demurring nod in the direction of James Barr) it is impossible to overread the connotations of *synesteken*.

On Hymnic Resonance and Community Cohesion

A reference to the hymnic character of Col 1:15–20 calls for a further investigation into a related set of resonant ligatures between the Old Testament and the New Testament. In an earlier footnote I referred to "other fraught occurrences of Hebrew *yaḥdaw*, 'together,' as expressing solidarity." In some of those passages, as we have seen, the LXX rendered this word with *hama*. Another frequent rendering is *homothymadon* (deriving from the Greek noun, *thymos*, "spirit, passionate animation"), a word connoting concerted and spirited action. The entry on this word in LSJ takes up less than four lines, with two of the textual citations being Exod 19:8 and Acts 15:25. An unhurried canvass of all its biblical occurrences will fill out the apparently prosaic "with one accord" of LSJ's definition, and recover its connotation of spirited, conjoint resonance. In Exod 19:8 the scene is the solemn assembly at Sinai where God makes the preliminary offer of a covenant with Israel, and the people answer together (*yaḥdaw*; LXX: *homothymadon*), "all that the LORD has spoken we will do." The word connotes congregational solidarity again in Num 27:21. In Job 2:11, Job's friends, on hearing of his calamitous plight, are moved to "make an appointment together [*yaḥdaw*; LXX: *homothymadon*]" in order to "condole with him and comfort him." Their aim is to identify with Job in his grief, and to help him to undergo and recover from his grief in such a way as to modulate his grief from something that bogs his life down in a morass of lamentation into something that enables him to go on with his life. Thereafter, *homothymadon* occurs thirteen more times in Job (out of its twenty canonical occurrences in all), almost always connoting conflict or solidarity in the face of conflict. I note in particular the "furrows of the field" weeping "together" in outcry against Job (31:38); and, analogously, the ramparts and walls of Zion that, under the destructive hand of God, "lament, they languish together" (Lam 2:8). The term is picked up as a prominent motif deuterocanonically, in 1 Esd 5:47, 58; 9:38 (all in solemn sacral assembly); Jud 4:12;* 7:29;* 13:17;* 15:2, 5, 9*; and 3 Macc 4:4, 6;* 5:50–51;* 6:39. The asterisked passages in the last two books portray groups in a solidarity of lament, petition, praise or blessing.

These occurrences of *homothymadon* in the LXX provide, I suggest, a canonical context within which to listen for the resonances this word may carry in the New Testament. And the incidences are striking. Apart from Rom 15:6, the term occurs only in Acts, and there it might be said to constitute a signature trope on the dispensation of the Spirit as the aegis of the nascent Church. The trope appears first in the upper room where, following the ascension of the risen Jesus, the small band of his followers devote themselves to prayer *homothymadon* (Acts 1:14). Following the descent of the Spirit, they continue attending the temple *homothymadon*, and "breaking bread in their homes" (2:46). In 4:24 the group responds to what they have heard by raising their voices *homothymadon* in a paean of praise to God. Thereafer, the term marks an oscillation between concerted and conflicted scenarios (5:12; 7:57; 8:6; 12:20; 15:25; 18:12; 19:29) that brings the proleptic eschatology of the unification (one might say, harmonization) of languages in the upper room back down to the plane of history where such "caves of resonant signification" as do exist at times seem only like the lairs of primitive warring groups armed with clubs and spears.

PART FOUR: An Interlude

With its Scriptural context in the Old Testament in view, and especially its function as a trope in Acts, it is suggestive now to note the one occurrence of *homothymadon* in Rom 15:6. Paul has been attempting strenuously to hold together two factions in the church in Rome, factions divided over the question of the continuing interpretation, or even relevance, of aspects of the Scriptures formerly agreed upon as governing concrete aspects of individual and group practice—in particular, foods and observance of sacred days. At the end of a long attempt to hold both parties within speaking distance of each other, Paul comes to the point where he writes,

> May the God of steadfastness and encouragement grant you to live in such harmony with one another [*to auto phronein en allelois*], in accord with Christ Jesus [*kata Christon Iesoun*], that together [*homothymadon*] you may with one voice [*en heni stomati*] glorify the God and Father of our Lord Jesus Christ.

Here we see the appeal to *to auto phronein* that runs also through Philippians, an appeal introducing the hymn (!) in Phil 2:6-11. What does *to auto phronein* mean? And in what relation does it stand to glorifying God "with one voice"? Does it mean that one can only glorify God in worship with those of like mind? (So that, for example, certain archbishops in the Anglican Communion would absent themselves from a eucharist at which other archbishops with opposing views on issues of human sexuality are participants?) But what if the meaning of *to auto phronein en allelois* is to be found *in* the practice indicated by "with one voice [*en heni stomati*] glorify[ing] the God and Father of our Lord Jesus Christ"? What if the purpose clause with *hina* ("so that") + subjunctive mood ("you may glorify") stands in explicative parallelism with the wish expressed in the optative *ho de theos . . . doe hymin*, "may God . . . grant you . . ."? In that case, God "gives" the "living in harmony with each other" (as RSV and NRSV render *to auto phronein en allelois*) in and through their willingness, in the midst of their differences, to come together in conjoint actions in which "*together* you may *with one voice* glorify the God and Father of our Lord Jesus Christ."[138] The *mesothesis*, so to speak, that mediates and "holds to-

---

138. For the primal and eschatological significance of such hymnic references in the Bible, see further Janzen, "'More Joy in Heaven?'" and, quoted there, the following lines from Erhard Gerstenberger, "The Dynamics of Praise in the Ancient Near East, Or: Poetry and Politics" (paper delivered at the 2010 Annual Meeting of the Society of Biblical Literature, and used with the author's permission): "Hebrew *hallelujah*, 'extol (imperative) Yahweh' . . . and . . . 'Allah is the greatest' (*allahu akbar*) . . . really are hymns in a nutshell, often condensed expressions of power . . . [P]raise is not only an

Toward a Hermeneutics of Resonance

gether" their polarized positions on the issues at hand, is the act of singing together their common hymnic affirmations of Christ such as we have in Phil 2:6–11. This, I suggest, is the implication of the way Paul encloses this wish and this exhortation in what precedes and what follows. In Rom 15:3, Paul quotes "Christ" as uttering words from Ps 69:9 (!)—words the context of which resonate with the context in which Paul quotes them and which thereby beautifully exemplify Hays' central thesis. One may recall how that psalmist's conflicted situation, in which at the outset he can find no standpoint in the midst of those who taunt and reproach him, transmogrifies in the end into a paean of praise that embraces heaven and earth in its hope for God's rebuilding of Zion as an enduring dwelling-place for all who love God's name. Then, in Rom 15:4, Paul offers a generic comment on the function of the canonical writings as consisting in the generating and sustaining of a hope that has its ground in the God's steadfastness and encouragement. And, shortly after the call in 15:5–6, Paul, in effect, hands out to his addressees in Rome copies of the hymn texts in the singing of which they may find themselves, at least in this hymnic mode, "living in . . . harmony with one another, in accord with Christ Jesus." If they can situate their current sharp differences over issues of food and sacral days within the resonant matrix of such solidarity in praise of "the God and Father of our Lord Jesus Christ" (going so far, one may suggest, as adding the hymn in Phil 2:6–11 to their repertoire), the transformative—one may say, sublimating—energy of such resonance may so re-situate themselves in God's prevenient or *prothetic* grace as to draw their polarized positions into a polarity mediated by that grace as *mesothetic*, as the Spirit binding one to another, toward an eventual *synthetic* resolution. The resonance lies in the music; and, while poetry is distinguishable from music as a discrete art form, poetry is by definition the music of meaning in linguistic form.

---

aesthetic or stylistic speech form, with possible theological implications, but a real primordial force not to be tamed by modern interpretations . . . Praise language . . . stretches out into the past, wrestles with reality by acknowledging accomplishments, grasping actuality and probing into the future, and in all these regards also tries to mould reality according to desired well-being, peace and justice . . . Hymn-singing sustains all beings, conferring strength to everything which is in need of it . . . Hymnal language of old may . . . emphasize the unity of all existence and a common responsibility for order and justice by recognizing participation of all agents in the universal powerplay."

part five

New Testament Afterword

15

"Hid with Christ in God"

IN THIS ESSAY I want to explore some texts in the New Testament that I take to resonate with the theme of the "secret place" in which prayer "takes place." I shall criss-cross back and forth between these texts through a series of forays, each of which can be pondered in its own right as a mini-essay, but the cumulative effect of which, I hope, will be to suggest a richer, "thicker" picture than any one foray by itself. Dizzying as this procedure may be on first reading, it calls for slowing down and pondering the individual texts cited and reflecting patiently on their mutual resonances, and it should suggest something of the richness of the scriptural thesaurus the New Testament writers drew on in attempting to explore the relationship between "things new and things old."[1] I shall begin with Col 3:1–4, from which the title for this essay is taken. The passage reads as follows:

> If then you have been raised [*synegerthete*] with Christ, seek *the things that are above* [*ta ano*], where Christ is, seated at the right hand of God. Set your minds on *things that are above* [*ta ano*], not on *things that are on earth* [*ta epi tes ges*]. For you have died, and your life is hid [*kekryptai*] with Christ in God. When Christ who is our life appears, then you also will appear with him in glory.

This exhortation may strike readers today as alien and out of sync with the temper of contemporary life. One could be forgiven for taking it as just one more sponsor of the sort of this-world-deprecating focus on heaven and the afterlife that apparently preoccupied, not to say obsessed, much of Western Christendom in the Middle Ages but that has been

---

1. The reference is to Matt 13:52 where, after a collection of parables that in many respects may be seen as retellings of Old Testament themes, Jesus is portrayed as saying, "Therefore every scribe who has been trained for the kingdom of heaven is like a householder who brings out of his treasure [*thesauros*] what is new and what is old."

PART FIVE: New Testament Afterword

left behind in the secular (literally, "this age" or "this world") temper of modern times. Yet such a reading of the passage would mistake the point of its emphasis, and would deprive the present reader of its importance precisely for life in this world and in this age.[2] The passage is listed in many Christian lectionaries as the Epistle for Easter Day. As Paul for his part frequently emphasizes (for example, in Romans 6 and in Philippians 2 and 3), the resurrection of Christ bears not only on a person's post-mortem prospects, but very much on how one lives within the present world and among one's fellow creatures. Colossians 2:1–4, when attended to carefully, is moved by just such a concern.

At this point, I shall limit myself to a few preliminary observations, all of which revolve around the peculiar expression, "hid with Christ in God." (1) The opening "you have been raised with Christ" carries forward the thought in Col 2:12, "you were buried [*syntaphentes*] with him in baptism, in which you were also *raised with* [*synegerthete*] him." So, one may wonder if being "hid with Christ in God" is somehow related to having been "buried with him in baptism." (2) In Rom 6:4 (the only other place in the New Testament where the compound verb *synthapto* occurs), the theme of being "buried-with" Christ in baptism has, as the correlate of Christ's being raised, the Christian's walking "in newness of life." So we may suspect that the call to seek and set one's mind on *ta ano*, and not on *ta epi tes ges*, by its connection with Col 2:12, similarly bears on how one conducts one's life here and now. (3) That this is the case is suggested by "connecting the dots" between the following texts: (a) in Gal 4:26, "the Jerusalem above [*ano*]" is contrasted to "the present Jerusalem"; (b) in Phil 3:13–14 Paul speaks of the "upward" (*ano*) call of God as involving a forgetting of what lies behind and straining forward to what lies ahead—this in the context of calling upon his readers to fashion their *present behavior* according to the mindset (the noun *phronema*) displayed in Christ; (c) this mindset is

---

2. In *A Very Brief History of Eternity*, Carlos Eire offers an illuminating analysis of how different understandings of eternity in successive periods of Western culture have shaped social, economic, political, and individual human existence in the "here and now." What is most interesting is his analysis of current secular understandings of eternity as informed by modern astronomy—an impersonal, meaningless cosmic timelessness, or cyclical "eternal return" of cosmic processes (think Einstein), under the eyeless blank glare of which human existence in its breath-taking ephemerality and post-modern individualism becomes, in Whitehead's words, "a flash of occasional enjoyments lighting up a mass of pain and misery, a bagatelle of transient experience" (*Science and the Modern World*, 192). For all its superficial strangeness, then, Col 3:1–4 may offer a strangely relevant alternative, for life in this world, to current views on "things above"!

to be informed by the *politeuma*—the "polity"—of the risen Christ who is "in heaven" (Phil 3:20), in contrast to those (Phil 3:19) whose "*minds-are-set [phronountes]* on earthly things [*ta epigeia*]." (4) This last phrase, *ta epigeia*, a close equivalent of *ta epi tes ges* in Col 3:2, further supports the suspicion that in Col 3:1-4 the concern is not with escaping this world but with living within it in accordance with a possibility, an authorization, a power, grounded in the risen Christ who at present is "seated at the right hand of God/hid in God."[3] Specifically, I propose that it is appropriate to hear resonances between the word, "hid" (*kekryptai*), in Col 3:3 and the sorts of connotations traced in the Introduction for the Hebrew phrase, *beseter*, and related expressions in the Hebrew Bible.

To these observations, I add a heuristic suggestion on the basis of my threefold characterization of "place" in the Introduction: that to "set one's mind" on *ta epi tes ges* is to live on the assumption that one's standpoint in life is effectively "plotted," and sustained and determined primarily in interaction with factors and forces within the space-time grid of the natural world and human forces making up one's social world (compare our idiom, "maintain one's social position");[4] and that to "set one's mind" on *ta ano* is to live within these two realms as fully engaged with both of them but as having one's ultimate standing in a "place" that is established, circumscribed, and indwelt by the risen Christ seated at the right hand of God. I shall return later to Col 3:1-4 and comment on its relevance to prayer as a "taking place." At present I want to return to the point from which the Introduction proceeded—Jesus' instruction to his disciples to pray to God *en to krypto*, "in secret," or, as I have suggested, "in the secret place," which in the Psalms and elsewhere is repeatedly indicated as God's earthly or heavenly sanctuary, or as God's own presence as refuge and "hiding place."

---

3. Apart from Col 2:12 and 3:1, the compound verb *synegeiro* occurs only in Eph 2:6, "*raised us up* with him, and made us sit with him in the heavenly places [*en tois epouraniois*]." The latter clause associates Christians in some fashion with Christ's current "heavenly" rule such that they are to be delivered, in the present, from subjection to the "principalities and powers" that otherwise exercise such power over human life (Eph 3:10; 6:12).

4. This is, for many in our day, "under the eye" of an impersonal cosmic eternity incapable of even being indifferent, let alone hostile, to human interests or welfare, incapable of itself containing let alone displaying anything that would correspond to what we mean by "meaningfulness" or "purpose" in the sense of "point," as in "what's the *point* of it all?" The equivalent in ancient Israel would be the "practical atheist" discussed in chapter 8.

PART FIVE: New Testament Afterword

Praying to the Father Who Is "in Secret" (Matthew 6:6)

Matthew 6 opens with the general warning, "Beware of practicing your piety before men in order to be seen [*theathenai*] by them." Such persons Jesus calls hypocrites. The Greek word *hypokrites* carries a number of connotations relevant here. In Greek drama performed in a theater (*theatron*), it refers to an actor as wearing the mask of his or her role. Piety, for its part, as a *religious* act is properly directed toward God. Why would anyone engage in a Godward act to be seen (*theathenai*) by others? Presumably to gain their approval or esteem, and so to gain or sustain one's standing among them.[5] Such people, says Jesus, "have their reward" (6:2, 5, 16) in the social standing they enjoy. But in the Last Day—after the play is over—they will have no reward, no final standing, no place to stand (cf. Ps 1:5); they will be unmasked and shown to have taken their stand not on the rock of Jesus' words but on sand of conventional human worldly opinion (Matt 7:24–27).

Over against such pious "theater," Jesus counsels a *practice of piety* (the Greek expression in 6:1 is *ten dikaiosynen poiein*, literally, "doing righteousness") "in secret," before God who is "in secret" (both *en to krypto*). The counsel concerning the practice of prayer introduces the "Our Father," whose opening three-clause petition centering in "thy kingdom come" gives the whole prayer an eschatological orientation. Does the eschatological orientation of this prayer stand in implicit contrast to a similar orientation in the ostentatious prayer critiqued in 6:5?[6] Are these "hypocritical" prayers moved by a desire to exhibit an exemplary passion for the coming of God's kingdom? A clue may lie in the way the counsel in 6:6 closely echoes the LXX. Jesus' counsel on prayer reads as follows,

> When you pray [*proseuche*], go [*eiselthe*] into your room [*eis to tameion sou*] and shut the door [*kleisas ten thyran*] and pray to your Father who is in secret [*en to krypto*].

---

5. I hasten to register my conviction that "a decent respect to the opinions of mankind" is, so to speak, part of the fabric of God's creation, and functions as one kind of glue to bind people together in community. So important is it, as a good, that—like all goods—it can become one's highest good, and as such, an idol. In that way of looking at it, a prayer that is truly addressed to God "in a secret place" even as it is offered along with others in prayer, is an act of smashing idols, actual or potential, including one's own.

6. One may muse, wryly, that if the reward of hypocrites lies in the impression they succeed in making on others, their prayer is answered immediately. So much for instant gratification and immediately realized eschatology!

"Hid with Christ in God"

With this we may compare Isa 26:20, where God calls upon the community,

> Come, my people, go [*eiselthe*] into your chambers [*eis ta tamieia sou*], and shut your doors behind you [*apokleison ten thyran sou*]; hide yourselves [*apokrybethi*] for a little while until the wrath is past.[7]

The counsel is to Israel in exile and comes near the end of a long section (Isaiah 25–26),[8] charged with eschatological import and concluding with a sentence that begins, "for behold, the LORD is *coming forth* out of *his place* to punish the inhabitants of the earth for their iniquity." The apocalyptic ring of these words is reinforced by the concluding lines: "and the earth will disclose [*anakalypsei*] the blood shed upon her, and will no more cover [*katakalypsei*] her slain."[9] The interest of this long passage for eschatological expectation in the time of Jesus is suggested by its numerous quotations and echoes in the New Testament.[10] While the passage opens with a first-person singular affirmation of praise for God's "wonderful acts," arising out of "plans formed of old, faithful and sure" (Isa 25:1), and, while it expresses hopes that at times enlarge the boundaries of eschatological hope to embrace even the dead (Isa 25:6–9; 26:19), its latter part in 26:8–18 consists in a lengthy expression of yearning and impassioned prayer elaborating on the opening words, "O LORD, we wait for thee." Isaiah 26:20 comes immediately after that long prayer, before the announcement of God's eschatological "coming forth." Given how Matt 6:6 is followed immediately by the "Our Father" with its prayer for God's *coming* kingdom, one wonders whether the prayers excoriated in Matt 6:5

---

7. One may also compare Matt 6:6 with 2 Kgs 4:33. But Matthew's *eiselthe eis to tameion sou* more closely echoes *eiselthe eis ta tamieia sou* in Isa 26:20 than it does *eiselthen eis ton oikon* in 2 Kgs 4:33. Not only does Isa 26:20 employ the same word, *tam(i)eion*, but also, as in the Matthean text the sentence functions as an exhortation and is not presented simply as a narrator's observation (as it is in 2 Kings).

8. The exilic origin and bearing of Isaiah 25–26 are generally recognized.

9. For a prayer of vindication against such "covered-up" violence, see Job 16:18–21. For the theme in Matthew of the eschatological vindication of such violence, see the other passage excoriating religious hypocrisy, with its ending, in Matt 23:35.

10. The following passages in Isaiah 25–26 are listed in the Nestle-Aland *Novum Testamentum Graece* (26th ed.) as cited (marked here with *) or echoed in the New Testament: Isa 25:4 in 2 Thess 3:2. Isa 25:8 in 1 Cor 15:34*; Rev 7:17*; 20:14; 21:4*. Isa 25:29 in Luke 21:35. Isa 26:11 in Heb 10:27. Isa 26:13 in 2 Tim 2:19*. Isa 26:17 in John 16:21; Rev 12:2. Isa 26:19 in Matt 11:5*; 27:52; Luke 7:22*; John 5:28. Isa 26:20 in Matt 6:6*. Isa 26:21 in Rev 8:13.

are to be taken as ostentatious repetitive public prayers for the coming of God's kingdom in the spirit of passages like Isaiah 25–26.

The petitions in the "Our Father" end with the plea to be spared temptation or the time of trial, and to be delivered *apo tou ponerou*—"from evil," or better, "the evil one."[11] This theme of "the evil one" crops up again at Caesarea Philippi, as Satan, in Jesus' sharp rebuke of the conventional messianic mindset that underlies Peter's vehement objection to Jesus' words about his vocation as a "crucified messiah." And it is implicit in Gethsemane, not only in Jesus' call to Peter and the others to "watch and pray that you may not enter into temptation [*peirasmon*]" (26:41), but in his own struggle in prayer over God's will (26:39, 42). But Jesus' temptation, or trial, appears most fully in Matt 4:1–11 where, after his baptism, in which he is designated God's Son (echoing Psalm 2), he is "led up by the Spirit into the wilderness to be tempted by the devil." Given how Jesus is more than once described as withdrawing to a solitary place for prayer (Matt 14:23; 26:39, 42, 44; cf. Mark 1:35), one may take it that Jesus' aim in resorting to such a deserted *place* is for such a time of prayer and reflection on what his announced vocation as messianic Son may entail. Let us suppose that a good part of that prayerful reflection would involve pondering and praying the Psalms—beginning with Psalm 1 and its *torah* orientation vis-à-vis the "wicked, sinners, scoffers," and Psalm 2 as echoed in his baptism. The psalms explicitly focusing on the king would come in for especial attention. But insofar as the whole Psalter would by Jesus' day be heard as "psalms of David,"[12] all would bear on his mission. Insofar, then, as Psalm 91 becomes a vehicle for one of his temptations, or trials, and insofar as that psalm concerns one who "dwells in the shelter [*beseter*] of the Most High, who abides in the shadow of the Almighty," and who says to the LORD, "my refuge [*maḥsi*] and my fortress; my God, in whom I trust," one may appreciate the close relevance of this psalm for an exposition of the security of the Son of Psalm 2 which ends, "blessed are all those who take refuge [*ḥose*] in him." In what follows, I want now to engage in a thought-experiment, of a sort that might trace a hypothetical train of reflection that suddenly finds itself on the precipice of a temptation or trial. I shall begin with a brief comment on some salient features in Psalm 91.

11. With RSV text note, NRSV, NIV, and in accordance with the Greek phrase's meaning in Matt 5:37 (RSV text note; NRSV, NIV); 13:9; 13:38.

12. On the question of whether we may legitimately hear the voice of David, as a literary construct, in all the psalms regardless of who actually wrote this or that psalm, see, affirmatively, Mays, *The Lord Reigns*, esp. 85–116 (Part 4, "David as Psalmist and Messiah").

"Hid with Christ in God"

God's "Secret Place" as Temptation and as Reality

Psalm 91 presents perhaps the most elaborate expression in the psalms of prayerful trust in God's "secret place" (*seter*) or "shadow" (*ṣel*) (v. 1), a place characterized also as a refuge (*maḥseh*) and fortress (v. 2) where, under God's "pinions" or "wings" one may "find refuge" (*teḥseh*, v. 4). These last images correspond to the implicit image, in 91:3, of the psalmist as a (young) bird vulnerable to "the snare of the fowler." The bird image for the object of God's care goes back to Exod 19:3–6 and Deut 32:10–11, and appears in connection with the Jerusalem temple in Ps 84:1–3. When Psalm 91 is read, then, within the context of the completed Psalter, anyone using it as an aid to prayer may be said to "take refuge" in the God whose rule is expressed in the Son of Psalm 2. But in the completed Psalter the quintessential voice heard in Psalm 91 is the implied figure of David, who as Israel's ideal king is the exemplary offerer of prayer.[13] Given how v. 9 parallels vv. 1–2, and v. 10 parallels v. 3, the image of the pinions and wings of v. 4 may be taken as paralleled by the image of the "angels" that *bear one up* in vv. 11–12.

At this point I want to suggest one way in which a faithful, prayerful meditation on this psalm could inadvertently segue into an insidious temptation. The segue may be imagined as turning on the connotations of the word, "wings," in v. 4, rendered in Greek as *pteryx*. Jesus, led into the wilderness by the Spirit that descended on him at his baptism (Matt 4:1), may be imagined as pondering, there, the implications of that baptism with the help of the Psalms—a natural recourse, given how the voice at his baptism had echoed a line from Psalm 2. In the course of his meditations he comes to Psalm 91. Having reflected on the "refuge" theme in v. 2 and recollected its first appearance in Psalm 2, Jesus goes on to ponder the image of the angel ministry in Ps 91:11–12. But that word *pteryx*, evoking the image of God as liberating and nurturing and protecting eagle, in the exodus, the wilderness (!), and the sanctuary, continues to resonate in the back of his mind; and in doing so it segues, through the "bear you up" language of Ps 91:12, into its close cognate, *pterygion*, connoting a "wing" or

---

13. Compare 1 Kings 8, where Solomon offers the prayer dedicating the newly completed temple in Jerusalem as the place where prayer is to be offered (but heard "in heaven [God's] dwelling place"); and 1 Chronicles 29, where David's prayer is even more foundational, in connection with the freewill offerings he makes and calls for toward the building of the temple. In Jesus' day, then, David and Solomon, as successively holding the office of God's anointed "son," were the royal founders of the temple as the place of prayer, with David (in 1 Chronicles 16) as founding patron of the use of the Psalms for prayer and praise in that place.

PART FIVE: New Testament Afterword

"pinnacle" of the temple.[14] Carried up in imagination to that height, Jesus finds himself momentarily enticed by the thought that, were he to jump off that "wing," God's angels would "bear him up" like the wings of the eagle in Exod 19:4 and Deut 32:11 and assure him a "soft landing." But the very thought is answered with a word that God had spoken through Moses to Israel in the wilderness, "You shall not put the LORD your God to the test." Tellingly, it is only *after* the period of testing is over, according to Matthew, that the angels (of Psalm 91?) come and minister to Jesus (Matt 4:11).

The above scenario is only an imaginative construct (the mode, of course, in which all temptations come!). But the possibility of a resonant play between *pteryx*, "wings," in Ps 91:4 and *pterygion*, "wing," in Matt 4:5 is an invitation to enter imaginatively into the interplay of the text. This much, at any rate, I take from the involvement of Psalm 91 in Jesus' experience of trial and temptation: The temptation is not only to make a (hypocritical?) public display of his confidence in God's assurance of protection in this psalm, but, deeper than that, to construe the *nature* of that protection as offering immunity from all physical and social harm and danger. In the terms of John's Gospel, this would be a docetic or only "seeming" incarnation of the Word.

With the imagery of God's sheltering wings, and of Israel as young birds seeking refuge under them from the snare of the fowler, we may now compare Jesus' lament in Matt 23:37-38 (concluding the other Matthean excoriation of hypocrites), "O Jerusalem, Jerusalem, killing the prophets and stoning those who are sent to you! How often would I have gathered your children together as a hen gathers her brood [*ta nossia*] under her wings [*tas pterygas*], and you would not!"

In the image of the sheltering wings one should hear resonances not only with the wings of Deut 32:10-12 (where the LXX renders Hebrew *qen*, "nest," with *nossia*), and Isa 40:31, and the wings of the cherubim, but also with the protecting "shadow [*ṣel*] of thy wings" so often sought by the psalmists as a refuge from their enemies (Pss 17:8; 36:7; 57:1; 61:4; 63:7; and, above all, 91:4).[15] Coming at this point in this Gospel—after

---

14. In the LXX, the word *pteryx* refers to God's "wings" in Exod 19:4; Deut 32:11; and often in the psalms, possibly in reference to the cherubim. The word *pterygion* refers to a cherub's wingtip in 1 Kgs 6:24; to the "ends" of a breastplate in Exod 39:19; and to the hem or skirt of a garment in 1 Sam 15:27; 24:4, 11; as also in Ruth 3:9 where the image of Boaz's "skirt" (*pterygion*) appears as the human complement to the image of God's "wings" (*pterygion*) in Ruth 2:12. Matthew 4:5 is the only place in the Bible where *pterygion* refers to a feature of the temple's external architecture.

15. In the Old Testament, the theme of "refuge," whether imaged as protecting

"Hid with Christ in God"

the temptation story and especially after the scene at Caesarea Philippi in Mathew 16—the "you would not" takes on a suggestive connotation. The "would not," like Peter's rebuke of Jesus' conception of messianic vocation, may in part be a refusal to accept the sort of "refuge" that a non-idolatrous appropriation of Psalm 91 would imply—a refuge that would leave one vulnerable to all the vicissitudes of a life fully participating (fully incarnate) in the physical and social world, but would provide assurance that, in words of Paul that I shall return to, no such vicissitudes are able to separate one "from the love of God in Christ Jesus our Lord."

"Your Life Is Hid with Christ in God"

Considered solely by itself, Col 3:1–4 seems not to have anything to do with "prayer taking place." But this impression overlooks the subtle ligature by which the passage is connected to other biblical passages involving prayer. It makes clear that to be "hid with Christ in God" is to be connected in some way with one who is "seated *at the right hand of God,*" the image, as James Dunn says, "clearly echoing Ps 110:1."[16] And in that psalm, "David" is celebrated as not only king but "a priest after the order of Melchizedek." So it is most likely that, whether Colossians was written by Paul or a member of his school, the connotations in Col 3:1–4 include what is said in Rom 8:34, where Paul writes, echoing Psalm 110, "who is to condemn? Is it Christ Jesus, who died, yes, who was raised from the dead, who is *at the right hand of God*, who indeed *intercedes* for us?"[17] The

---

"shadow" or "wings," typically is applied to God; so its application to Jesus here is striking. But see also Lam 4:2: "The breath of our nostrils, the LORD's *anointed*, was taken in their pits, he of whom we said, 'Under his *shadow* [ṣel] we shall live among the nations.'" In Judg 9:15 a human figure says, in a parable, "If in good faith you are *anointing* me *king* over you, then come and take refuge [ḥasu] in my shade [ṣel]." And compare the thematic interaction between God's "wings" (*kenapayim / pterygas*) in Ruth 2:12 and Boaz's "skirt" (*kanap / pterygion*) in Ruth 3:9. Most commentators on the last line in Psalm 2 take the "him" in whom the blessed take refuge as God rather than the anointed Son. If so, the clear implication of the psalm is nevertheless that God's refuge is concretely to be sought in God's anointed Son. See further below.

16. Dunn, *Epistles to the Colossians and to Philemon*, 203.

17. I follow RSV and some recent commentators (against KJV, NRSV, and several recent commentaries) in taking Rom 8:34b as a rhetorical question, having perhaps in the background the possibility that the Christ, or Messiah, when he appears, will come chiefly to indict and condemn his enemies. The possibility is here raised only to be dismissed—if *he* does not condemn, but intercedes for us, then no other condemnation against us can prevail. I would also draw attention, yet again, to David's function

import of this latter passage for the present essay calls for a foray into the many-dimensioned portrayal of prayer in Romans 8.

Prayer in Romans 8 as the Nexus of the Solidarity of Heaven and Earth

The intercession of the heavenly Christ in Rom 8:34 is intimately linked with Paul's earlier references to prayer in this chapter. To begin with, it resonates with the intercession of the Spirit in 8:26–27.[18] And, given how Paul speaks in 8:9–11 of the Spirit as both the Spirit of God and the Spirit of Christ, we may venture that the intercession of the Spirit arising from within the depths of the human heart is the presence in the human heart of the intercession of Christ at the right hand of God. This may even be suggested by 8:15–16: "When we cry [*krazomen*], 'Abba! Father!' it is the Spirit personally[19] bearing witness with our spirit that we are children of God, and if children, then heirs, heirs of God and fellow heirs with Christ, provided we suffer with him in order that we may also be glorified with him."

I take the expression, "fellow heirs with Christ," to resonate with the motif of the anointed Son's "heirship" in Ps 2:8, as a participation in his messianic vocation (compare already Isa 55:3–5 where the addressee is Israel in the plural), with the proviso of "suffering with" on the way to "glorification with." This close identification with Christ in his suffering may also be signaled in the word through which the Spirit (of Christ) "bears witness with our spirit that we are children of God." For, in the only New Testament instance outside of Paul's use here and in Gal 4:4, the Aramaic word transliterated as *Abba* comes on the lips of Jesus in Mark's version of his travailing prayer in Gethsemane, "*Abba*, Father, all things are possible to thee; remove this cup from me; yet not what I will, but what thou wilt." The Spirit, then, that bears witness with our spirit, is the Spirit evidenced most clearly in that Gethsemane prayer. Now, it is not insignificant that almost always in the LXX, and most frequently in the New Testament, the verb *krazo* connotes a cry of distress or other extremity. In any case, that stressful connotation of the cry in Rom 8:15–16 is surely implied in

---

as royal intercessor in 1 Chronicles 29, and especially to Solomon's in 1 Kings 8, on which see further below.

18. The verb, *hyperentygchanei*, in Rom 8:26 is matched by the verbal phrase, *entygchanei hyper*, in 8:34.

19. The Greek phrase is *auto to pneuma*—precisely as in 8:26.

"Hid with Christ in God"

the proviso that "we suffer with him." As such, this cry in 8:15-16 arises in solidarity with the groaning of creation (v. 22), of those who have the first-fruits of the Spirit (v. 23), and of the Spirit itself (v. 23); and with the seven-fold extremities in v. 35 in relation to which the heavenly Christ intercedes in v. 34.

If, then, we may connect the many-dimensioned portrayal of "solidarity in prayer"[20] in Romans 8 specifically with the prayer of Christ in Gethsemane, this may throw some light on what Paul means in 8:26 in saying "we do not know how to pray as we ought (*katho dei*)." James D. G. Dunn asserts that, as generally recognized, "*katho dei* here is more or less equivalent to *kata theon* ['according to the will of God']" in 8:27."[21]

In view of Paul's language in 8:29, "For those whom he foreknew [*proegno*] he also predestined [*proorisen*] to be conformed to the image of his Son," it is easy to take *katho dei* as reflecting a "high divine determinism" here, referring simply to the divine predetermination as to the shape of worldly events in accordance with God's ultimate purposes. But three considerations lead me to a more nuanced reading of *katho dei*, considerations that bear directly on prayer as lodging the situational concerns of earth in the bosom of the situational providence of heaven.

In the first place, the verb *proorisen* in 8:29 refers to a specific "pre-horizoning"—conformation to the image of God's Son. This resumes the point in 8:17, and it suggests to me that all of 8:15-28 is to be read through the lens, as it were, of Gospel passages such as Mark 8:34//Matt 16:24 and such Pauline passages as Phil 2:12-13 after 2:5-11 (also 3:10-11). The spiritual and ethical pattern, the mindset (*phronema*) with which one engages all specific, variable situations in the world, is to be conformed to the pattern of Christ in his crucifixion and resurrection/glorification.

Second, despite the appeal to such Stoic writers as Epictetus, it is not necessary to read *dei* with strictly deterministic connotations. Even in Epictetus the term often means simply, "if you want to achieve result x, pattern of life y is *dei*." It is all too clear from Epictetus' discourse that,

---

20. Greek *syn*, "with," occurs free-standing or as a prefix ten times in Rom 8:15-39, tracing an anatomy of solidarity in vicissitude and redemption: *symmartyrei*, "bears witness with" (8:16); *sygkleronomoi*, "co-inheritors" (8:17); *sympaschomen*, "we suffering together" (8:17); *syndoxasthomen*, "glorified together" (8:17); *systenazei*, "groans together" (8:22); *synodinei*, "travails together" (8:22); *synantilambanetai*, "joins in assistance with" (8:26); *synergei*, "works together" (8:28); *symmorphous*, "conformed with" (8:29); and, finally, free-standing *syn*, "with" (8:32), returning to the theme of co-inheritance in 8:17.

21. Dunn, *Romans 1-8*, 477.

despite his logical and rhetorical efforts to persuade them, many of his desultory hearers who might desire the result in question are not willing to apply themselves to the life-discipline called for. I take *dei* in Rom 8:26, then, as carrying a connotation that brings it into closer semantic relation with its cognate prayer word, *deesis* (as in Phil 4:6, "supplication"), referring to "what is needed or wanted." The verb *dei* is a third person singular present indicative form of *deomai*, "to be lacking, to need, want," and in this "impersonal" usage means, "what is needed or wanted," or "what the situation calls for." To the degree that *dei* carries connotations of "divine determinism," it refers to what God determines the general or specific situation to call for. One factor in the difference between a human and the divine estimate of what the situation calls for is that the "horizon" within which humans experience a pressing sense of need is often confined to the immediate or proximate present, or the span of one's own lifetime, whereas the "horizon" within which God determines what is called for attends not only to such an immediate or proximate present, but also to the wider and history-long redemptive scenario through which God will bring all things "unto himself" (Rom 11:36).

A third consideration in construing the connotation of *dei* in Rom 8:26 arises from the way in which the heavenly intercession of Christ as God's Son resonates with the programmatic intercession of Solomon as God's Son (so called in 2 Sam 7:12–14) at the inauguration of the Jerusalem temple in 1 Kings 8. In that intercession (which for its part has its royal precedent in 2 Sam 7:18–29 and 1 Chron 29:10–19), Solomon offers a seven-fold petition covering seven types of exigency in which Israel may in years to come find itself. Beginning with a recollection of God's promise to David in 2 Sam 7:4–17, Solomon asks that God "have regard to the prayer [*deesis*] of thy servant and to his supplication," and that God's eyes "may be open *night and day* toward this house, *the place* [*maqom*; LXX: *topos*] of which thou hast said, 'My name shall be there,' that thou mayest hearken to the prayer which thy servant offers toward this place" (1 Kgs 8:28–29). For the the biblical historical theologian, writing in a Deuteronomistic vein, from now on the earthly "place" for prayer is the temple. When, then, in future years, Israelites pray "toward this place" (8:30 and repeatedly), Solomon asks that God hear "in heaven *thy dwelling place* [*meqom šibteka*]" (8:30 and repeatedly).

I pause here to note the correlation between where earthly prayer "takes place" and where God's hearing of it "takes place." While God's "ears," so to speak, are in heaven, God's "name" is in the temple, in the

"Hid with Christ in God"

form it takes when people there "call on God's name." Following the sevenfold petition, Solomon blesses the people (8:54–58); and then he says,

> Let *these words of mine*, wherewith I have made supplication before the LORD, *be near to the LORD* our God *day and night* [*yomam walayelah*], and may he *maintain the cause* [literally, "do the *mišpaṭ*"] of his servant, and the cause [*mišpaṭ*] of his people Israel, *as each day requires* [*debar-yom beyomo*], that all the peoples of the earth may know that the LORD is God; there is no other.

This conclusion contains a theologically bold inversion of one of God's concluding words to Israel through Moses. In Deut 30:14 God says to Israel, concerning all that God has said by way of commandment and statute and ordinance, "the word [*dabar*] is very near [*qarob meʾod*] you; it is in your mouth and in your heart, so that you can do it." The only other verse in the Bible where the words "word" and "near" occur together is here in 1 Kgs 8:59. If God's word is "very near" to Israel, so that they are to meditate on the book of the *torah* "day and night" [*yomam walayelah*]" (Josh 1:8), so in return Solomon asks that "these words" of his programmatic intercession, offered before God "this day" (8:28), may be "near" God "day and night." That is, Solomon asks that his programmatic prayer be a "standing prayer," providing the intercessory agenda for God's merciful dealing with Israel in whichever of the seven typical exigencies Israel may at any given time find itself.

Solomon drives home the quotidian specificity of his request with the idiomatic expression, *debar-yom beyomo*. This idiom (something like "the matter/concern of a day on that day") occurs 13 times in the Hebrew Bible. Seven times it refers to a daily task imposed by Pharaoh or cultic duty mandated by God.[22] Six times it refers to daily allotment of food by God, a king, or all Israel.[23] In the first sort, *debar-yom beyomo* involves a claim on human activity—what is *dei*, "called for," in accordance with the decree of God (one could say, *kata theon*) or of Pharaoh. But in the second sort, the idiom involves a certain need of certain people (*katho dei*), a need to be met by God or a king or the community as a whole. Clearly, in 1 Kgs 8:58 the idiom functions in the second sense: God is besought to respond to Solomon's standing prayer in accordance with the specific need in the people's exigent situation on any given day. It goes without saying, of course, that such divine response, of "doing *mišpaṭ*," is in

---

22. Exod 5:13, 19; Lev 23:37; Ezr 3:4; 1 Chr 16:37; 2 Chr 8:14; 31:16.
23. Exod 16:4; 2 Kgs 25:30; Jer 52:34; Dan 1:5; Neh 11:23; 12:47.

PART FIVE: New Testament Afterword

accord with the divine governance of the world in general. But that *mišpaṭ* will take on specific form depending on the specifics of the exigency. In one sort, God's *mišpaṭ* (LXX: *dikaioma*) relates to Israel's military battles, and connotes victory (8:44–45); but in another sort—the seventh and climactic sort—Israel's sin, consequent judgment in exile, repentance, and confession there (8:46–48)—God's *mišpaṭ*[24] consists in God's forgiveness and granting compassion on Israel in the eyes of their captors (8:49–50). If one were to use Paul's phrases in Rom 8:26–27, Solomon would be asking God's *mišpaṭ* to take a form *katho dei*—as Israel's situation called for.[25] And the extreme exigency would be Israel's captivity in exile because of its radical covenant infidelity and sin.

To this point, the picture in 1 Kings 8 is only suggestive of how one might construe the correlation between *katho dei* and *kata theon*. But a number of features suggest a strong resonance between the two passages. There is the fact of Christ as a royal intercessory figure in Rom 8:34, whose fullest Old Testament predecessor is Solomon in 1 Kings 8. There is a

---

24. The LXX here lacks "and maintain their *mišpaṭ*," probably by scribal error, the copyist's eye jumping in Greek from "their" to "their."

25. I am intrigued by the possibility that the use of the term *mišpaṭ* three times in Solomon's temple-related intercesion in 1 Kings 8 may be related somehow to its occurrence in the Priestly tradition in Exod 28:29–30. In the latter passage, Aaron, as High Priest, is to "bear the names [by tribe] of the children of Israel in the breastplate of judgment [*mišpaṭ*] upon his heart, . . . to bring them to continual remembrance before the LORD." The last clause identifies Aaron's action here as intercessory. The precise nature and function of the "Urim and Thummim" that are to be placed in this breastplace are not clear to scholars, who generally take them as some sort of binary system for determining God's *mišpaṭ*, taken to mean the divine "judgment" in response to the intercession. I think, rather, that *mišpaṭ* here refers, as in 1 Kings 8, to the people's "cause" as the priest holds it up before God "continually," that is (as the Hebrew term *tamid* can mean in cultic contexts), "day by day." In reflecting on such a possible connection between the intercessor and the people's *mišpaṭ* or cause as it takes specific form day by day, I find the following comment of Samuel Taylor Coleridge (clearly interpreting this biblical passage) remarkably apt. In a notebook entry reflecting on the Trinity, and the question of the procession of the Spirit from the Father *and the Son* (in Western Christian tradition), Coleridge wrote, "the *clearness . . .* in the Idea [of procession through the Son] is a full compensation for the *twilight* in the written Word, for as many as rightly comprehend our Lord's assurance, that he will always be with his Church, & as the *spiritual and inward Urim and Thummim*, giving the truth *according to the needs, circumstances, and capabilities of the Individual Believer, and of the Church and the Age*" (*Notebooks of Samuel Taylor Coleridge*, 5:no. 6499; the last two italics added). Coleridge's implicit reference to the indwelling Spirit of Christ as "inward Urim and Thummim," and his concluding words, connect thematically both with 1 Kgs 8:59 and with—as I take it—the earthly locale, in and through the Spirit's intercession in Rom 8:26–27, of the Son's heavenly intercession in Rom 8:34.

congruence between the way in which ongoing Israel's ad hoc prayers in any given exigency will participate in Solomon's standing prayer "before God in heaven," and the way in which Christian ad hoc prayers (Rom 8:15-16, 26a) participate in the intercession of Christ in heaven (Rom 8:34) and Christ's Spirit in their hearts (Rom 8:15-16, 26b-27). And a definite if tiny echo of Solomon's prayer may be detected in the Greek word *dikaioma* in Rom 5:16, an echo that reinforces the homology that I see and the resonance that I hear between Solomon's standing prayer in 1 Kings 8 and Jesus' heavenly intercession in Romans 8.

Hearing the echo between Rom 5:16 and 1 Kings 8 turns on recognition of the following factors. (1) In Rom 5:12-21, which begins the second major section of Romans stretching to the end of chapter 8, the major question concerns the "reign" or effective Power under which people live: whether the "reign" of sin and death (5:14, 21) or the "reign" of grace through Christ (5:21). (2) The imagery here is of Sin and Death as transhuman "Principalities" (as in 8:38-39) or Powers exercising their reign in and through earthly powers and institutions. (3) For Israel prior to Christ, the human embodiments of these Powers at the political level have been Egypt, Assyria, Babylon, and subsequent imperial powers from Persia down to Rome. In 5:12-21 Paul begins to argue that, through his death and resurrection, Christ has victoriously delivered his followers from the transhuman power of Sin and Death. (4) In the context of these thematics of captivity to the reign of sin, Paul speaks of liberation through grace by using the Greek terms *dikaiosis* (v. 18) and *dikaiosyne* (vv. 17, 21)—words in which the modern reader may most readily hear juridical connotations, but which, in the case of *dikaiosyne*, often carries connotations of God's deliverance from alien power in the LXX. (5) In Rom 5:16 (as in 5:18), the cognate noun *dikaioma* is generally taken as a synonym of *dikaiosyne*, "justification" (so, e.g., KJV, RSV, NRSV, NIV), despite the failure of commentators to produce any evidence for such a connotation; this general interpretation has recently been challenged, in favor of the term's usual meaning of "judgment, penalty, or reparation."[26] (6) But the general interpretation in fact receives support from one overlooked passage—in Solomon's intercessory prayer! In 1 Kgs 8:45, 49, 59, Solomon prays that God will, "do *mišpaṭ*," "maintain the cause," for the people in battle against their enemies (v. 45), in exile and oppression because of their sin (v. 49), and in whatever situation of need they find themselves in (v. 59). As already observed, the idiom is missing in the LXX of v. 49,

---

26. See Kirk, "Reconsidering *dikaioma* in Romans 5:16."

PART FIVE: New Testament Afterword

probably owing to scribal error; but in vv. 45 and 49 the idiom is rendered, "do *dikaioma*." Here—and only here in extant Greek literature—the latter Greek word carries exactly the connotation that fits its use also in Rom 5:16, 18; and the two passages in which it uniquely carries this connotation, 1 Kgs 8:44–59 and Rom 5:12–21, are exactly homologous in thematic profile.

The point of all this is to give an illustration of the gossamer tones of echo and resonance that can connect passages of scripture to one another. In the present instance, it suggests that the prayer of Solomon—climaxing in a concern for the eschatological deliverance of Israel envisioned as fallen into captivity and oppression because of covenant infidelity and sin—operates as a Scriptural context for Paul's whole argument beginning in Romans 5 and ending in Romans 8, with the prayer dimension of the Davidic Son in 1 Kings 8 surfacing explicitly in the intercession of the Son in Rom 8:34.

This brings me to a question that is not suggested by the New Testament texts themselves, but by the way Solomon asks, in 1 Kgs 8:59, that his prayer on this day of temple-consecration stand before God as a "standing prayer" for the people, down through the generations, presumably undergirding their own prayers as offered in any one of the seven sorts of exigencies he has already held up before God. My question is this: What is the relation between the heavenly intercession of Christ in Rom 8:34, as undergirding the prayers in 8:25–26, and the *abba* prayer of Jesus in Gethsemane according to Mark 14:6? Is the heavenly prayer in part a prolongation of the Gethsemane prayer? In that case, the four-way nexus between Mark 14:36; Rom 8:15–16; Rom 8:25–26; and Rom 8:34 would become a four-dimensional nexus uniting "time" and (earthly and heavenly) "space" in a solidarity that holds all secular times and spaces in a matrix of potential reconciliation and redemption. The "place" that grounds all such prayer, all such *groaning* and *crying out*, and delivers it from mere captivity *to* the realm of space and time, setting it free for agential participation *in* the realm of space and time, is, as Rom 8:34 and Col 1:1–4 both have it, "at the right hand of God."

If, then, the counsel of Jesus in Matthew 6 to pray "in secret," involves a participation, through the Spirit, in the intercession of Christ who now sits at the right hand of God, it should be evident, through the creation-embracing emphasis on solidarity connoted through the repeated prepositional element *syn*, "with," in Romans 8, that such prayer "in secret," and such setting one's mind on such "heavenly things" (Col 3:1–4), far from being individualistic other-worldly escapism, stands in profound solidarity with the whole creation in its unfathomable passage "from God and through God and unto God" (Rom 11:36). It is only an apparent paradox, that genuine entry through prayer into such a "secret place" at the same time places one at the heart of this world.

Garrisoned in Prayer

This understanding of the "place" of prayer is congruent with another classic counsel on prayer, when, in Phil 4:6–7, Paul writes,

> Have no anxiety about anything [*meden merimnate*],[27] but in everything by prayer and supplication [*deesis*] with thanksgiving let your requests be made known to God. And the peace of God, which passes all understanding [*nous*], will keep [*phrouresei*] your hearts and your minds [*noemata*] in Christ Jesus.

To start with, I take the noun *noema*, in the plural, to refer to the specific "thoughts" arising in one's mind in relation to those issues in one's circumstances that generate a spirit of anxiety, issues that (like those confronting the woman in Psalm 131), prove too much for one's powers, baffle one's understanding, and threaten to make a mockery of the notion that "in all these things *we more than conquer* [*hypernikomen*] through him who loved us" (Rom 8:37). It is not that Paul's readers are forbidden to be anxious (a futile counsel), but that those anxieties are to be gathered up in prayers and petitions in which one's specific requests are leavened with thanksgiving. In saying that "the *peace* of God will *keep* your hearts and minds in Christ Jesus" (RSV), one could take Paul to be invoking his readers' familiarity with the High Priestly Prayer of Num 6:24–26 whose first line ends with "keep you" and whose second line ends with "give you peace." But if that echo were intended, why would Paul not use the Greek verb *phylasso* as in the LXX of Num 6:24? Even apart from this blessing, the verb *phylasso* occurs regularly in the Old Testament to render the Hebrew verb *šamar*, including all twenty times that the latter verb occurs in the Psalms with God as subject. (Relevant to our general topic, one may note especially Pss 17:8; 25:20; 41:2; and five instances in Psalm 121.) Why

---

27. With Phil 4:6–7 we may compare John 14:27, "Peace I leave with you; my peace I give to you; not as the world gives do I give to you. Let not your hearts be troubled [*tarassestho*], neither let them be afraid."

PART FIVE: New Testament Afterword

does Paul resort to the highly unusual verb, *phroureo*, which NRSV and NIV render "guard"?

Greek words built on the stem *phrour–* carry the general connotation of guarding or protecting, and in many instances refer to a garrison. In the LXX, words from this stem appear only infrequently, but their connotation there is most suggestive. The data are these: (1) The nouns *phroura* and *phrourion* occur only in 1 Esdras (1x), Judith (1x), and 1 and 2 Maccabees (11x), and in these 13 instances always refer to a "garrison," having the same sense as that English word, "*a body of soldiers* stationed in a fortress or other place for purposes of defence, etc.," or "*a place* in which troops are quartered for defensive or other military purposes" (*OED*; italics added). In these passages, the region under consideration is contested, so that the garrison is stationed there to "pacify" it. (2) The verb, *phroureo*, occurs literally, of a garrison, in 1 Esdras 4:56 and Judith 3:6; and it is used metaphorically, in Wisdom of Solomon 17:15–16, in the context of that book's description of the exodus plague of darkness that terrified the Egyptians, instilling such terror in them that they were "now paralyzed by the betrayal of their own minds, for sudden and unexpected fear poured over them. Thus, whosoever fell down there was *held bound* [*ephroureito*], locked [*katakleistheis*] in a prison without bars."[28]

All the above occurrences come in the deutero-canonical LXX; they illustrate copiously—and consistently—the meaning, "garrison," for that literature. But they give no indication as to why Paul might have chosen such a word for use in reference to prayer and the peace of God in Christ Jesus. But the LXX contains one more passage where the noun *phroura*, "garrison," occurs, and here—the only instance in the Hebrew Bible—the context is highly suggestive. The context is this:

David, having established his royal headquarters in Jerusalem, and having brought up the ark to Jerusalem from its Philistine exile, proposes to build a sanctuary for the LORD (2 Sam 7:1–3). But he is informed through Nathan that, instead, God will build *him* a house, in the form of a dynasty such that "your house and your kingdom shall be made sure forever before me; your throne shall be established forever" (2 Sam 7:4–17). This dynastic promise evokes a long prayer in which David asks that this promise be actualized in David's line and for Israel. (The prayer, especially as arising out of the existential question, "who am I, and what is my house?" [7:18] surely forms the template for David's prayer in 1 Chronicles 29.) After this oracle and this prayer, consolidating David's reign internally,

---

28. The translation is that of Winston, *Wisdom of Solomon*, 303.

"Hid with Christ in God"

2 Samuel 8 narrates David's extension of power over the surrounding regions of "Edom, Moab, the Ammonites, the Philistines, Amalek, and ... Zobah" (8:12). To secure these regions to his royal rule from Jerusalem, David then "put *garrisons* [*phrouran*] in Aram of Damascus; and the Syrians became servants to David and brought tribute. And the LORD gave victory [*wayyošaʻ*; LXX: *esosen*, literally, "saved"] to David wherever he went" (2 Sam 8:6 // 1 Chr 18:6). Also, "he put *garrisons* [*phrouran*] in Edom; throughout all Edom he put garrisons, and all the Edomites became David's servants. And the LORD gave victory to David wherever he went" (2 Sam 8:14 // 1 Chr 18:13).

The scenario here is a Davidic analogue to the later *pax Romana* in which Rome kept order throughout its empire through garrisons established in key colonial centers like Philippi in Asia Minor. The citizens of the latter city, we recall, enjoyed all the benefits of Roman citizenship, including governance under its laws and protection by its garrisoned troops. What I am suggesting is that the Paul who could in Phil 3:20 analogize the Christians of Philippi as a colony of "the Jerusalem *ano*," here in 4:7 invokes on those who pray the benefits of what we might call the *pax Hierusalem*.[29] If for the registered citizens of Philippi, living in that city is tantamount to living in Rome, then analogously, having one's *politeuma* "in heaven" is tantamount to living there,[30] tantamount to being, as Colossians will put it, "hid with Christ in God." The "safety" of that *politeuma* is, then, the safety of one who has "died" and been "buried" with Christ in baptism and raised to a new life. This is the basis of a *pax* that may be known in the very midst of one's participation in—indeed, solidarity with—the exigencies and vicissitudes of the physical and social world.

Two questions remain. First, it is not a far reach to suppose that the Paul who in Phil 3:20 could refer to the Philippians as having their "commonwealth" in heaven could a few verses later elaborate the image with the verb, "garrison." But could he also intend an echo of the word,

---

29. *šalom* is, of course, intrinsic to the city's name (and see especially Psalm 122). And there is the ark with all its symbolism as God's throne-seat between the wings of the cherubim (1 Sam 4:4), which in the wilderness had gone before Israel to scatter the enemy (Num 10:35), which for Israel's sin had been taken into captivity by the Philistines (1 Sam 4:11; Ps 78:60–61), but under David was "brought up" to Jerusalem (2 Sam 6:15) where eventually, "hidden" behind the curtain, it became the invisible focus of Israel's worship and the place from which God's protection and help was sought (see, e.g., Psalm 80). How could such a drama not resonate with Paul ransacking his scriptures for means of understanding the heavenly reign of Christ in such places as his Philippian "garrision" after his crucifixion?

30. Compare the thought in Eph 2:6.

"garrisons," in 2 Samuel 8? Whether or not he intended it, the resonance between the Davidic situation, the political status of Philippi within the Roman Empire, and Paul's conception of the standing of the Philippian Christians under the present reign of Christ, makes for a remarkable resonance between Phil 4:7 and 2 Samuel 8. But the further, practical question is then, *in what experiential form* is God's garrisoning peace present with them? If Romans 8 is any guide, we may say that the Spirit of the Christ of Phil 3:20, the Christ who "in heaven" intercedes at the right hand of God for them—this Spirit dwelling in the hearts of the Philippian Christians and interceding for them, is the garrisoning presence.[31] Paul more than once associates the gift of peace closely with the gift of the Spirit (Rom 8:6; 14:17; 15:13; compare Eph 4:3); so that one may view the list of the Spirit's fruit in Gal 5:22—23 as, one might say, a calling of the roll of those garrisoning virtues that fortify the Christian vis-à-vis the hostile forces of this world. It is through prayerful hosting of such spiritual energies (cf. Phil 4:8) that God's peace will be with them (Phil 4:9).

Some words of Samuel Taylor Coleridge are pertinent here. Apparently following a conversation with a "well-intentioned convert-hunting young Lady" who equated salvation with a constant tranquility, and conscious enjoyment of the peace it affords, Coleridge identified her "two fundamental Mistakes":

> [F]irst, that the Peace that passeth all understanding, promised to the true Believer, must be at all times the object of the Believer's own immediate <and direct> Consciousness. In opposition to this consult Hooker's admirable discourse on the Perseverance of Saints, and Leighton's Works passim—or rather read Job and the . . . lamenting Psalms.—Not what we see or imagine that we see, but what God sees in us, is . . . [the] Ground of our Salvation.
>
> The second mistake is consequent on the first—viz. that a . . . man or woman is no Christian *at all*, unless he or she is a faultless Christian. Of *the growth* of a Christian Life these Enthusiasts seem to have no conception.[32]

---

31. Compare Phil 2:1 where the expression, "*koinonia* in the Spirit" is complemented in 3:10 by the expression, "*koinonia* of his [Christ's] sufferings."

32. Coleridge, *Notebooks*, 5:no. 6494, written when he was 60. In *Aids to Reflection*, 78–79, Coleridge offers a theological interpretation of Rom 8:26-27 that reflects more fully on the objective presence of the Spirit of God deep within the human spirit at "an unknown distance" from conscious awareness.

A Postscript to Be Read in Retrospect

The following remarks are offered as a postscript to my comments on Paul's teaching on prayer in Phil 4:6–8. But, since I came to write them only after completing the final essay in this volume, and since their full intelligibility presupposes a retrospective reading in light of that essay, it is recommended that this postscript be read only then. Once again, Frost's words cited in the Introduction apply when he invited his readers to "circulate" among his poems, reading each in the light of all the others so as to gain deeper understanding of each with every "pass" through all the others. It is in this sense that we may appreciate the old saying, "Scripture is its own interpreter."

The element that I want to pick up in Phil 4:6–8 is Paul's use there of the verb, *gnorizo*, "make known." The comments of O'Brien are so helpful here, both in their positive content and as laying the basis for my remarks, that they deserve to be quoted in full.

> The Philippians are urged, as a corrective to being anxious, to let their specific requests be made known to God . . . This expression is unusual, for *gnorizestho* suggests that God is unaware of their petitions or lacks information about them. The clause is even more surprising when it is recalled that in the very section of Jesus' teaching that Paul is here thought to echo, namely Mt 6:32, our Lord urges his disciples not to worry about everyday things "because (*gar*) your heavenly Father *knows* (*oiden*) that you need them." Why then are the requests of the Philippians to be made known to God? Not because he is unaware of either the petitions or their content. Rather, by bringing to him their *aitemata* ["requests"], which reflect every possible cause of anxiety, they are laying all their troubles before him . . . or casting all their cares upon him (cf. 1 Pet 5:7). In doing this, the Philippians acknowledge their total dependence upon God, and at the same time *they are assured that he knows* their earnest desires. They have told him of them.[33]

All of this is very much to the point, and very helpful. But more remains to be said, involving what I would call a stronger meaning of the verb, "make known." I may illustrate what I take to be this stronger meaning from the verb's frequent occurrence in Theodotion's revision of the LXX of the book of Daniel. There, the issue in the first instance, in Daniel 2, is *what* Nebuchadnezzar has dreamt and what it *means*. For it seems

---

33. O'Brien, *Epistle to the Philippians*, 493 (his italics).

PART FIVE: New Testament Afterword

to be the case that he is troubled by a dream even the substance of which he cannot remember, let alone its possible meaning; so he asks his court interpreters to tell him the dream and then interpret it for him. When they cannot do this, he sentences them to death. But Daniel, upon praying with his friends, is able both to tell Nebuchadnezzar his dream and to interpret it for him. In Theodotion's revision of Daniel, then, the verb *gnorizo* occurs frequently (where RSV reads, "make known") in reference to God making things known through revelation of a divine mystery, and Daniel making that mystery known to rulers like Nebuchadnezzar. As argued in Chapter 17 in the present volume, Daniel's "making known" the mystery of God's revelation is also a revelation to Nebuchadnezzar of his own thoughts as they trouble him without knowing what their content is.

The phenomenon is, after all, not uncommon. "What's bothering you?" we say, to someone we overhear muttering or groaning quietly off in a corner, to which the response not infrequently is, "Oh, I don't know ... I'm just feeling so (sad, uneasy, anxious)." "About what?" "That's just the problem, I don't *know*. All I know is that I'm just so sad (or ...)." We may say, "Well, do you want to tell me what sorts of things are going through your mind when you feel this way? Any images, thoughts, past events, or future plans that seem to be in your mind when you feel this way?" And the person begins to talk. And as they talk, a picture begins to form in our minds, and eventually we say, "it sounds to me as though you are concerned about such-and-so," and the person says, "That's it!"

Psalm 32 gives one sort of example of this in prayer. There, the psalmist says, "When I kept quiet, my body wasted away / through my groaning [*ša'agah*; LXX: *krazein*] all day long. / For day and night thy hand was heavy upon me; / my strength was dried up as by the heat of summer" (32:3–4). The Hebrew verb suggests the sounds that a lion makes, in this instance perhaps a deep internal inarticulate grumbling as of an animal in pain. The Greek verb, as it happens, is the same verb we find in Rom 8:15 where the only articulate words seem to be, "Abba, Father," the rest being a sort of groaning like that of creation generally—a groaning over the fact that "we do not know what to pray for *katho dei*"; a groaning like the unutterable groaning, too deep for words, of the divine Spirit within; a groaning whose "mind" is known only to God (Rom 8:26–27). So long as the Psalmist keeps this internal groaning suppressed within, it works to drag down even the body. And, given the topic of this groaning (when it eventually comes out), it is likely that the Psalmist is attempting to keep the matter below the level of explicit consciousness. (Who is not familiar with this dynamic?) But then we read, "I made known [*'odia'*; LXX: *egnorisa* (!)] my sin to thee, and I did not hide [*lo' kissiti*; LXX: *ekalypsa!*] my iniquity; I said, 'I will confess my transgressions to the LORD;' then thou didst forgive the guilt of my sin" (Ps 32:5). Here we have the themes of hiding one's internal groanings from God, but also in a way from oneself; and then of "making them known" as a form of "uncovering" or "revealing" them both to God and to the self.

Of course, we may suppose, on general grounds or on the basis of a passage like Psalm 139, that the notion of "making known" or "revealing" to God what had been hidden is, after all, only a figure of speech, and cannot be supposed to show to God what God had not already seen. But such a supposition may obscure what is being articulated, or at least intimated here. As a heuristic strategy, we may ask, What might it mean, that there is something that we can hide from God, and that God cannot see or know until we make it known? The primal form of the issue is God's question to Adam, "where are you?" and the shamefaced answer, "I heard the sound of you in the garden, and I was afraid, because I was naked; and I hid myself" (Gen 3:9–10). Is the hiding futile in the face of God's penetrating omniscience? Or does the hiding veil something from God that God seeks, and that God can see and know only if the veil is removed through owning up to the truth? What if the "knowing" here is not simply informational but relational? What if what God seeks here is the open face returning the shining of the divine face, as when the psalmist says in Ps 34:5, "They looked unto him, and were radiant: and their faces were not ashamed [*yehparu*]"?[34] (For this verb to connote the darkening or "veiling" in shame of a "face" that by nature should shine brightly, see Isa 24:23. The "face-to-face shining" implied in Ps 34:5—with which compare Ps 27:4—is also implicit in Gen 2:25, "and they were not ashamed [*welo' yitbošašu*].") In that case, the "opening up" to God in confession is a genuine making known of what until then is not available to God to be known because it does not exist until that moment of turning to God. I recur to comments made in connection with David's prayer of amazement in 1 Chr 29:14 (see Chapter 3): "who am I, and what is my people, that we should be able [*na'ṣor koaḥ*, literally, 'retain the power'] thus to offer willingly?" To "retain this power" is to have the power either to give willingly—as God desires according to Exod 25:2—or to withhold such a gift. It is the same with the face. To hide one's inner truth from God, when that inner truth is a lie or a wrong, is to withhold from God what God seeks.

---

34. KJV; RSV and NRSV use second person verbs that have no basis in the Hebrew.

PART FIVE: New Testament Afterword

For even when the inner truth is a wrong or lie, when it is freely disclosed to God, "a broken and a contrite heart, O God, you will not despise" (Ps 51:17). But it is the same with one's deepest desires and wants. Kept locked up within one's bosom, nursed perhaps in self-pity or as motivations for calculating ambition or revenge, they have the same effect of thwarting the open relation God seeks. To "make them known" to God—even in passionate, angry complaint—is to honor God as the one acknowledged to be in ultimate charge of things. Anything short of such a strong reading of the verb, "make known," drains the dynamic vitality and existential truth of what these passages are talking about in favor of a desiccated conceptual abstraction that leaves prayer feeling like so much play-acting.[35]

Just so, I propose, we may appreciate the "strong" connotations of *gnorizo*. On the basis of this particular sort of unburdening, the psalmist generalizes (Ps 32:6) to any sort of "time of distress" that may be likened to being overwhelmed by the onrush of "many waters," and encourages others in any and all such situations to pray to God who has "encompassed" him (compare the verb, "garrison," in Phil 4:8) with deliverance. In Phil 4:6–8, then, I take it that *gnorizo* means to "open up" to God in such a way that behind the specific wants and needs initially voiced, the deeper concerns at the root of one's "anxiety" (4:6) may be "revealed."

Such a reading of *gnorizo* in Phil 4:6–8 may shed light also on the occurrence of the same verb in Phil 1:22–24 where Paul writes, "to me to live is Christ, and to die is gain. If it is to be life in the flesh, that means fruitful labor for me. Yet which I shall choose I cannot tell [*ou gnorizo*]. I am hard pressed [or 'torn'] between the two." Commentators debate the meaning of what they take to be the unusual use of the verb here. Some suggest that it may be taken to mean, "I do not know" in the sense of not knowing which to choose. But it never has that sense in biblical Greek. I suggest that Paul is here using the verb elliptically in the same sense in which he uses it in Phil 4:6–8 and in the spirit of Rom 8:26a—that is, of not knowing what to pray for—not knowing which "request" to "make known" to God. He simply finds himself pulled in two opposing directions, and, not making either or even both "known" to God as petitions, he falls back on his general "standing orders" from God (Phil 1:24–25), and concludes that he is to continue on his mission unless and until God indicates otherwise. In the spirit of Daniel, and of the Paul of Ephesians, it is enough to be privy to the general apocalyptic scenario which that mission is to serve, and to leave the deeper matters of his own immediate fate to the deep

---

35. Compare the thought in Eph 2:6.

intercession of the Spirit who, according to Rom 8:27, prays within him, in wordless groans below the level of his knowing what to pray for, *kata theon*. In such a case, declining to "make known" is not a withholding of one's inner depths, but rather an entrusting of them to God within the context of continuing faithful apostolic labor.

16

Faith as a Foothold "within the Veil"

Afterwords in the Letter to the Hebrews

IN THIS CHAPTER I shall offer some forays into selected aspects of the letter to the Hebrews when they are read as afterwords to Old Testament themes considered in the preceding essays. Such an approach will lead at times to interpretations at variance with the magnificent commentaries chiefly consulted for this study—namely, those of Westcott, Attridge, and Johnson.[1] There is no space here to offer extended support for my interpretations or close interaction with these or other commentaries. Instead, I offer the following comments simply as heuristic signposts, starting-points for the reader's own testing and further reflection. The aim, as stated in the introductory chapter, is to entice the reader to "linger lovingly" over the passages of scripture so sign-posted.

The letter to the Hebrews may be said to revolve elliptically around two locations with their respective contexts, locations that are nicely connected in 12:2 where the writer calls upon readers to look "to Jesus the pioneer and perfecter of our faith [*tes pisteos*], who for the joy that was set before him endured [*hypemeinen*] the cross, despising the shame, and is seated at the right hand of the throne of God." The first location is heavenly, where Jesus is "seated at the right hand of the throne of God." The second is the implied earthly location of the author's readers, who are being exhorted to "look unto Jesus," as the pioneer and perfecter of their *pistis*, their "faith(fulness)"[2] as latter-day successors to the gallery of

---

1. Westcott, *Epistle to the Hebrews*; Attridge, *Epistle to the Hebrews*; Johnson, *Hebrews*.

2. The Greek adjective *pistos* means "faithful, trustworthy." The noun *pistis* in the New Testament is frequently translated simply, "faith." Given the history of the Christian use of this word, where it is often associated with the word, "belief," I here and often render it "faith(fulness)" to bring out the dynamic sense the word still often

the faithful presented in chapter 11. This "spatial" contrast is bridged by Jesus' one-time sharing of their earthly situation, when he "endured the cross," an enduring that is offered as an interpretive key and encouragement to them where they (still) are. (One may note the resonance between "looking unto Jesus ... seated at the right hand of the throne of God" and the imagery and exhortation in Col 3:1–4, Romans 8 and related Pauline passages.) In this Afterword I shall oscillate between these two loci and their contexts, beginning with the readers and the call for them to endure and persevere in faith.

Faith as a Foothold on Things Hoped For

From Heb 12:2 we may gather that to have faith is to "endure" or "persevere" (*hypomeno*)[3] under present sufferings, in light of the joy that is held out before one's hope. So understood, faith, while distinguishable from hope, is inseparable from it. This is clear from 11:1, perhaps the classic text on *pistis* in the New Testament; but it is also clear from the way "faith" and "hope" are paired in texts like 3:6 and 10:23 where hope is grounded in the fact that Jesus is faithful (*pistos*).[4] These considerations, and especially the usage of *pistos* in 3:5–6, help to suggest the resonance of *pistos* and *pistis* with the connotations of "faithfulness" that they carry in the Old Testament. If in 3:5–6 Jesus as "faithful" is related to Moses who was "faithful," then the latter reference is probably to Num 12:7 where Moses is termed *pistos*, in Hebrew *ne'eman*, which is related to *he'emin*, "to believe" in the sense, "to rely on, to trust."[5] The scriptural grounding and resonance of "faith" in this letter is reinforced by the Old Testament text whose quotation in Heb 10:37–38 immediately precedes 11:1—namely, Hab 2:4, where *pistis* translates *'emunah*.

But the connection between Heb 11:1 and Hab 2:1–4 goes beyond the keyword *pistis*. In Heb 11:1, *pistis* is characterized in two ways: in 11:1a faith(fulness) is said to function as *hypostasis*, and in 11:1b as *elegchos*. The latter word, occurring often in the LXX, appears only here in the New Testament, and is subject to various interpretations such as "evidence"

---

carries in the New Testament and typically in the Old Testament where it renders *'emunah*, "faithfulness."

3. The verb is used in Heb 10:32; 12:3, 7; and noun, *hypomone*, in Heb 10:36; 12:1.

4. Compare also the easy alternation between "full assurance of hope" (6:11) and "full assurance of faith" (10:22).

5. See the discussion of *he'emin* in relation to Ps 27:13 in Chapter 10.

PART FIVE: New Testament Afterword

(KJV) or "conviction" (RSV, NRSV). If the interpretive principle that sees quotations or echoes of specific Old Testament texts often implying the relevance of the wider Old Testament context for interpretation of the New Testament context is correct,[6] then it is highly suggestive for our interpretation of Heb 11:1 that *elegchos* occurs in Hab 2:1 where the prophet, initially dismayed at the spectacle of the advance of unstoppable Chaldean forces, says, "I will *take my stand* [*'e'emodah*; LXX: *stesomai*] to *watch*, and *station myself* on the tower, and *look forth* to *see* what he will say to me, and what I will answer concerning *my complaint* [*tokahti*; LXX: *elegchon mou*]."

The Hebrew noun *tokahat* is from a verb that means, "to decide, adjudge, approve," that is, to adjudge and approve or disapprove the merits of someone or something. In chapter one Habakkuk's complaint to God is offered (as generally recognized) in the spirit and tone of the complaint psalms.[7] Now the prophet determines to "take my stand" and "station myself" before God. The language is metaphorical, and resonates with the thematics of "standpoint" within society and before God that I discussed in relation to Ps 69:3 in Chapter 9.[8] The determination to "look forth and see" finds resonance in the "looking to Jesus" of Heb 12:2. And God's counsel, in Hab 2:3, to "wait" (*hypomeinon*) for the envisioned "end" finds resonance in the "endurance" language of Hebrews canvassed above. Moreover, we may wonder whether the image of "running" in Hab 2:4 is echoed in the "running" image of Heb 12:1.[9] If Heb 12:1 in this way carries

---

6. I have in mind principally Hays, *Echoes of Scripture in the Letters of Paul*; and, more recently, Hays, Alkier, Huizinga, eds., *Reading the Bible Intertextually*.

7. The noun *elegchos* also occurs in Job 13:6; 16:21; 23:4, 7; and the verb *elegcho* in 13:3, 15, in reference to Job's stance of persistent complaint before God over the issue of his unmerited suffering. The verb *hypomeno* occurs 6x on Job's lips and the noun *hypomone* once, all translating Hebrew words for "hope," indicting the latter's dynamic quality. In 6:11 Job laments, "What is my strength, that I should wait? / And what is my end, that I should be patient?" Yet he does persist and endure—an overlooked prototype of the gallery of the faithful in Hebrews 11.

8. In that chapter I have indicated a close connection between the "vision," "end-time" and "stand(point)" language in Hab 2:1-4 and eschatological visionary texts in Daniel. Does *hypostasis* in Heb 11:1 resonate with the *stesomai* of Hab 2:1? See further below.

9. For an interpretation of the "running" in Hab 2:2 as referring to a faithful response to reading the vision, see Janzen, "Habakkuk 2:2-4 in the Light of Recent Philological Advances." To be sure, the Greek verb in Hab 2:2 is *dioko* (for Hebrew *ruş*, "run") and Heb 11:1 has *trecho*; but the latter is a close synonym of the former in metaphorical application to faithful living. (Cf., e.g., Paul's self-description by the former in Phil 3:12, 14, and by the latter in Phil 2:16.) Given the importance of Habakkuk for

forward an echo of Hab 2:2, then the whole of Hebrews 11 may be thought of as cradled in the interpretive frame of Hab 2:2–4. But this suggests that in Heb 11:1 the word *elegchos* may carry the connotation of testing and approving, under affliction and suffering, the merits of the unseen hope (or joy, as in 12:2) that the Gospel holds out before one.

"Faith," then, in Heb 11:1, may be taken to involve, in part, the "adjudging and approving" of things unseen as a *hypostasis* (11:1a) of things hoped for, a *hypostasis* that will enable one to endure. The word *hypostasis*, which carries a variety of connotations in classical (nonbiblical) Greek and in the LXX, is in Heb 11:1 variously interpreted by commentators. Luke Johnson comes closest to my interpretation in observing, but dismissively as "less likely" for this passage, that *hypostasis* can refer to "a foundation or substructure, as of a temple (Nah 2:7; Ezek 43:11)."[10] But Johnson has overlooked the related, metaphorical use of *hypostasis* (Hebrew *mo'omad*)[11] in Ps 69:2, in an existential setting that is closely resonant with the social setting of Hebrews and especially chapters 10–13. So I shall return briefly to that psalm and others like it, as a scriptural background for Heb 11:1a.

Reproach and "Standing" in Hebrews and in Psalm 69

Two aspects of the situational resonance between Hebrews and Psalm 69 may be highlighted: First, the theme of "reproach," conveyed in the verb *oneidizo* and its cognate noun *oneidismos*, "reproach" (KJV) or "abuse" (RSV, NRSV) in Heb 10:33; 11:26; 13:13, resonates with *oneidismos* in Ps 69:7, 9, 10, 19, 20 (the most dense repetition of this word in any Old Testament passage). Second, the exhortation to "look to Jesus" for inspiration to endure under suffering (Heb 12:2) identifies, in 12:3, the aspect of his

---

eschatological speculation, as evidenced by the Habakkuk Commentary (1QpHab) among the Dead Sea Scrolls, we may note further how the image of "running" and of the "joy set before Jesus" in Heb 12:1–2 resonates with the language in Hab 3:18–19 of "rejoicing" and finding "footing" through a vision of God's expected action.

10. Johnson, *Hebrews*, 277. I may note that the English New Testament scholar C. F. D. Moule writes, "Hebrews 11:1 is enough to remind us how inseparable hope is from trust in God—in the God whose love is constant: 'It is trust [in God] which forms the foundation to our hopes'" (*Meaning of Hope*, 20). Though Moule does not elaborate the grounds for his construal of *hypostasis* as "foundation," it is clear that he assumes its lexical and contextual plausibility. One function of the present essay is to indicate those grounds.

11. This word *mo'omad* is related to the verbal form *'e'emodah*, "I will take my stand," in Hab 2:1, and is used in an analogous figurative sense.

PART FIVE: New Testament Afterword

crucifixion that is to be held especially in view. Here, RSV and NRSV are distinctly unhelpful in rendering the expression, "he endured such *antilogia* of sinners against himself" as referring, generically, to "hostility." The word *antilogia* refers to a specific *form* of hostility—"contradiction" (KJV), literally, "speaking against." The theme of linguistic abuse—the slander, accusation, scorn, reproach, and ridicule heaped on the innocent, devout psalmist—is ubiquitous in the Psalms. The word *antilogia* occurs with special force in Ps 18:43, where God delivers "David" from "*strife with the peoples*"; in Ps 80:6, where the people complain, "thou dost make us the *scorn* of our neighbors"; in Ps 55:9 where the psalmist, resonating with Jeremiah against his opponents, says, "O Lord, confuse their tongues; for I see violence and *strife* in the city"; and especially in Ps 31:20, where the psalmist affirms, of those who "take refuge in thee" (echoing Ps 2:11), "*In the covert* [*beseter*; LXX: *en apokrypho*] of thy presence thou *hidest* [*tastir*; LXX: *katakrypseis*] them from the plots of men; thou holdest them safe under thy shelter [*besukkah*; LXX: *en skene*] from the strife [*rib*; LXX: *antilogias*] of tongues."

As the Hebrew and Greek indicate, the "shelter" in Psalm 31 is specifically the sanctuary viewed as a *booth* or *tent*. (See further below, on the tabernacle in Hebrews.) The "strife of tongues" in Psalm 31:10 resonates with the theme of "reproach" in passages like Psalm 69. In Heb 12:2 the *antilogia* against Jesus on the cross carries forward this psalmistic theme of verbal abuse or reproach. The Psalm's conclusion (31:23-24) suggests what sort of action constitutes "finding refuge" in God's "covert" where one may be "hidden" from such contradiction or reproach: "The LORD preserves the faithful [*'emunim*], but abundantly requites him who acts haughtily"—as if to suggest that "finding refuge" in God involves acts of *'emunim*, acts that, in the act, find God to be such a refuge. The character of such acts of "faith(fulness)," is then elaborated in the concluding verse, "Be strong [*ḥizqu*], and he shall strengthen [*ya'ameṣ*] your heart, all you who wait for the LORD!" (Ps 31:24?). As in Ps 27:13-14, the act of faithful trusting is elaborated as a self-reflexive act, "be strong," which is at the same time an experience of divine action in and through that self-reflexive act: "he shall strengthen [*ya'ameṣ*] your heart."[12] Corresponding to the reference to the "covert" and the "tent" in Ps 31:20, the act of trusting belief in Ps 27:13-14 occurs within the interior space which is the internalization of God's "house / temple" (27:4), God's "shelter" (*sukkah*) / "the cover

---

12. For my argument concerning this translation of *ya'ameṣ* as referring to divine action in and through the psalmist's "taking courage," see Chapter 10.

(seter)" of God's tent (27:5). In that interior space, the act of "belief" or "trust" includes the reflexive act of "self-address" that we have seen in the Prayer of Hannah. It is as though, in that space, God's strength is present to the psalmist as the strength to call upon himself or herself to be strong. Moreover, it is in this interior space, within God as sanctuary and within the psalmist, that the psalmist recovers the standing or "foothold" that human society does not provide.

This scene of interior action, I suggest, provides a scriptural frame of reference within which we may appreciate what is meant when Heb 11:1 says, "Faith is the *hypostasis* of things hoped for, the *elegchos* of things unseen." But "frame of reference" is too conceptual, too conscious a way of putting what I mean to imply by proposing to read Heb 11:1 against the background of psalms like these. Their scene of interior action, I want to propose, itself becomes internalized in generations of readers who have found these psalms to resonate with and speak to their situation—and, once internalized in this way, how they shape a deep structure of what one may call a "scripturalized psyche" within such readers.[13] In the letter to the Hebrews, then, the writer makes appeal to this deep structure in his or her readers. Bereft of any foothold or standpoint in what is visible to the physical or social eye—one's latter standing being undercut by accusation and reproach, and thereby one's access to the physical structures where social standing finds its symbolic space—*pistis* takes the promises of God as its *hypostasis* or standpoint;[14] and that standpoint sometimes has no other evidence—nothing by which to test itself, and against which to (ap)prove itself, than this conviction itself.[15] That conviction, that standpoint,

---

13. Compare Mark George's use of the analytic categories of "conceptual" and "symbolic" space in reference to the Israelite tabernacle in his *Israel's Tabernacle as Social Space*, summarized below.

14. When Martin Luther allegedly said, "Here I stand," *where was he?* Physically, he was in the town of Worms in the Spring of 1521. Socially, he stood before a formal assembly answerable to a papal inquiry. Existentially, however, he took his stand upon "the Word of God"—*sola scriptura*—as affirming one's standing before God solely on the grounds of faith—*sola fide*—in, and enabled by, God's grace—*sola gratia*. In that posture I take him to exemplify the standpoint, on the one hand, of the speaker in Psalm 69, and on the other hand, of person of faith as generalized in Heb 11:1 and particularized in the Jesus of Heb 12:2. (If Luther's words are apocryphal, they nevertheless represent the stance of his followers in the following decades of religious strife, and they epitomize his hymn, *Ein' feste Burg*.)

15. Compare Coleridge's comment, in the context of his defense of prayer against Kant's dismissive or reductionist characterization of that practice. Following his opening characterization of "faith" as at heart "fidelity," and in the midst of an abstruse philosophical argument, Coleridge invokes the biblical references to "a God

PART FIVE: New Testament Afterword

is both the psalmist's act and the evidence of God's faithfulness. As is said in Ps 63:8, "My soul *clings* to thee; / thy right hand *upholds* me"—these two poetic lines referring to two dimensions of a single dynamic event, in which the strength to cling is experienced as the upholding of God's right hand. Here divine and human faithfulness are displayed in distinguishable but inseparable union. With this comment, it would be logical to shift the focus of this Afterword on to the characterization of Jesus as "the author and finisher of (our) faith." But in order gain a fuller appreciation of this characterization, it will be helpful to explore further the import of Jesus as Son and High Priest.

Jesus as Son on the Throne / High Priest in the Tabernacle

Jesus is presented in Hebrews in two chief aspects: as Son of God and as High Priest. These appear to connect him to two distinct spheres of activity as portrayed in the Old Testament: (1) as God's royal son in the political sphere (e.g., Psalm 2), and (2) as the chief cultic functionary as established by God in Exodus 25–31. But the royal and the priestly are in fact two aspects of one sphere, and this is demonstrated in Hebrews in a number of ways.

To begin with, Jesus is introduced in Hebrews 1 as Son and as "heir of all things" (a royal theme, as in Ps 2:8), and then through quotations from three royal psalms (Ps 2:7 in Heb 1:5; Ps 44:7 in Heb 1:8; Ps 110:1 in Heb 1:13). And reference to him as Son recurs eight more times throughout the letter. That Jesus is High Priest *as* royal Son, is indicated in several ways. First, the initial reference to activity in cultic terms comes in 1:3 where he, *as Son*, makes "purification for sins." Second, in three places Jesus' high priestly function is presented as embedded in his status as Son (Heb 4:14; 5:5; 7:28). Third, Jesus' function as High Priest is grounded in Ps 110:1, which, as a royal psalm, focuses chiefly on the royal figure as ruling over

---

that heareth Prayer" (Ps 65:2) and "the Prayer of a righteous Man availeth Much" (Jas 5:16), and then observes, "the only possible *proofs* of the Position [implied in these references] are so incapable of being raised into an outward *evidence* of the truth for others, rest so almost wholly in the . . . Individual's own secret persuasions, *the Faith itself being the main evidence of the truth of the Faith*" (*Notebooks*, 5: no. 6666; italics added) Kant himself, after characterizing prayer as a "superstitious illusion [*abergläubischer Wahn*]" and a "fetish-making," goes on to characterize praying "in words and formulas" as a means of "conversing within and really *with oneself*, but ostensibly . . . speaking the more intelligibly *with God*" (*Religion within the Limits of Reason Alone*, 183–85). Hannah and her like would presumably not find Kant's distinction amounting to much of a difference (see Chapter 5).

all his enemies from the throne at God's right hand. Fourth, Jesus' High Priestly function is not to be interpreted in reference to Aaron and sons, but to Melchizedek (Ps 110:4, quoted or echoed in Heb 5:6, 10; 6:20; 7:10, 15, 17); and in Heb 7:1, as also in Gen 14:18, Melchizedek (whose name refers to God as "my king") is identified first as "king of Salem" and then "priest of the Most High God."

Finally, we may note the present location of Jesus as Son and High Priest. It is at God's "right hand" (Heb 1:3, 13; 8:1; 10:12; 12:2), which is a royal location (8:1; 12:2). At the same time, this location is prominently portrayed in terms of the holy of holies in the tabernacle or tent sanctuary that Moses built under God's direction. But, as Mark George points outs, the tabernacle itself is a royal structure, its pattern comparable to royal dwellings in the ancient Near East, with its holy of holies in fact a throne room.[16] That earthly tabernacle is declared to have been "the example and shadow of heavenly things" (8:5), that is, of the "true" (8:2), "greater and more perfect" (9:11) tabernacle or sanctuary within which the enthroned Son functions as High Priest. Again, whereas the sacrificial, purificatory activity of the Son as High Priest was carried out "once for all" (1:3; 10:10, 12), his present "standing," or status, within the sanctuary is not that of a priest ministering *before* God (and entering the holy of holies only once a year for atonement) but at God's right hand, enthroned *within* the holy of holies.[17]

This throne is characterized as "a throne of grace" where "we may receive mercy and find grace to help in time of need" (4:16). The "spatial" relation between Jesus as enthroned Son and High Priest, and his followers, is represented in complex fashion. On the one hand, "we" are bidden to "draw near" (4:16, 10:22), having "confidence to *enter*" through "a new and living way which he opened for us *through the curtain*" (10:20). On the other hand, this drawing near is spoken of as taking place "through hope" (7:19), and this hope is imaged as "a sure and certain anchor of the soul, a *hope* that *enters into* the inner shrine behind the curtain" (6:19). One could say that this hope has objective and subjective poles, the objective pole consisting of Jesus within the heavenly holy of holies and the subjective pole consisting of the earthly person hoping. But the text implies that

16. George, *Israel's Tabernacle as Social Space*, e.g., 59, 62, 66 n. 47, 80.

17. The Christological import of this symbolism is staggering, in view, e.g., of Manasseh's action, in 2 Kgs 21:7, of "setting" (*wayyaśem*; LXX: *etheken*) the graven image of Asherah in the house of the LORD, an action presented as utter apostasy in defiance of God's saying to David and Solomon, "in this house . . . I will *put* [*'aśim*; LXX: *theso*] my name forever."

PART FIVE: New Testament Afterword

it is the subjective hope itself of such a person that "enters" the heavenly space and anchors itself there.

That hope is at one and the same time *subjective*, in reference to the followers of Jesus, and *objective*, in reference to Jesus "there already" as the object and ground of his followers' hope, for, as 6:20 goes on to say, the hope that "enters into the inner shrine" attaches to Jesus who is a "forerunner on our behalf." The "on our behalf" again, like the word "hope," signals distance between Jesus and "us." The complex character of this "spatial" imagery—of access and presence, and of distance and representation in behalf of—is indicated also in respect to prayer as it "takes place." On the one hand, prayer as petition for "grace to help in time of need" is spoken of as our "drawing near" (4:16). On the other hand, such prayer is bosomed in Jesus as High Priest who now "always lives to make intercession for" those who "draw near to God through him" (7:25). Here, prayer as a "drawing near to God," takes place—finds its space to enact itself—within the intercession that the Son and High Priest makes to God on "our" behalf.[18] The passage resonates with the words of Solomon in 1 Kings 8 (considered more fully in Chapter 15), where that king prays that the seven-fold prayer he has just offered may be "near" God as a standing prayer for the people, such that the people, in any and every vicissitude, may have their specific prayers heard in like fashion. We may also compare the complex picture of prayer presented in Romans 8, where the prayers of those who cry "*abba*, Father" are, as it were, bosomed in the intercessions of the Spirit within their depths and the Son at God's right hand, the latter image (in Rom 8:34) resonating with the picture in Hebrews through the common echo of Ps 110:1.

I want now to draw this brief discussion of the "spatial" imagery in Hebrews into resonance with the *beseter / en to krypto*, "in secret" imagery explored in earlier chapters, by summarizing relevant aspects of Mark George's study of *Israel's Tabernacle as Social Space*. This study came into my hands literally days after I completed work on Chapter 15, and some time after I had written the earlier pieces. I take it to support, in a more

---

18. I note the comment of Mark George, *Israel's Tabernacle as Social Space*, 86, that "spatial practice is performed space, in the sense of actions that must be taken to create it." The central idea of the present volume may be put in these terms from George: Prayer "takes place" in/as "performed space." But if prayer is a faithful action that creates the space within which it occurs, and as such creates a space for God within the one praying, that action itself occurs within the space created by the faithful—including prayerful—action of Jesus as "author and finisher of *pistos*." This is another way of putting the central argument in Chapter 4.

Faith as a Foothold "within the Veil"

systematic way, the impressionistic three-dimensional analysis of "spatial experience" that I sketched in my introductory chapter and that I explore more fully in Chapter 9.

George takes the tabernacle in the Old Testamemt to be a material structure whose significance lies in the way it both symbolizes and in turn *in*forms Israel's understanding and experience of intra-human and human-divine relations as spatial. He draws his analytical categories primarily from three recent studies of space, Gaston Bachelard's phenomenological study, *The Poetics of Space*; Henri Lefebvre's Marxist analysis of "critical spatiality" in *The Production of Space*; and the "New Historicist" approach of the literary theorist Stephen Greenblatt to representations of space in Shakespearean theatre.[19] George's succinct summaries of these works are themselves so rich in implication—let alone his analysis of the tabernacle within their terms—that my further summaries and use of them here do them a vast injustice. Without, I hope, fatal distortion, I siphon off the following insights for present purposes.

George summarizes Lefebvre as identifying space as a social *product* (the Marxist preoccupation with humankind as *producer* is evident here).[20] As a social product, space may be viewed under three aspects: *practical, conceptual,* and *symbolic*. *Practically*, social space consists in the way humans take material objects and (re-)shape them for human purposes.[21] *Conceptually*, social space consists in the "abstract, theoretical understandings" a society shares, a "conceptual space" it creates to "explain and conceive of space." Already the notion of "conceptual space" takes the literal meaning of the second word to a metaphorical level.[22] *Symbolic* spaces "are the socially significant meanings ascribed to space."[23] Here the literal and the preceding metaphorical meaning of "space" are both

---

19. Bachelard, *Poetics of Space*; Lefebvre, *Production of Space*; Greenblatt, "The Circulation of Social Energy," and Greenblatt, "Towards a Poetics of Culture." The markings and marginalia that I entered into my copy of Bachelard in 1972 indicate the impact it made on me then. I suspect that my subsequent reflections on space in the Bible—in the present volume as well as my discussion of the tabernacle in my *Exodus*, have all been informed by Bachelard's work in one way or another.

20. This preoccupation resonates with, for example, the concluding prayer in Psalm 90 concerning "the work of our hands," though that preoccupation, devoid of the transcendent reference, runs into the biblical strictures against the self-idolatry overcome in the confession of Ps 100:3.

21. George, *Israel's Tabernacle as Social Space*, 23-25.

22. Ibid., 25.

23. Ibid., 27.

taken to another metaphorical level.[24] The importance of symbolic space vis-à-vis material/practical and conceptual space is that, as George puts it later on, "[s]ymbolic space is the space in which a society and individual human beings emotionally and affectively live. It is socially meaningful because it creates a sense of excitement, fear, peace, disquiet, reverence, passion, pain, joy, awe, or other effect. Thus it is the space of social energy or cultural force."[25]

The terms "energy" and "force" lead me back to two other elements in the Lefebvrian/Greenblattian analytic schemes that George employs. The first is the rigorously materialist mode of analysis and explanation derived ultimately from the materialist epistemology and metaphysics implicit in modern science from the seventeenth century onward.[26] Though the phrase is not used, everything is analyzed and explained in terms of "efficient causation," that is, of unilinear energies and forces interacting upon one another in a material-social universe conceived as a closed system. As George puts the matter (following Greenblatt),

> [o]ne of the most important assumptions of New Historicists is that literature, as well as any and every cultural expression or expressive act, be it a painting, a Shakespearean play, a Donne poem, or a legal ruling, is enmeshed within a large, complex network of material practices and realities. No cultural

---

24. In a generic way I note the similarity of this multi-dimensional experience of space to the one I traced in my introductory chapter. George himself gives a wonderfully apt illustration of the usefulness of this tripartite analytic schema in his application of it to Yosemite National Park: the surrounding National Forests, the park itself, and the innermost "privileged zone of publicly demarcated Nature" (*Israel's Tabernacle as Social Space*, 29). (The similarity to the tripartite division of tabernacle space is clear if implicit.)

25. George, *Israel's Tabernacle as Social Space*, 141. While George uses the term "effect," in psychoanalytic discourse the listed descriptors, and others like them, would be given the label "affect." And in fact George himself earlier in the quotation uses the adjective "affectively." To speak of "affect," however, simply as an "effect" is to skew "where we live," however inadvertently, in the direction of a materialistic determinism, Marxist or scientific, that would explain all "affects" simply as "effects." As I understand it, "affect" at least potentially has both passive and active dimensions, identifying both the felt-emotional color and tone of our experience of *being acted upon* by what or who is not ourselves, and the felt-emotional color and tone of our *acting toward* what or who is not ourselves. Despite my tremendous appreciation for the fruitfulness of George's analysis of the tabernacle as social space, this goes to the heart of my following critique.

26. For a critique of modern science as materialistic in this sense, and therefore as insufficiently radical in not accounting for volitional freedom, see Whitehead, *Modes of Thought*, especially the last two chapters.

expression emerges as a gift from the gods or as a spontaneous act of genius, something without any connection to the world around it.[27]

Such an assumption is an understandable reaction to one-sided idealism. But it erases any possibility of free, responsible, creative action in the interval between *being acted upon* by such energies and forces and *acting upon* them in return. There is both literally and metaphorically no room here for human responsive/responsible freedom, what Whitehead terms *causa sui*, self-causation, or what he in one place calls the "elbow room" in the universe.[28] The power of the materialist mode of analysis lies in the fact that such forces and energies *do* constitute a large part of our awareness both of the world around us and of our own bodily and psychic and emotional dynamics. As Whitehead himself puts it, we are "never very free"; but there is always a margin of freedom, however slender—even if that margin be no more than, as Coleridge the high-functioning drug addict put it, the power to cry out to God in our bondage.[29]

The second element in the Lefebvrian/Greenblattian analytic scheme is the "default" view of society (and nature for that matter) as adequately analyzable in terms of a "conflict" model of interaction. This comes to expression within Lefebvre's analysis chiefly in reference to symbolic space. The issue is of such importance for the present book that another lengthy quotation is in order. As George summarizes Lefebvre,

> The same space may mean a variety of things to different people. This is the result . . . of different social meanings being overlaid on the physical objects of space. There are socially dominant meanings associated with certain spaces, coming from those with the most social power. But other socially significant meanings remain attached to those spaces, even though they may not be obvious, because they are coded, hidden, or imaginative in

---

27. George, *Israel's Tabernacle as Social Space*, 34.

28. Whitehead's phrase, "elbow room," comes just before his sentence, "our claim for freedom is rooted in our relationship to our contemporary environment" (*Adventures of Ideas*, 195). I develop the image of "elbow room" in Janzen, "Modes of Presence and the Communion of Saints." Whitehead's "dipolar" alternative to purely idealistic and purely materialistic systems of thought comes to pithy expression in *Adventures of Ideas* where, in what is in part an implied riposte to Marxist materialism, he says "the well-marked transition from one age to another can always be traced to some analogues to Steam and Democracy, . . . senseless agencies and formulated aspirations" (6).

29. See Whitehead, *Process and Reality*, 201–2; and Coleridge, *Aids to Reflection*, 157–60.

character. This may be the only effective means whereby the dominant symbolic meanings in symbolic space may be challenged. *Symbolic space is* socially *contested space, because it is oversignified space.*[30]

This materially-grounded analysis is so largely on target that it itself threatens to drive underground into a hidden (*beseter*?) "imaginative space" more empirically-radical analyses inclusive of a freedom, however marginal, for interaction, even with outsiders and aliens, in a spirit of cooperation. The code of hospitality as "sanctuary" within traditional societies is a case in point, whether that hospitality is honored in the family home or in the community's sanctuary properly so-called; and the erosion, and eventually the erasure, of that code in modern societies, with the diminution in meaning of "hospitality" to what is exchanged between cordial friends and neighbors, testifies to the capacity of modern materialistic secularism to "mystify" consciousness with a false sense of what is fully real.

But the above critique is aimed primarily at the implications of the conceptuality that George draws on, and only secondarily at what, from my point of view, are some reductionist comments that this conceptuality leads him to make at certain points. In the main, his analysis of Israel's tabernacle as embodiment of space is subtle and richly fruitful. On the basis of my own, largely intra-textual work on the tabernacle, and biblical themes related to it, I find our perspectives highly congenial. In the present chapter there is no space to do more than make a few comments relevant to the images, motifs, and themes traced in George's book as they appear in the letter to the Hebrews.

George draws on previous modern studies of the tabernacle that have shown how it resembles the architecture of ancient Near Eastern temples and, like them, stands as a microcosm of the cosmos. But his discussion of the details of ordinary house or apartment holdings as reflected in Assyrian documents of purchase adds an "on the ground" dimension of appreciation for how (dwelling in) one's own building locates one in social space—and, implicitly, this discussion then resonates with God's request[31] to Moses in Exod 25:8, "let them make me a sanctuary, that I may 'tabernacle' [*wešakanti*, "tent"] among them." Further, like those temples the

---

30. George, *Israel's Tabernacle as Social Space*, 28 (italics added).

31. One might take my choice of this verb as softening a more realistic biblical action of commanding or instructing. My choice of verb is informed by the emphasis, in the tabernacle narratives in Exodus as well as the temple narrative in 1 Chronicles 29, on Israel's response as voluntary.

Faith as a Foothold "within the Veil"

tabernacle is a royal dwelling; among other things, God is present in its innermost room as enthroned above the cherubim. But in several respects, as George shows, the tabernacle, as a material, social, and symbolic space, is counter-cultural vis-à-vis the fixed stone or brick temples of neighboring nations and states. Three elements of this contrast are of particular relevance to the present discussion. First, the tabernacle is a moveable space, not predictably plottable, we might say, with a GPS. Second, there is no mention of a human king's special function or place in the tabernacle.[32] The only chief functionary is the high priest, usually ministering *before* the curtain and only once a year entering the holy of holies. The only king, in other words, is the LORD, enthroned *beseter* on the cherubim. Third, while typically temples are built of materials seized, negotiated for, or purchased and fashioned by forced labor, the materials and the work for the tabernacle are voluntarily given.

At this point I want to identify two ways in which George's analysis reflects the materialist influence of the analytic schemes he draws on. First, in what a materialist exegesis might be called a "demystifying" move, at a number of points George "demythologizes" the text, and converts statements about divine word or action into evidence for human perspective and motivation. For example, in characterizing "what happens within the walls of the tabernacle complex and tabernacle proper" as "part of a self-contained (social) world," he notes that "this sense of separateness" from the world "beyond" is "enhanced by the narrative's theological claim that it is a space that originated and came from the deity . . . of divine origin, rather than human, and therefore not of the human world."[33] That this is a human "claim," and not in any sense or degree a divine request or authorization is implicit from George's earlier uncritical presentation of Greenblatt's New Historicist (i.e., social-materialist) denial of "divine gift" or "human genius" as factors.[34] It becomes explicit when, in reference to the "space" that priests occupy and act within, in this tabernacle, George contrasts the Priestly tabernacle with the "rival" eschatological temple in Ezekiel.[35] Whereas Ezekiel retained a (reduced) role for the human king (demoted to "prince") in the temple, the tabernacle left the king—if such there be—with the rank of a mere commoner. "There might be conflict

---

32. See, contrastingly, the reference in 2 Chr 34:31 to the king's "place" ('*omed*) in the temple, and my discussion in Chapter 9.
33. George, *Israel's Tabernacle as Social Space*, 146.
34. Ibid., 34, 137.
35. Ibid., 132.

341

PART FIVE: New Testament Afterword

in the [exilic] community over Israel's identity," George writes, "but the Priestly writers attempt to overcome all conflict and objection by putting their views into the mouth of the deity."[36] So, theological statements are re-characterized as ideological backing for human motivations and purposes in conflicted social situations. Such a move leaves no room for God to dwell, either in the tabernacle or in the human heart and imagination. This materialist methodology itself results in a hermeneutical Egyptian bondage and Babylonian captivity under which some exegetes may wish to cry out as in Exod 2:23–25 or Ps 137:1, 4.

This is not to say that any and all claims to divine inspiration or authorization are without an alloy of human motivation and perspective, or even at times mask purely human machinations for the sake of human power. But a hermeneutic that leaves no room for God within its categories or within the human imagination also leaves no room for freedom within the human heart, reducing all human perception, affect (sic!), motivation, and action to a quantifiable reaction-response "exchange" of social-material energies, a deterministic wheel of existence from which there is no release. (In psychoanalytic terms, this turns the discussion of sublimation from Loewald back to Freud.) In the main, my critique is overly harsh as applied directly to George's discussion. It is directed primarily at the Marxist and New Historicist conceptualities that he employs, and only secondarily at the way those conceptualities at times infect his own analysis. But another indication of this infection comes when he implicitly (and, surely, inadvertently) undercuts his own contrast between the voluntary basis for tabernacle construction with the typically enforced construction of temples.

Early in his analysis George makes statements such as the following:

> [The Israelites'] participation is *willing* and their skills *divinely bequeathed* . . . Moses commands only those with a generous heart (*kol nedib libbo*; Exod 35:5, 22) to bring an offering . . . [a]ll whose hearts are stirred (*neśaʾo libbo*; 35:21; 36:2; see also 35:26) and spirits are willing (*nadebah ruḥo ʾoto*; 35:21).[37]

The first rendering, "those with a generous heart," is unexceptionable. But the expression *kol . . . neśaʾo libbo* means, literally, "everyone whose heart lifts him up," and its idiomatic force is suggested in 2 Kgs 14:10 // 2 Chr 25:1 as referring to a reflexive or "internally transitive" action where one's

---

36. Ibid., 134.
37. Ibid., 65 (italics added).

own heart is the actor and the origin of how one takes oneself into the world around. "Whose hearts are stirred" flattens the "voluntary" connotations here. Similarly, in *nadebah ruḥo 'oto*, one's "spirit" is the subject of the verb *nadab*, "to incite, rouse to action"—hence: "everyone whose spirit incites him."[38] My comments here might be taken as a typical grammarian's quibble, —especially in view of George's conclusion that

> The willingness and generosity of those who contribute raw materials for the tabernacle . . . is a social practice that stands in marked contrast to common building practice in the ancient Near East. It produces a social space, the tabernacle, quite different from the social spaces elsewhere in the ancient Near East that are created as a result of conscripted or corvée labor.[39]

But later George's language slips to "all those whose hearts *were stirred* or *compelled* (*ndb*)."[40] The people "perform this work not because they are coerced into doing it but because their hearts *are stirred* within them, *compelling* them to do it."[41] But, again, their hearts are not stirred (in the passive voice); their hearts are *doing the stirring*. And they are not compelled; they consent, as it were, to what their hearts prompt them to do, and promptly do it.

The critical point, here—particularly for the resonance of the tabernacle traditions with the place and dynamics of faith(fulness) in Hebrews—is the portrayal of the interaction of divine Spirit and the human heart, such that *free human response to God* (—and human freedom as a social energy, enabling human participation in society and the wider natural world as responsible and responsive rather than merely reactive and determined by material and biological and emotional and psychic forces)—*is the effective symbol-making sign*, both *hypostasis* and *elegchos* within creation, of the presence of the transcendent divine ground of creation. The awareness of this mystery informs David's prayer of amazement in 2 Chr 29:14, as sketched in Chapter 3, where he exclaims, "But who am I, and what is my people, that we should be able thus *to offer willingly*

---

38. For an analogously subtle portrayal in the Hebrew text of the various states of the human heart as acting freely or under compulsion, see my discussion (Janzen, *Exodus*, 69–79) of Pharaoh in Exodus, who at first "hardens his own heart," whose heart then later simply "becomes hard," until, in the end, God "hardens his heart." The progression is from (ir)responsible freedom to unfree compulsion.

39 George, *Israel's Tabernacle as Social Space*, 66–67.

40. Ibid., 165 (italics added).

41. Ibid., 166 (italics added).

PART FIVE: New Testament Afterword

[*hitnaddeb*]? For all things come from thee, and of thy own have we given thee."

In this prayer, David's reference to the amazing power humans have before God to "offer willingly," that is, "freely," refers in the first instance to the material gifts that he and others are voluntarily providing for the construction of the temple. But the prayer itself is also such a free action. Indeed, the material sanctuary will become the outward and visible sign of the inward and spiritual acts of worship that will go on within it and that are inaugurated in this prayer of David. Those acts of worship and covenant fidelity (1 Chr 29:18-19) are acts of free response, grounded in who God is and what God has done (1 Chr 29:10-12). To put this in different terms, the shaping of human lived space in response to God's creative action of providing space for humanity in creation, arises through acts of faithfulness enabled by and responsive to the faithfulness of God. This leads us to the testimony in Heb 12:2 that Jesus is *tes pisteos archegon kai teleioten*, "the pioneer and perfecter of faith(fulness)."[42] Having come around finally to this expression, I now want to explore some of what I hear to be its freighted resonances.

Jesus as *archegon kai teleioten* of Faith(fulness)

In speaking specifically of *Jesus* as *archegon kai teleioten*, the writer has in focus the earthly race in which he shares our common humanity.[43] But given the high Christology of this epistle, we may wonder whether the range of the *pistis* in question is to be confined to his earthly race. Surely it extends to his heavenly intercession as a faithful (*pistos*) High Priest. But where does his faith(fulness) begin?

I start with the claim in Heb 1:2-3 that it is through this "Son" that God "made the world" (*epoiesen tous aionas*). This affirmation is repeated

---

42. KJV, RSV, NRSV, and NIV translate *tes pisteos*, "of our faith." But before the *pistis* is "ours," it is Jesus'. Westcott (*Epistle to the Hebrews*, 397) writes, "in Jesus Christ Himself we have the perfect example . . . of that faith which we are to imitate, trusting in Him." Attridge (*Epistle to the Hebrews*, 356-57) writes, "the 'faith' . . . that Christ inaugurates and brings to perfect expression is not the content of Christian belief, but the *fidelity and trust that he himself exhibited* in a fully adequate way and that his followers are called upon to share . . . It is precisely as the one who perfectly embodies faith that he serves as *the ground of its possibility in others* . . . and the model they are to follow" (italics added).

43. Johnson (*Hebrews*, 317) writes, "the simple name of Jesus . . . emphasiz[es] in a manner characteristic of Hebrews the shared humanity of Jesus and those who follow after him (Heb 2:9; 3:1; 6:20; 7:22; 10:19; 12:24; 13:12)."

Faith as a Foothold "within the Veil"

in 1:10 through a boldly Christological application of Ps 102:25 which in the LXX reads, "Of old [*kat' archas*] thou, Lord, didst lay the foundation of the earth, and the heavens are the work of thy hands." This divine work of creation is said in Ps 33:4, 6 to be "done in faithfulness" [*'emunah*; LXX: *en pistei*]. If, then, God "made the world" through the "Son," it is surely not a stretch to identify the beginning of the faithfulness of Jesus as *archegon* at least as early as this *kat' archas*.

This Son, moreover, "upholds all things [*pheron te ta panta*] by the word of his power," and, that "when he had made purification for sins, he sat down at the right hand of the Majesty on high"—this last action, insofar as it was "once for all" and needed never to be repeated, is taken as an act of perfecting or completing at least that aspect of the Son's work. Moreover, in connection with this High Priestly aspect of his work, the Son, by the foreshadowing found in Melchizedek, is implied to have "neither beginning [*arche*] of days nor end [*telos*] of life," and so "continues a priest *forever* [*eis to dienekes*]" (Heb 7:3).

We may note the peculiar expression, *eis to dienekes*, where we should have expected *eis ton aiona* as in Heb 6:20; 7:17, 21, 24, 28 following Ps 110:4.[44] In the LXX the word *dienekes* occurs only adverbially, in Est 13:4 and 3 Macc 3:11, 22; 4:16. In classical Greek the adjective means, "continuous, unbroken"; but its verbal cognate gives a richer sense of its connotations. The verb *diaphero* (*dia* + *phero*), "carry over or across," is often used with temporal reference, for example, in Herodotus, "to *go through* life." In the middle voice it means "to live, continue," with the connotations, "bear with, bear to the end, go through with, bear the burden of, endure, support."[45] That the writer should, at Heb 7:3, abandon the expected phrase and choose this word, suggests a connotation that should not be missed. It pertains to the function in which a priest continually bears the people upon his heart before God.[46] If I am not reading too much into the etymology of *dienekes*, the expression in Heb 7:3 resonates with the verb *phero* in 1:3, suggesting that the "word of power" by which the Son "upholds" or "bears" "all things" is the royal/high priestly word of intercession. And this is in fact explicit in 7:25.

---

44. In the New Testament, *dienekes* occurs again only in Heb 10:1, 12, 14, all with reference to priestly function.

45. For these meanings, see LSJ, 417.

46. Compare the occurrence of *phero* in Exod 28:30, where it is said of Aaron as High Priest that he "shall *bear* the judgment [*mišpaṭ*] of the people of Israel upon his heart before the LORD continually [*tamid*; LXX: *dia pantos*]." (As argued in Chapter 15, the word *mišpaṭ* here may mean, as in 1 Kgs 8:59, "cause.")

PART FIVE: New Testament Afterword

But this "continual bearing" is characterized for us in Heb 2:17 (and see 3:2, 5–6), where Jesus is called a "faithful" (*pistos*) High Priest. If this High Priest/Son "bears all things by his word of power," this means that the work of the Son/High Priest from initial creation to ultimate completion is marked by faithfulness. And this implies that Jesus is *archegon kai teleioten* in the first instance by the Son's embodying faithfulness in creation and bearing of all things through to the end. And, given the exalted Christology in this letter, one may wonder whether we might hear embedded in the words A*rchegon and Teleioten* an inter-lingual play on the initial and final letters of the Hebrew alphabet, *aleph* and *tau*, which of course in Greek become Alpha and Omega, which in the Book of Revelation are linked with Old Testament attestations to God as First and Last, Beginning and End. These latter attestations, in Second Isaiah, attest to God's faithfulness from creation, through all of cosmic history and until the envisioned end, a faithfulness that, in Isa 46:3–4, is said to involve God's bearing and carrying Israel as a people "from birth ... to old age." If in Isa 40:31 the exiles who "wait for the LORD" are able to "mount up with wings as eagles," to "run and not be weary," to "walk and not faint," it is because their strength derives from the God who in creating the ends of the earth "does not faint or grow weary" (40:28). So it is with human faith(fulness) in the letter to the Hebrews. It is grounded in the faith(fulness) of God (Heb 10:23; 11:11) as embodied and manifest in Jesus Christ "who is the same yesterday, and today, and forever" (13:8).

Such a reading of the call to "look unto Jesus the *archegon kai teleioten* of faith(fulness)" illuminates the call, in Heb 12:12–13, to "lift your drooping hands and strengthen your weak knees, and make straight paths for your feet, so that what is lame may not be put out of joint but rather be healed." The first verse is a loose quotation of Isa 35:3 which, interestingly for the present study, employs the verb pair, *ḥazaq // 'amaṣ*, studied closely in Chapter 10. Isa 35:4 follows with the exhortation, "Say to those who are of a fearful heart, 'Be strong [*ḥizqu*], fear not!'" The way in which Heb 12:12–13 weaves the inter-personal and intra-personal multi-dimensionality of the latter word-event into the call to look unto Jesus as pioneer and perfecter of faith(fulness), helps further to explicate the inter-personal and intra-personal dimensions of faith(fulness) in following this Jesus.

All of this comes to spatial expression in reference to this Royal High Priest's everlasting intercessory reign in the heavenly sanctuary (analogous to the imagery in Col 3:1–3), and his followers' following him into that "place" in and through their own hopes and prayers. This, I suggest, is one way of taking Heb 11:1b: the faith(fulness) mentioned there is the *elegchos*, the assessing and affirming, the ultimate *hypostasis* of one's own faith(fulness), of "things unseen," above all, the "unseen" heavenly intercession of the Son as faithful high priest. It is into the bosom of this unseen heavenly activity *beseter / en to krypto* that the prayers of Jesus' followers are invited to "take place."

17

Redeeming the Expression "Redeeming the Time"

THE FOREGOING ESSAYS HAVE for the most part turned on images of space, and what happens when prayer "takes place," especially in circumstances where such a place is not granted or supported by one's community. In one of its articles, the American Constitution has enshrined certain protections, commonly referred to as articulating a doctrine of "the separation of church and state," of the right of individuals and groups to congregate for prayer without encroachment by the state. Conversely, however, this doctrine has increasingly been construed to mean that vocal prayer is "out of place" in such places as public schools. Even in such places, of course, nothing can prevent one from offering prayer under one's breath. Such an act may signify only a personal or private aspiration. Or, in the current culture wars, it may be intended as a reclaiming, however *sub rosa*, of a public space rendered spiritually naked by what is taken to be a secularity never intended by the Founding Fathers.

In the present essay I want to explore what it might mean when prayer in this last sense takes place in a slightly different sense of the word, "place." This different sense is conveyed, in the English language, in one fundamental meaning of the word, "space"—indeed, the first meaning listed in the *OED*—namely, as an interval or duration of time, as when the KJV translates Rev 8:1, "there was silence in heaven about the space of half an hour." It is also reflected when one says, colloquially, "the jobs he gave me for the afternoon kept me so busy I hardly had space to breathe." When prayer, either vocal or in Quaker silence, takes place in this sense, it may do so for durations or within intervals for which provision is made by the community. Where no such provision is made—or where the official or de facto powers-that-be claim one's time totally for other purposes—*sub rosa* prayer may also intend a reclaiming of time, a "redeeming of the time," from what is experienced as a threat to turn all time into purely secular time monitored by the relentless eye of the clock ticking on the wall of the powers-that-be. I shall explore this topic through a re-examination of Eph 5:16. As I hope will become apparent, the issues here transcend one's identification with one side or the other in the current culture wars.

Ephesians 5:16 reads, in Greek, *exagorazomenoi ton kairon, hoti hai hemerai ponerai eisin*. In the KJV, this is rendered, "Redeeming the time, because the days are evil," but more recently the first clause has been rendered, "making the most of the time" (RSV, NRSV) or "making the most of every opportunity" (NIV). The latter expressions suggest a picture in which the time or opportunity that lies at hand is to be exploited before it passes and goes to waste. As N. T. Wright has put it, "Christians are to see every day, every hour, every minute, as an opportunity for serving the Lord, for understanding what his will is and getting on and doing it."[1] But the identical expression occurs in Dan 2:8 where Nebuchadnezzar accuses his court interpreters, *kairon hymeis exagorazete*, and here commentators, and the above translations, agree on a meaning in which the interpreters are accused of trying to "buy time" *beyond* the time (as in, "now!") given them by the king for responding to his order.

Entirely apart from questions of any dependence of Eph 5:16 on Dan 2:8, how can the same Greek expression—found nowhere else, but only in these two passages—convey two diametrically opposite meanings? An obvious answer might be that meanings of words and even of longer expressions depend upon their contexts. As James Barr has argued with great effectiveness, the unit of meaning is not the word but the sentence, and an individual word in a particular instance, whatever its connotations in other sentence-contexts, conveys a connotation appropriate to its sentence in that instance.[2] And, one could say in extending this principle, the meaning of a sentence depends on *its* larger context, and so on in expanding contexts until one has arrived at a community's or a culture's worldview or "imagined world."

---

1. Wright, *Paul for Everyone*, 63. As if catching himself in the act of sounding like the time-and-motion efficiency experts Frank Gilbreth and Lillian Gilbreth Carey, whose book, *Cheaper by the Dozen*, was made into a famous movie, Wright quickly cautions, "verse 16 can, of course, lead people to an obsessive lifestyle, calculating and counting every minute and giving oneself and everyone else no peace." Quite! I shall argue that the expression, "the days are evil," refers to a world in which precisely such a "quantifying, commodifying" approach to time (a fundamental aspect of what Max Weber referred to as the "iron cage" of an overly bureaucratized society) is to be "redeemed."

2. See Barr, *Semantics of Biblical Language*.

PART FIVE: New Testament Afterword

In the present study I want to reexamine this Greek expression, first in respect to its component parts, then in respect to its contexts in Daniel and in Ephesians, and then, briefly, in respect to its bearing upon life in our own time with its fondness for expressions like "time is money" and "the sixty-second minute," or, under a contrasting economic regime, "the five-year plan," and then, again, under a different domestic regime, the parental adoption of the language of clock-driven sports in saying to "naughty" children, "time out."[3] I shall begin with a brief sketch of the verb, *exagorazo*, considered by itself.

Exagorazo in Classical Greek

The semantic genealogy of the verb goes something like this.[4] A verb, *ageiro*, "gather together," produces a noun, *agora*, which basically means "place of assembly." Often the noun indicates a marketplace, or the activity of people gathered there such as public speaking and especially commerce. Interestingly, the word can also signify in expressions marking times. The idiom, *agora plethousa*, literally, "the market being filled," refers to "the forenoon," it being the case, apparently, that Greeks typically did their shopping in the early part of the day; and the expression, *agores dialysis*, literally, "the breakup or disbanding of an *agora*," in the case of market-day refers to the time past mid-day when people begin to head home. By itself, then, *agora* can carry the temporal connotation, "market-day." The usage is similar to that in, say, the England of Thomas Hardy's novels when people would congregate in central locations for commerce and for social events between friends, but also for official events conducted by the government such as "assizes," trials, and hangings. In these latter events, while the public might come as spectators (hangings being a special drawing card), official parties attended *sub poena*, however much they might wish—especially in the case of those convicted of capital crimes—to "buy more time." (For the condemned, needless to say, market days were "evil

---

3. A three-year-old grandchild, Lucy—generally of a lively, bouncy disposition—is observed sitting quietly on a chair in the corner of the kitchen. Asked what she is doing, she says, "I'm giving myself a time out." As I shall suggest later, she is *wiser* (in Daniel's and Paul's sense of the word) than she knows. In passing, let me say that while the majority of commentators may be right in taking Ephesians as pseudo- or deutero-Pauline, for simplicity in writing, and actually as a kind of counter-cultural practice of second naiveté within a "willing suspension of disbelief," I shall respond to the invitation of Eph 3:1 by referring to the implied author of this letter simply as Paul.

4. The following information is gleaned from LSJ, 13.

days.") In the words of Anthony Powell's twentieth-century epic novel, the *agora* specifies the space-time coordinates of one prominent public and official *Dance to the Music of Time*.

This noun, *agora*, spawns a verb, *agorazo*, which means "frequent the market," or more specifically, "buy (for oneself) in the market." When the preposition *ek* augments a verb, "the sense of *removal* prevails; *out, away, off*"; but it can also "express *completion* like our *utterly*."[5] The augmented form, *exagorazo*, means, then, "buy *from*," or "buy *off*." One could, then, take *exagorazo* as meaning, in some contexts, "buy up," as when a land speculator "buys up" all the lots in a newly opened subdivision—in that sense, taking advantage of the opportunity provided in the *agora*. (To be clear here: I am trying to grant as much semantic room as possible for the current construal of Eph 5:16 as meaning, "making the most of the time/opportunity at hand" for the sake of argument.[6]) But the verb can also mean, "redeem," as in one instance where a Roman knight "redeems" a slave. Here the reference is to individuals who, as slaves or servants, are at the unceasing beck and call of their masters, but who, as "redeemed," or "manumitted," become freedpersons. When Paul, then, says in Gal 3:13, "Christ redeemed [*exegorasen*] us from the curse of the law," and in Gal 4:4–5, "when the fullness of time had come, God sent his Son, born of a woman, born under the law, in order to redeem [*exagorase*] those who were under the law," the verb here apparently carries the connotation of manumission from slavery, having as its immediate literary context Paul's binary contrast between existence as "slave" in "bondage" and as "son" in "freedom." This "bondage" is to what he twice (Gal 4:3, 9) calls *ta stoicheia tou kosmou* (see also Col 2:8, 20), the "elementary principles" or "elementary spirits" of the cosmos. As examples of these "elementary principles" to which, Paul fears, his Galatian readers want again to submit in bondage, he immediately says (Gal 4:10), "You are observing special days, and months, and seasons, and years." They are dancing, as it were, to the music of times set for them by these *stoicheia*, structuring elements of "this evil age" (Gal 1:4), principles or "spirits" presumably expressing the hegemony of what Paul elsewhere will term "principalities and powers"—as when he avers, in Rom 8:38–39, that "neither death, nor life, nor angels, nor rulers,

---

5. LSJ, 499.

6. But why express such a notion with an expression that nowhere else means that, and that in Dan 2:8 means just the opposite! Such a notion could quite easily be expressed by any one of a number of classical (nonbiblical) Greek idioms such as *harpazein ton kairon*, "to seize the opportunity." For other idioms expressing this idea, and further references, see Hoover, "The Harpagmos Enigma."

PART FIVE: New Testament Afterword

nor *things present*, nor *things to come*, nor powers, nor height, nor depth, nor anything else in all creation, will be able to separate us from the love of God in Christ Jesus our Lord." The way the two time references come between the words "rulers" and "powers" simply underscores rhetorically what is the case existentially: that—whether one is thinking of the institution of the Julian Calendar, the attempt to institute a new calendar in the aftermath of the French Revolution, or conversely the Jewish practice of indicating year-dates not only according to the Common Era but also according to the traditional reckoning from creation—those who have the power to determine the times have the power to "call the tune" to which life is to be danced. Freedom from such hegemony comes only in "marching to a different drummer," or in dancing, *at one and the same time*, like Jews with their double notation, to a different tune.

But I am getting ahead of myself. I want to propose an approach to the relation between the expressions in Dan 2:8 and Eph 5:16 through the method of intertextual interpretation proposed by Richard B. Hays.[7] To summarize again, this method turns on the following basic idea: Where a word or longer textual unit in Paul—or presumably elsewhere in the New Testament—may be said to echo a word or similar unit in the Old Testament, the context for the item in the Old Testament stands in some kind of meaningful correlation (whether direct or inverse) with the context for the item in the New Testament. My way of analogizing this is to think of buttons and buttonholes. I may attach one side of my overcoat to the other by affixing only a single button to its corresponding hole, but when that button is attached to the correct buttonhole, all the other unattached buttons and buttonholes stand in correct if implicit correlation with each other. Just so, when a textual item in the New Testament that is not an explicit quotation or clear allusion, but seems to this or that reader to echo a corresponding item in the Old Testament, one test of whether it is a genuine echo is the way in which other elements in the Old Testament context resonate with other elements in the New Testament context. The correlation need not be exact—every item in the one context matching every item in the other context, button for button-hole; but where the inter-textual match-up is felicitously identified, prominent features of the Old Testament context will shed light on how the echoing term in the New Testament functions in relation to the chief features of its context. As I shall argue, so many individual elements in the context of Dan 2:8 match up with elements in Ephesians—along with general features of the

---

7. Hays, *Echoes of Scripture in the Letters of Paul*.

Redeeming the Expression"Redeeming the Time"

worldview in each book—that an interpreter is hard-pressed to defend the notion that in Eph 5:16 means "make the most of the opportunity." Rather, one should make every effort to construe the expression in Ephesians in a fashion that is at least generically similar to its connotation in Daniel.

Buying Time in Daniel 2

Nebuchadnezzar, we read, "dreamed dreams; and his spirit was troubled [*wattitpa'em*], and his sleep left him." (I shall return later to the indicated verb.) He summons assorted practitioners of the interpretive arts "to tell [*lehaggid*] the king his dream." Is he asking them to intepret it? Or is he asking them to first tell him what he had dreamt? The verb *higgid*, depending on the context, could mean either. As BDB puts it, the verb means "1. *tell, announce, report,*" and "2. *declare, make known, expound*, esp. of something before not understood, concealed or mysterious."[8] Yet, as the passage unfolds, the repeated occurrence of the verbs "tell" and "interpret" in sequence, together with Daniel's eventual actions of first telling the dream (with no indication that he is repeating what the king had first narrated to him) and then interpreting it, suggests that the king has forgotten the substance of his dream and remembers only that he is troubled by it. (Or does the verb give him a subliminal clue?) Given the importance of dreams in the ancient world, a ruler would want to know a dream and its meaning especially if it troubled him. So he repeats to the assembled interpreters, "I dreamed a dream, and my spirit was troubled [*watippa'em*] to know the dream." They respond, respectfully and—we might suppose—reasonably (the chapter now shifting from Hebrew to Aramaic), "O king, live forever! Tell [*'emar*] your servants the dream, and we will show the interpretation [*pišra'*]." At this the king responds peremptorily that they had better make known the dream and its interpretation or they will face certain death. They respond a second time that if the king will tell them the dream they will show its interpretation. At this the king announces,

> I know with certainty that *you are buying time* [*'iddana' 'antun zabenin*; LXX: *kairon hymeis exagorazete*], because you see that the word from me is sure that if you do not make the dream known to me, there is but one sentence for you. You have agreed to speak lying and corrupt words before me till the times change [*'ad di 'iddana' yištanne'*; LXX: *heos an ho kairos alloiothe*].

8. BDB, 616.

PART FIVE: New Testament Afterword

> Therefore tell me the dream, and I shall know that you can show me its interpretation.

I find the interplay between the king's "you are buying time" and "until the times change" fascinating. What does he mean by the latter expression? Until what times change? What is it about this dream that he cannot recall, that suggests to him that it has to do with "the times changing"? Is it something that is conveyed by the narrator's, and his, use of the verb, *paʿam*?

The Verb *paʿam* as a "Beating of Times"

This verb, which occurs very seldom in the Hebrew Bible, appears to mean, at root, something like "beat, strike, hit." It is related to the noun, *paʿam*, which can refer to a (repeated) hoof-beat or footfall, an anvil (on which a hammer beats), a foot, or, in the Bible, most often, to "a time" as though counted off by a stroke or beat. (Think of the sounds of Big Ben at twelve noon.) I hasten to say that I do not mean to claim that all these connotations are present in the noun or verb in every one of its occurrences, but only to explore the word's semantic range with a view to ascertaining why the writer in Daniel 2 chose this word rather than, say, the idiom in 1 Sam 24:5 and 2 Sam 24:10 where, as KJV and RSV translate literally, "David's heart smote him." There, the impact, so to speak, is of a single blow. Or why does the writer not use the verb *nibhal*, "be troubled, in anxious haste," which is quite at home with such a connotation in visions like Nebuchadnezzar's (see Isa 21:3; Dan 4:5; 5:6; 7:15)? Does *paʿam* signify something more than a single blow—perhaps a series of palpitations, for example? For in Dan 2:1, in the narrator's report, the verb occurs in the Hithpael form, which often refers to reiterated action. And does such a repetitive beat carry additional connotations concerning the passage of time?

The expression in Daniel 2, about Nebuchadnezzar's "spirit" being "troubled by a dream," has as its biblical prototype Pharaoh's dream of the seven fat and seven lean cows in Genesis 41. And as it turns out, that dream is about time sequence: seven repeated fat years followed by a "change in economic seasons" to seven lean years. Is the "troubling [*paʿam*] of spirit," here, specifically a sense—as Freud or Jung might say, a symbolic representation deep in Pharaoh's dream-self—that the times are about to change?

The verb occurs again in Ps 77:4, where the psalmist says, "I am so *troubled* that I cannot speak." The "trouble" concerns the psalmist's present plight. Considering "the days of old" and "the years of long ago," and searching in his spirit, the psalmist can only conclude concerning "the *day* of my distress [*ṣarah*]" that "the right hand of the Most High has changed [*šenot yemin 'elyon*]." It is the psalmist's sense of being caught within the crushing constraints of his present *day* as contrasted with the "days of old" as informed by God's promises—a sense that his times have changed, and that God has changed them—that gives rise to his grief. His solution is to call to mind God's redeeming "wonders of old" in the Exodus, where Israel was led out of bondage; and, as he recalls its epic scenes, ending with the scene of God leading Israel like a flock by the hand of Moses and Aaron, the psalm ends on a note of quietness that may evoke for the reader the line, "he leads me beside the still waters." In this way, through prayerful and meditative reflection on God's past acts, the psalmist finds alleviation from the *ṣarah* with which his present "evil day" hems him in.

What of the remaining occurrence? In Judg 13:25 we read that when young Samson has grown up "the spirit of the Lord began to *stir* him." How? On a jaunt down into Philistine country he spots a young Philistine woman, and returns home to tell his parents that he wants her for a wife. When God "stirs" him (*pa'am*) to action through the sight of this young woman, how can one not be reminded of Adam who, seeing the animals pass by before him, recognizes in none of them a "peer helper," but who, when he lays eyes on the one made out of a rib taken from his side (near his heart?), exclaims, "*zo't happa'am*—this time!—bone of my bones and flesh of my flesh," the *pa'am* here connoting not simply a particular item in a temporal series, but that this figure before him palpably "strikes a note" in him, quickens his pulse, in a way that his own body attests. In Samson's case, the hidden agenda in this apparently natural internal "stirring" within him is God's intent to deliver Israel from Philistine hegemony—a changing of the times, as Nebuchadnezzar would put it. Not to put too fine a point on it, but the intimation is of a correlation between this change in the "seasons" of Samson's life, from his tutelage to his manhood, and a change in Israel's fortunes from Philistine hegemony to a regaining of autonomy. All this is by way of suggesting that, for the reader deeply immersed in the Scriptures and closely attuned to its words, Nebuchadnezzar's recognition that his dream "troubles" (*pa'am*) his spirit is his dim apprehension that, as Bob Dylan puts it, "the times they are a'changin." But he, as a ruler generally preoccupied with governing in the waking world, cannot even recall the content of the dream, let alone what it means, only that something he dreamt has to do with changing times that threaten his rule which, like that of his predecessors and counterparts, has been ritually if hyperbolically assured will endure forever.

To the king's reiterated demand, the professional interpreters respond that what he is asking lies beyond human capacity: only the gods "can show it to the king." When Daniel hears of the sentence—one which will engulf him and his three friends in the general execution—he addresses the captain of the king's guard "with prudence and discretion," asking, "Why is the king's decree so *mehaḥṣepah*?" The LXX's term here, *pikros*, takes the Aramaic word to mean "sharp, bitter, vindictive." But John J. Collins takes Daniel's subsequent successful bid for time (2:16) to indicate that "the hastiness was the issue," and he translates the Aramaic adjective, "peremptory."⁹ This rendering underscores the implicit issue throughout this chapter as the king's concern to retain control of time through his commands. When Daniel then asks the king to "give him a time" [*zeman yinten-leh*] to show the king the interpretation, the king accedes. Noting that, from a compositional point of view, some have argued that "Daniel's intervention and the report of revelation are secondary elaborations of the narrative," Collins refers, as supporting such an argument, to the fact that "in v 16 Daniel goes to the king and requests a delay of the executions, although the king's anger was aroused precisely by the attempt of the Chaldeans to 'buy time' (v 8)."

But a different explanation may be offered from within the dynamics of the issue implicitly running through the chapter. When the king's interpreters respond to the king's demand that they recount the dream and then interpret it simply by saying, "tell us the dream and we will interpret it," the king construes this as a challenge to his authority to order as he pleases, and as a lying ploy to "buy time." Daniel's implicitly prudent and discreet request that the king grant him (a) time to show the interpretation is an implicit recognition of the king's authority to orchestrate others' actions in time. Given what will ensue, Daniel's request is also subversive;

---

9. Collins, *Daniel*, 149, translating in Dan 2:15. Though there is no indication that Collins is aware of this, the choice of this word is, so to speak, inspired, given that—as the Latin expression, *caveat emptor* indicates—it derives from a verb, *emere, emptum*, originally "to take," and then "to buy." This connotation is, of course, not present in either the Aramaic word or its LXX translation, but it underscores the way the king responds to what he perceives as an attempt to wrest control of time from him by tightening his control of it.

but, veiled under his prudence and discretion, it is a "hidden" subversion of this king's control of time that will come to light only with the interpretation.

Daniel then does what the other interpreters don't do. He convenes a meeting of his three friends for prayer "to seek the mercy of the god of heaven concerning this mystery [*razah*; LXX: *mysterion*]." When this mystery is revealed (*gali*; LXX: *exephanthe*; Theodotion: *apekalyphthe*) to him in a night vision (2:19), his response takes the form of what W. Sibley Towner has identified as a hymn of individual thanksgiving.[10] Specifically, it takes the form of what is called a *berakhah*:

> Blessed be the name of God forever and ever,
> to whom belong *wisdom* and *might*.
> He changes times and seasons
> [*wehu' mehašne' 'iddanayya' wezimnayya'*];
> he removes kings and sets up kings;
> he gives *wisdom* to the wise
> and *knowledge* to those who have *understanding*;
> he reveals deep and mysterious things
> [*hu' gale' 'ammiqata' umesatterata'*; LXX: *anakalypton ta bathea kai skoteina*; Theodotion: *autos apokalyptei bathea kai apokrypha*];
> he knows what is in the *darkness*,
> and the *light* dwells with him.
> To thee, O God of my fathers, I give thanks and praise,
> for thou hast given me wisdom and strength,
> and hast now made known [LXX: *egnorisas*]
> to me what we asked of thee,
> for thou hast made known [LXX: *egnorisas*]
> to us the king's matter.

This *berakhah* deserves close analysis in its own right. Here I note only that it resonates with terms and themes that we encounter throughout Ephesians; while the term *mesatterata'*, literally "hidden things," is related to the other Hebrew words built on the root *str*, including the noun *seter*, "secret (place)," and forms an echoing link with the vision in Second Isaiah where words built on *str* occur repeatedly as a leitmotif,[11] most especially in connection with the travail of the suffering figure in chapter 53 and that of Zion in chapter 54, in implicit reference to both of which God says, in synonymous fashion in Isa 48:6–7, "From this time forth I make you hear

---

10. Towner, "The Poetic Passages in Daniel 1–6," 319.
11. See Isa 40:27; 45:15, 19; 48:16; 49:2; 50:6; 53:3; 54:8.

PART FIVE: New Testament Afterword

new things, hidden [*neṣurot*] things which you have not known. They are created now, not long ago; before today you have never heard of them, lest you should say, 'Behold, I knew them.'"

When Daniel reports to the captain of the guard that he has an interpretation of the king's dream, and the order for the execution of the other interpreters should be suspended, the captain brings Daniel before the king "in haste" (*behitbehalah*; LXX: *kata spouden*)—another indicator of the fact that any suggestion of interruption in events ordered by the king must immediately secure the king's sanction. When the king asks if Daniel can make known to him both his dream and the interpretation, Daniel's response, after a suitable prologue, focuses on the dream re-characterized as "your thought [*ra'yonak*] of what would be hereafter" and "the thoughts of your heart [*ra'yone libebak*]." The texture of these thoughts, as "troubling his spirit," is suggested by the character of the "thoughts" as referenced elsewhere in Daniel. In 4:19, Daniel, upon hearing the report of another of Nebuchadnezzar's dreams, is dismayed, and "his *thoughts* alarmed him [*yebahalunneh*]." The same expression occurs in 5:6, 10 and 7:28, the verbal component in each case being the same as in 2:24 where the captain runs to the king "in haste." This implies that the king's thoughts "of what would be hereafter" are likewise tumultuous and alarming—exactly the picture suggested, in my view, by the verb *pa'am* in the opening scene where Nebuchadnezzar's dream is portrayed as "troubling" his spirit. What Daniel will report, now, is both the content of these troubled "dream thoughts" and their "objective correlative"[12] in the changing times and seasons of overlordship on the international scene. The dream concerns a succession of world empires that in the end will give way to an empire represented by "a stone cut from a mountain by no human hand" and "standing" forever.[13]

This is the context within which the court interpreters attempt to "buy time" from one whom they in fact—despite their reference to the gods (2:11)—evidently take to be ultimately in charge of things, and within which Daniel, for his part, seeks and gains enough time from the king to seek and gain wisdom from God for telling the king his dream. I find it intriguing that Daniel characterizes the king's dream here not as a

---

12. Compare the implicit correlation, noted above, between God's "stirring" Samson at the sight of a young woman to a new season in his psycho-social life (from tutelage to manhood where, according to Genesis 2, he will "leave his father and mother and cleave to this woman"), and a change in the seasons of Israel's existence (from subjection to the Philistines to autonomy).

13. Could the "mountain" here be Zion, and the stone cut from it to become itself a great mountain to fill the whole earth an echo of Isa 2:1–5?

message from God to him, but as the king's own thoughts—thoughts that are so deeply veiled from his conscious recall that Daniel must make them known to him. In such a way of viewing the matter, the "revelation" of the "mystery" is at one and the same time a "making known" of the king's own thoughts hidden deeply within him and a "making known" or "revealing" of God's word to the king. (As so read, Nebuchadnezzar's dream discloses the vestigial awareness, buried deep within the heart of even such a king as he, of the primordial vocation to which humankind was called in Genesis 1 and 2. In such a reading, the famous "hanging gardens" of Babylon would be another vestigial, veiled symbolic expression.) As I shall now seek to show, this context and these respective buyings and requestings of time bear directly on the interpretation of Eph 5:16 in the context of Ephesians as a whole.

Redeeming the Time in Ephesians 5:16

As noted earlier, the verb *exagorazo* occurs twice in Galatians where Paul speaks of Christ as coming to "*redeem* those who were under the law" (Gal 4:5) or "to *redeem* us from the curse of the law" (Gal 3:13); and these statements function as specific applications of the programmatic statement in Gal 1:3 that "our Lord Jesus Christ . . . gave himself for our sins to *deliver* [*exeletai*] us from the present evil age." The latter programmatic verb typically refers, in the LXX, to a rescue of one party from the oppressive power of another party; and the two verbs in these three passages, while not strictly synonymous, are closely related in meaning, both here operating within an apocalyptic frame of reference. Given Paul's reference in Gal 4:9–10 to the observance of "days, and months, and seasons, and years" as turning back and becoming slaves once again to the "elemental spirits," one may assume that Christ came to redeem—*exagorazein*—from these elemental spirits, including at least a freeing of those who were previously under the rule of these calendrical elements. Is it a far semantic leap from such a conception to "redeeming the time, because the days are evil"? It may be objected that this usage is Pauline whereas the usage in Eph 5:16 is pseudo- or deutero-Pauline. Even if so, it is at any rate deutero-*Pauline*; and methodologically it is worth asking whether the verb in Eph 5:16 carries a meaning similar if not identical to that in Galatians, referring in some sense to the setting free—for those who take themselves to be beneficiaries of Christ's "rescuing" and "redeeming" activity—of time that otherwise is under the power and control of "this present evil age." I am proposing that the congruence between the apocalyptic drama in Daniel 2 and the apocalyptic drama in Ephesians is such that we should consider more closely the bearing of the usage in Dan 2:8 on that in Eph 5:16.

I shall start with Peter O'Brien's comments on the circumstantial clause, "because the days are evil." He writes

> In apocalyptic literature evil was dominated by rulers or demonic powers which were doomed to pass away (1 Cor 2:6, 7). The notion that "the days are evil" appears to be similar to the idea of "this present evil age" in Galatians 1:4 (cf. "the *evil day*," Eph 6:13). These "evil" days are under the control of the prince of the power of the air (Eph 2:2) who is opposed to God and his purposes. He exercises effective and compelling authority over men and women outside of Christ, keeping them in terrible bondage (2:1–3). But the Ephesian Christians have already participated in the world to come, the powers of the new age have broken in upon them, and they have become "light in the Lord" (5:8).[14]

The very word *aion* ("age") connotes a temporal frame within which one lives, and the statement, "once you were darkness, but now you are light in the Lord" connotes not only a shift in moral and spiritual character (cf. Rom 12:2, "be not conformed to this age [*aion*]") but a *temporal* shift, from one time period to another. (Compare the shift from primordial darkness to the first day of creation, in Gen 1:2–5.) The age in question, for the writer, correlates with the final, everlasting "season" in Nebuchadnezzar's dream as Daniel reports it in Dan 2:34–35 and interprets it in Dan 2:44–45. (Note also the "darkness/light" reference in Dan 2:22). One's experience of time, I submit, is shaped in significant measure by the ruling power that defines an *aion* as an *aion*. (Think of the connotations of "the Augustan Age" or "the Elizabethan Age.") For all intents and purposes Daniel's times are in the hand of Nebuchadnezzar, so that he must ask this king for time to tell and interpret the king's dream. But his behavior, in contrast to those other interpreters who do not pray or sing to God, but simply attempt to "buy time" by their own devices, suggests that in Daniel's case, as Ps 31:15 puts it, "my times are in [God's] hand."

At this point I want to explore further connections between Daniel and Ephesians, through Eric Lincoln's and Peter O'Brien's references to

---

14. O'Brien, *Letter to the Ephesians*, 383. With reference to Eph 2:2 and the prince of the power of the air, one should note that in Wis 13:2 some idolaters are said to have "supposed that either fire or wind or swift air [*aera*], or the circle of the stars, or turbulent water, or the luminaries of heaven were the gods that rule the world."

Daniel in their commentaries on Ephesians.[15] (1) Noting that in recent scholarship the use of the word, *mysterion*, "mystery," in Eph 1:9 (and 3:3, 4, 9; 5:32; 6:19) is taken to reflect a Semitic background, Lincoln writes, "The usage of *mysterion* in LXX Daniel, where it translates the Aramaic *raz*, forms a major aspect of its Semitic background (cf. Dan 2:18, 19, 27–30, 47) and provides some parallels with its use in Ephesians, since in Daniel it refers to God's purpose, which is a unified plan with eschatological and cosmic dimensions."[16]

(2) Commenting on Eph 3:4, where the writer speaks of "my insight into [*synesin . . . en*] the mystery of Christ," Lincoln writes, "The notion of insight into mysteries being given to those who have received revelations or visions is a frequent one in apocalyptic literature (cf. Dan 10:1; 4 Ezra 5:22; 14:40, 47; *T. Levi* 2:3; 18:1, 7)."[17] O'Brien, more precisely, notes that "the expression *synesis* (*syniemi*) *en* ('insight into'), which is not classical, is probably drawn from the Greek Old Testament, especially Daniel, where it is used of the understanding of dreams and visions, and by implication associated with revelations (1:4, 17; 9:13, 23; 10:1, 11)."[18] O'Brien's listing of passages from Daniel needs to be nuanced as follows:

> Dan 1:4: "Skillful in all wisdom";
> LXX, *epistemonas* (Theodotion: *synientas*) *en pase sophia*;
>
> Dan 1:17a: LXX and Theodotion, *synesin en*;
>
> Dan 1:17b: LXX, *synesin en*;
> Theodotion, *syneken en*;
>
> Dan 9:13: LXX, *dianoethenai ten dikaiosynen sou*;
> Theodotion, *synienai en pase aletheia sou*;
>
> Dan 9:23: LXX, *dianoetheti to prostagma*;
> Theodotion, *synes en te optasia*;
>
> Dan 10:1: "a word *was revealed* to Daniel";
> LXX, *edeichthe*;
> Theodotion, *apekalyphthe*;
>
> Dan 10:1: "he had understanding of the vision";
> LXX, *dienoethen auto en horamati*;
> Theodotion, *synesis edothe en te optasia*;

---

15. Lincoln, *Ephesians*; O'Brien, *Letter to the Ephesians*.
16. Lincoln, *Ephesians*, 30.
17. Ibid., 176.
18. O'Brien, *Letter to the Ephesians*, 230 n. 20.

PART FIVE: New Testament Afterword

> Dan 10:11: "give heed to the words";
> LXX, *dianoetheti tois prostagmasin*;
> Theodotion, *synes en tois logois*.

A number of things are to be noted in these data. First, though the LXX displays the "nonclassical" Greek construction twice, in Dan 1:17, Theodotion routinely revises the LXX in using the classical construction, beginning at 1:4. If Eph 3:4 echoes the usage in Daniel, it likely echoes that book in Theodotion's recension. Second, in Dan 10:1, Theodotion's revision of the LXX *edeichthe* to *apekalyphthe* indicates that the non-classical construction, "used of the understanding of dreams and visions," is not only "by implication" but explicitly in Theodotion associated with revelations. In fact, whereas LXX never translates Hebrew *galah* or Aramaic *gale'* with *apokalypto*, Theodotion revises LXX's renderings to *apokalypto* in Dan 2:19, 22, 28, 29, 30, 47; 10:1; 11:35. Especially noteworthy is the collocation of terms in 2:28–30:

> there is a God in heaven who reveals mysteries [*apokalypton mysteria*], and he has made known [*egnorisen*] to King Nebuchadnezzar what will be in the latter days. Your dream and the visions of your head as you lay in bed are these: To you, O king, as you lay in bed came thoughts of what would be hereafter, and he who reveals mysteries [*apokalypton mysteria*] made known [*egnorisen*] to you what is to be. But as for me, not because of any wisdom that I have more than all the living has this mystery [*mysterion*] been revealed [*apekalyphthe*] to me, but in order that the interpretation may be made known [*gnorisai*] to the king, and that you may know the thoughts of your mind.

As with *apokalypto*, the verb *gnorizo* is in Daniel peculiar to Theodotion where it appears 19 times, ten of which occur in chapter 2. It appears in Ephesians five times with reference to the *revelation* and/or the preaching of the *mystery* of the Gospel. Indeed, given the repeated collocation of the words in Dan 2:28–30, the collocation in Eph 3:3 strikes me as a resounding echo where Paul writes, "the mystery [*mysterion*] was made known [*egnoristhe*] to me by revelation [*apokalypsin*]." Finally, *optasia* in Theodotion 9:23; 10:1, 7, 8, 16 (but never in the LXX)[19] appears only once

---

19. The noun, *optasia*, occurs also in Ezek 1:1 (Theodotion); elsewhere in the Greek Bible only 4x in the LXX, of which one, Mal 3:2, is relevant: "But who can endure the day of his coming, and who can stand when he appears [*en te optasia autou*]?" Given the eschatological tenor of this passage—together with the question of "who can stand" and the depiction of Daniel in Daniel 10 as at first falling before being called to

in the New Testament: in 2 Cor 12:1, where Paul, in defense of his apostleship, says, "I will go on to visions [*optasias*] and revelations [*apokalypseis*] of the Lord." If Paul truly received the Gospel, and/or was subsequently illumined as to its meaning and significance, by revelation through a visionary experience (Gal 1:12, 16),[20] it is hard to imagine him not studying Daniel closely in an attempt to understand that "inexpressible" experience. And if someone else was the immediate author of Ephesians, that person must have been "channeling" Paul so closely that it becomes easier to practice the "willing suspension of disbelief" and, at least for the duration of reading the letter, imagine it as coming directly from Paul. As for the difference in style—the pleonasms and the quasi-lyrical, almost quasi-rhapsodic rhythms—one might almost imagine the Paul of 2 Cor 12:4, who "heard things that cannot be told, which man may not utter," finally now, caught up in recollection into heaven (or, as he now says in Ephesians, "the heavenlies"),[21] venturing to speak, if not directly of what

---

stand up—it seems likely that Theodotion in Daniel revises the LXX to *optasia* under the influence of Mal 3:2.

20. Alan F. Segal's thorough discussion of "Paul's Ecstasy," within a framework of first-century and earlier Jewish apocalyptic mysticism, bears on my point in several ways, including Segal's several references to visionary reports in Daniel. See Segal, *Paul the Convert*, 34–71.

21. As often observed, the set phrase, "in the heavenlies" (*en tois epouraniois*), or "heavenly places," is peculiar to Ephesians. O'Brien comments that it "is bound up with the divine saving events and is to be understood within a Pauline eschatological perspective" (*Letter to the Ephesians*, 97). On the basis of his close study, Lincoln concludes that "the origin of the expression is uncertain" (*Ephesians*, 20) But note the following considerations: (1) the word refers to "the heavenly God" (*ho epouranios theos*) in 3 Macc 6:28; 7:6, or "the one who has his dwelling in the heavens" (*ho ten katoikian epouranion echon*), as watching over the Jews since the time of the ancestors. (2) That the term is drawn from Psalm 68—itself a synopsis of Jewish history from the beginning to anticipated end in universal if not cosmic redemption—is suggested by the full ascription in 3 Macc 6:28, *tou pantokratoros epouraniou theou zontos*, where *pantokratoros* echoes one common translation of Hebrew *šadday* in Ps 68:15 but the latter epithet is there translated, *ton epouranion*(!)—"When *the one in the heavens* scattered kings . . ." (3) Ephesians 4:8 quotes Ps 68:18 in connection with divine victory over all enemies on behalf of God's people. (4) Ephesians 1:3, "who has blessed us in Christ with every spiritual blessing *en tois epouraniois*" may echo Paul's appropriation of Ps 68:18 in Eph 4:8. (5) In Ps 68:28, "Benjamin" is "the least" of the tribes, with whom we may compare Benjaminite Paul's self-description in Eph 3:8. (6) This Benjamin is shown "leading them" [*rodem*] in triumphal procession; but the LXX construes *rodem* as an otherwise unattested Qal stem of the usual Niphal form, *nirdam*, "be, or fall into, heavy sleep or trance" (as in Dan 8:18; 10:9!), and translates, *en ekstasei*, which, even if we discount the portrayal of Paul *en ekstasei* in Acts 22:17, is a translation that would be of interest to the Paul of the revelatory vision in 2 Corinthians 12

he then saw and heard, at least of earthly things as viewed from that exalted standpoint. But this comment is only offered by the way, as I return now to matters more strictly intertextual.

(3) On Eph 3:14, "for this reason I *bow the knee* [*kampto ta gonata mou*] to the father, from whom every family in heaven and on earth derives its name," Lincoln notes that the usual Jewish and early Christian practice was to stand for prayer, but that kneeling is found in Theodotion's version of Dan 6:11 (*kamptein epi ta gonata*); 1 Kgs 8:54 (*oklazein epi ta gonata*); and 1 Chr 29:20 (*kamptein ta gonata*). I note:

a. that the prayer in Eph 3:14–21 resumes, as Lincoln observes, an anacoluthon in Eph 3:1 that reads, "I, Paul, a prisoner for Christ *on behalf of you Gentiles*";

b. that the digression beginning in Eph 3:2 ends in 3:13 with a request that "you not lose heart over what I am suffering *for you*";

c. that Eph 3:1 and 3:13 thus form an *inclusio* in reference to Paul's situation;

d. that Eph 3:4 through 3:12 presents a compact restatement of essential elements in the *berakah* or "blessing" with which the letter opens (1:3–14), with several elements resonating with themes in Daniel and especially Daniel 2;

e. that, as Lincoln observes the doxology in Eph 3:21 ends with a formulation *eis pasas tas geneas tou aionos ton aiono*, whose concluding terms, *tou aionos* (singular) *ton aiono* (plural) depart from the standard plural-plural pattern and are paralleled *only* in Dan 3:90 (LXX) and 7:18 (LXX, Theodotion);[22]

f. that Dan 3:90 comes as the concluding doxology of the Greek text of the Song of the Three Children in the fiery furnace, while Dan 7:17–18 gives assurance that "these four great beasts are four kings who shall arise out of the earth; but the saints of the Most High shall receive the kingdom, and possess the kingdom forever, forever and ever [*heos tou aionos kai heos tou aionos ton aionon*]."

---

where he was caught up "into the third heaven." These "dots," as "connected," might suggest the origin of "in the heavenlies" in a psalm of intense personal interest to Paul the Benjaminite visionary "least" among the people of God; and in so doing, might go toward identifying an authentic Pauline voice in Ephesians.

22. Lincoln, *Ephesians*, 217.

Is Paul's confidence in prayer, and the ground for his assurance to his readers—no doubt rooted in the resurrection of Jesus—also backed up by the (Greek) story of the young men who survived the fiery furnace, and then by the story of Daniel who, after bending the knee to pray in defiance of the king's orders to petition to no god or other power but only to him for thirty days (Dan 6:6–10), is thrown into the lions' den only to be delivered?

(4) Lincoln notes that the term, *apolytrosis*, "redemption" (Eph 1:7, 14; 4:30), occurring in the LXX only in Dan 4:34, occurs ten times in the New Testament, seven "in the Pauline corpus."[23] Engaging the issue of whether this term refers to some kind of "ransom," Lincoln favors a general connotation of "deliverance, liberation" like that in the deliverance from Egypt, and he compares such a connotation to the function of *exagorazein* (!) in Galatians. Of the occurrence in Dan 4:34 he says only, "the use [there] does not contain this idea," the latter passage reading, in Hebrew, "my reason returned to me," and the LXX of Dan 4:34 reading (in Collins' translation), "the time of my redemption [*apolytroseos*] came, and my sins and my ignorance [*hai hamartiai mou kai hai agnoiai mou*] were complete [*eplerothesan*] before the God of heaven."

But what are people being redeemed *from*, and redeemed *to*, in Ephesians? Are they not being redeemed from the so-called wisdom of the principalities and powers? Do those powers not rule through their own "wisdom," and through their capacity to determine what counts on earth as wisdom and right action in response to their hegemony? Granted, in Eph 1:17 the *apolytrosis* is through Christ's blood, for the forgiveness of sins, but greater than the sins of transgressing the torah or any Gentile moral code—as serious as those sins are—is that "you once walked, following the course of this world [*kosmou*], following the prince of the power of the air, the *spirit* that is now at work in the children of disobedience" (Eph 2:2).

The phrasing here is so compact that it calls for unpacking, first of all by clarifying the grammar of the sentence. There is only one verb, "the trespasses and sins in which you once *walked*." This walk was not simply a matter of following one's own willful and wayward propensities and predilections, but rather "*in accord with* [*kata*] the course [*aion*] of this world." The basic meaning of the word *aion* is a period of time. If translated as "course" (KJV, RSV, NRSV; cf. NIV: "ways") the word is to be understood as carrying the meaning which the *OED* gives under "III, *fig.* of time, events or action," sub-meaning "19, habitual or ordinary manner

---

23. Ibid., 27.

PART FIVE: New Testament Afterword

of procedure; way, custom, practice." The waywardness thus consisted in walking in accord with a customary pattern larger than the individual and indeed transcending the community. More specifically, then, one had walked "*in accord with* [*kata*] the prince [*archon*] of the power [*exousia*] of the air." The *aion* of this world is given its shape, its "customary manner of procedure," by the transcendent power here called a "prince." The latter term answers to the transcendent "prince of Persia" or "prince of Greece" in Dan 10:13, 20, 21 over against Michael as transcendent "prince" and protector of Daniel and his people (Dan 10:13, 21; 12:1). (Note, that, in these passages, Theodotion revises the LXX's *strategos* or *aggelos* to *archon*, in closer conformity to Hebrew *śar*.) The term *exousia* occurs some two dozen times in Daniel, sometimes referring to a human regional office and authority under the imperial king, and by implication under the transcendent "prince."

The phrase, "the prince of the power of the air" is then followed immediately, without a conjunction, almost like a piece of Hebrew parallelism, by the phrase, "the spirit that is now at work in the children of disobedience." This phrase identifies the immanent dimension of "the prince of the power of the air" as the spirit animating and conforming individuals to disobedience. The phrase "children of disobedience" can be taken simply as a way referring to members of a class. But the "genetic" image implicit in "children" is of individuals conforming to a culture into which they are born and in which they are raised. That culture—to complete the circle—is "the *aion* of this world." In contrast to this "prince" stands the one whom God has

> at his right hand *in the heavenly places*, far above all princely rule [*arche*] and authority [*exousia*] and power and dominion, and above every name that is named, not only in this age [*aion*] but also in that which is to come; and he has put all things under his feet and has made him the head over all things for the church, which is his body, the fulness of him who fills all in all.

The juxtaposition of Eph 1:20–23 and 2:1–3 provides, in a nutshell, the frame of reference within which 5:16 is to be interpreted: As still living within "this age," but participating already in the age to come in the power of the Spirit of the one enthroned above all transcendent princely powers, one "walks" in accord with this new age by redeeming the time, by dancing in time to the music of eternity.

In Eph 4:18 Paul speaks of those who "are darkened in their understanding, alienated from the life of God because of the ignorance [*agnoian*] that is in them." As for Nebuchadnezzar's confession that God's "redemption" has come when "my sins and my ignorance were complete [*eplerothesan*] before the God of heaven," does that "fullness" of sin in ignorance not come, despite Daniel's dream interpretation in chapter 4, in the king's utter hubris expressed in Dan 4:29–30? With this we may compare Paul's saying in Gal 4:4–5 that "when the time had *fully* come [*hote de elthen to pleroma tou chronou*], God sent forth his Son, born of woman, born under the law, to redeem [*exagorase*] those who were under the law." If we may interpret what Paul meant here by "the fullness of time" by recourse to 1 Cor 2:7–8, the "fullness" of the time is measured, for Paul, by the fact that the "rulers of this age," in accordance with their conceptions of ruling wisdom and power, "crucified the Lord of glory," and in this fashion displayed and enacted their sinful ignorance in all its "fullness." As the running theme of wisdom and insight in Ephesians makes clear, as well as the theme of light versus darkness in Eph 5:8–14, the redemption in Eph 1:7 includes a transformation (as in Rom 12:1–2) through the renewal of one's mind.[24] But a transformation of mind is *exactly* what the term means in the LXX of Dan 4:34.

(5) Noting the "much discussed feature" of an epistolary thanksgiving in Eph 2:15–22 following the introductory *berakah* in 2:3–14, Lincoln adduces epistolary practices that support the sequence, and then adds that "in terms of Jewish liturgical background a sequence of eulogy followed by thanksgiving can often be found (e.g., Dan 2:20–23; *Jub.* 22:6–9).[25] But the Daniel instance is not simply one among many examples because, as we have seen, it forms part of a series of echoes between this book and Ephesians. In particular, Daniel's *berakah* follows immediately the narrator's report that, in response to his prayer "the *mystery* was *revealed* to him" concerning the *wisdom* and *might* by which God "changes times and seasons" and "gives wisdom to the wise and knowledge to those who have understanding." In the present case, the "wisdom and strength" that God has given to Daniel (cf. Eph 3:3–5) concerns the way in which a succession of human imperial epochs is brought to an end by a divinely-instituted epoch that will have no end. Given the general apocalyptic frame of reference of both passages, one cannot think of a scriptural passage that

---

24. Note also the collocation of terms descriptive of Jesus in 1 Cor 1:30: first, "our wisdom," and *then* "our righteousness, sanctification, and *apolytrosis*." Concerns for these last three are not foreign to the respective parties indicated in 1 Cor 1:18–25; the issue for Paul there is what *wisdom* informs those concerns, vis-à-vis the "hidden" wisdom of God "revealed" in the crucified Jesus.

25. Lincoln, *Ephesians*, 53.

PART FIVE: New Testament Afterword

restates it more dramatically than Eph 1:9–10, "he has made known to us *in all wisdom and insight* the *mystery* of his will, according to his purpose which he set forth in Christ as a plan for the fulness of time, to unite all things [*anakephalaiosasthai ta panta*] in him, things in heaven and things on earth."

(6) At this point I would like to venture a possible echo, or at any rate a resonance, between Ephesians and Daniel that has not previously been suggested. Apart from the verb, *anakephalaiosasthai*, in Eph 1:10, the noun, *kephale*, "head," occurs in several passages. First of all it occurs in 1:20–23:

> He [God] raised him [Christ] from the dead and made him sit at his right hand in the heavenly places, far above all rule and authority and power and dominion, and above every name that is named, not only in this age but also in that which is to come; and he *has put all things under his feet* and *has made him the head [kephalen] over all things* for the church, which is his body, the fullness of him who fills all in all.

The connotation of "head" here is clear: it has to do with Christ as royal figure ("at his right hand," echoing Ps 110:1), not only over the nations (cf. 2 Sam 22:44 // Ps 18:43; also, in a figure, Ps 118:22) but also over all powers in heaven—that is, as *kephalen hyper panta*. The theme of "head" is picked up in Eph 5:23: "The husband is the *head* of the wife, even as Christ is the *head* of the church: and he is the savior of the body." The move in scope, here, is from the cosmos and its ruling powers earthly and heavenly, to the church, to the basic human social unit, the family. (Note the reference, in cosmic scope, to "every family in heaven and on earth" in 3:15.) Finally, the term occurs in 4:15: "Speaking the truth in love, we are to grow up in every way into him who is the *head*, into Christ."

It is commonly recognized that, while modes of discourse that we distinguish as "theological" and "ethical" run in varying proportions throughout Ephesians, chapters 1–3 are weighted on the theological side and chapters 4–6 on the ethical. In that case, one may suggest that 4:1–16 functions as an introduction to the specific counsels that follow, in a fashion parallel to the way the *berakah* in 1:3–14 function in relation to 1:15—3:21. Now, with the magnificent verb, *anakephalaiosasthai*, in 1:10 still ringing softly in one's mind (one may imagine the rhetorical gusto with which a *lector* in a local church would hear this word read forth), one cannot but be struck by the way 4:1–16 *virtually exegetes the implications of that verb*, in its repeated emphasis, through a variety of verbal

Redeeming the Expression "Redeeming the Time"

and rhetorical devices, on the theme of unity-in-diversity under Christ as *head*. This unity-in-diversity is in one sense an accomplished fact, in virtue of the present reign of Christ "in the heavens." But in another sense it is an eschatological vision, to be realized in part by the response of the members of the church, as this verse puts it, through "speaking the truth in love" to "grow up in every way into him who is the head, into Christ." This introduction then ends with another grand rhetorical flourish, so typical of Ephesians, "from whom the whole body, joined and knit together by every joint with which it is supplied, when each part is working properly, makes bodily growth and upbuilds itself in love."

If Richard Hays is right, and echoing items in New Testament passages invite us to interpret their contexts against the backdrop of Old Testament contexts in which the echoed items occur, and if this means that one prominent Old Testament context for reading Ephesians is the Book of Daniel, then I propose that one echo of the *kephale* motif in Ephesians is the use of the same term in Daniel. There, Daniel describes Nebuchadnezzar's dream as containing "a great image, mighty and of exceeding brightness ... its appearance ... frightening." The image is of a body (nb), distinguished in terms of head, breast and arms, belly and thighs, and legs and feet. Suddenly "a stone cut out by no human hand" smites the image, its parts fall to pieces and become like chaff, and they are swept away by the wind (Dan 2:34–35a). "But the stone that struck the image became a great mountain and filled the whole earth" (Dan 2:35b). Daniel interprets this vision as follows: Nebuchadnezzar, unto whom God has given rule over all things, is "the head of gold." But his kingdom will give way to another kingdom, and that to another, and that to another, and the parts of the last of these "will not hold together" (Dan 2:43). But after this the "God of heaven" (a locution peculiar to Daniel in the Bible) will set up a kingdom that "shall stand forever" (Dan 2:44). I suggest that the portrayal of Christ's body, in Eph 4:16, may be read as resonating (at the least) with the portrayal of the bodily image in Daniel 2, the difference being that, whereas human universal kingdoms cannot sustain themselves nor hold together through time, the body that belongs to Christ as "head over all things" is destined to endure forever through the cohesion of its parts as it grows up into the one who is its head.

(7) Such a reading may even throw suggestive light on the verb in the expression, "speaking-the-truth [*aletheuontes*] in love." I note how Eph 4:15a and 4:16b form a sort of *inclusio* through the repetition of "grow(th)" and the phrase, "in love," thus suggesting that "speaking-the-truth in love

PART FIVE: New Testament Afterword

has something to do with this unifying process. This stands in contrast to what is portrayed in Eph 4:14 as "every wind of *doctrine*." Such doctrines would at least implicitly be grounded in stories of the gods or in "fundamental principles" such as Paul refers to more than once in Galatians. Over against such "false" doctrines, what would it mean to *aletheuo*?

The verb occurs five times in the LXX: Gen 20:16; 42:16; Prov 21:3; Sir 34:4; and Isa 44:26. The first four are contextually not relevant, but Isa 44:26 could not be more germinally germane, as the exilic visionary announces to the exiles,

> Thus says the LORD, your Redeemer,
> who formed you from the womb:
> "I am the LORD, who made all things [*syntelon panta*],
> who stretched out the heavens alone,
> who spread out the earth—Who was with me?—
> who frustrates the omens of liars,
> and makes fools of diviners;
> who turns wise men back,
> and makes their knowledge foolish;
> who confirms [*meqim*; LXX: *histon*] the word of his servant,
> and makes good [*yašlim*; LXX: *aletheuon*]
> the counsel of his messengers;
> who says of Jerusalem, 'She shall be inhabited,'
> and of the cities of Judah, 'They shall be built,
> and I will raise up their ruins.'"

This, of course, is a lead-in to the Cyrus oracle of Isa 45:1–7 which augurs the downfall of Nebuchadnezzar's dynasty. That this oracle in some fashion informs Daniel 2 cannot be denied, given the latter book's repeated echoes of later Isaianic materials as well as Habakkuk. The contrast between the diviners and wise men of Nebuchadnezzar, and Daniel, comes to nice expression in the deployment, in Greek Daniel, of cognates built on the root, *aleth*-. The data are as follows:

In the LXX of Dan 2:5 Nebuchadnezzar warns his interpreters, "The word from me is sure [*ep' aletheias*]" concerning their fate if they do not tell him his dream and interpret it. In 2:8 he asserts, "I know with a certainty [LXX, Theodotion: *ep' aletheias*] that you are trying to gain time." In 2:9 he reiterates his threat, and the LXX glosses his threat with *ep' aletheias*. In 2:45 Daniel concludes his report and interpretation by saying, "The dream is *certain* and its interpretation is *true*"—and where the LXX reads *akribes* and *piste*, Theodotion reads *alethinon* and *piste*. Implicitly, we have a contrast here between the supposedly "true" words of Nebuchadnezzar (words which will turn out not to be true, since Daniel saves the interpreters from their fate) and the "true" words of the dream as interpreted by Daniel.

In 3:14, the king, furious that Daniel's three friends refuse to serve his gods, asks them, "Is it true [*haṣda'*; LXX: *dia ti*; Theodotion: *ei alethos*], that you do not serve my gods?" In 3:23, the king rises up "in haste," astonished at what he sees in the fiery furnace, and exclaims, "Did we not cast three men bound into the fire?" and his men respond, "True [Theodotion: *alethos*], O King." In these two passages, "truth"—*aletheia*—is entered not only on the side of Daniel as intepreter over the court interpreters, but on the side of his friends in their faithfulness to Daniel's and their God.

In 6:12, King Darius reaffirms his original decree, that for thirty days no one should petition anyone else, god or human, but only him, on pain of death, and seals his reaffirmation by saying, "The thing stands fast [*alethinos ho logos*], according to the law of the Medes and Persians, which cannot be revoked." In 8:26, Gabriel says to Daniel of the vision just given, "The vision of the evenings and the mornings which has been told is true [LXX: *ep' aletheias*; Theodotion: *alethinos estin*]; but seal up the vision, for it pertains to many days hence." In 11:2, the heavenly messenger says to Daniel, in introducing another vision, "And now I will show you the truth [LXX, Theodotion: *(ten) aletheian*]."

Finally and strikingly, two passages call for juxtaposition. In 9:12–13 Daniel confesses,

> He has confirmed [*wayyaqem*; LXX and Theodotion: *estesen*] his words, which he spoke against us and against our rulers who ruled us, by bringing upon us a great calamity; for under the whole heaven there has not been done the like of what has been done against Jerusalem. As it is written in the law of Moses, all this calamity has come upon us, yet we have not entreated the favor of the LORD our God, turning from our iniquities and giving heed to thy truth [*lehaśkil ba'amitteka*; LXX: *dianoethenai ten dikaiosynen sou*; Theodotion: *tou synienai en pase aletheia sou*].

Here, one may detect in the opening and closing words, and in the opening sentence generally, a clear echo of Isa 44:26 in reference to that prophet's earlier words about the foretellings of the "former things" including God's judgments for covenant infidelity. "Thy truth," then, consists not only in "the law of Moses," but also in those prophetic words spoken "against us and against our rulers who ruled us." With this implicit call to "give heed to thy truth" as including prophetic proclamation—where Theodotion once again displays the non-classical usage found also in Ephesians—we may compare Dan 10:21, where the aforementioned heavenly figure says to Daniel, "I will tell you what is inscribed in the book of truth [LXX, Theodotion: *en (apo)graphe aletheias*]: there is none who contends by my side against these except Michael, your prince."

To sum up all these occurrences in terms of this last one, the upshot of the book of Daniel is that anyone seeking to interpret cosmic goings-on—seeking to "speak the truth," as it were—will do so in terms of the eschatological vision found in this book. If, now, we may read Eph 4:15, "speaking-the-truth [*aletheuontes*] in love" in *its* immediate context in 4:1-16 as well as its parallel context in 1:3-14 and the letter as a whole, against the backdrop of the "truth" motif in the Book of Daniel, such speaking carries the connotation of articulating the cosmic and eschatological vision of the reign of Christ as "head of all things for the Church," as conveyed in terms of the love of God embodied and manifest in Christ crucified, risen and now at the right hand of the Father. This brings me to a final proposal for an intertextual resonance between Ephesians and Daniel.

(8) One of the most intriguing sentences in all the New Testament is surely the statement in Eph 3:9-10, that "the plan of the mystery hidden for ages in God who created all things [is] that *through the church* the manifold wisdom of God might now be made known [*gnoristhe*] to the principalities and powers in the heavenly places." I have already noted that *gnorizo* in Ephesians is frequently related to the "mystery" or "hidden wisdom" of God's redemptive "plan" (1:9; 3:3, 5, 10; 6:19); and that this same verb, rare in the LXX elsewhere, and never in the LXX of Daniel, occurs repeatedly in Theodotion's version of Daniel with reference to the revelation of the mystery through (Daniel's interpretation of) Nebuchadnezzar's dreams. As Collins notes, one dramatic outcome of Daniel's "making known" of this "mystery," hidden in the dream but now revealed through his interpretation, is the conversion of this Gentile world-power to the praise of Daniel's God. Under the categories of "setting and function," Collins writes that the story of Nebuchadnezzar "ends with the repentance and virtual conversion of the pagan king.... [T]he story expresses a stubborn hope for the reclamation of even the most arrogant tyrant and for universal recognition of the Most High God."[26]

---

26. Collins, *Daniel*, 234.

Redeeming the Expression "Redeeming the Time"

To return, then, to Eph 3:9–10, the precise way in which the church is to be the means of this mystery being made known to the heavenly powers is debated. Lincoln takes the church's pattern of life, in reconciling Jews and Gentiles, and both to God, and so overcoming the hostility endemic in human existence, to be the effective and proclamatory sign to the heavenly powers "that their malign regime . . . has come to an end."[27] It is interesting that Daniel himself is just such a sign, both through his words in interpreting Nebuchadnezzar's dream to him in chapter 2 and through his actions in resolutely praying to God, and being delivered from the lions' den in chapter 6. One may suggest that the household code in Eph 5:21—6:9—concerning the basic earthly form of the *patria*, or "family," of which the writer speaks in 3:15—provides one model of how the church "makes known" the mystery of the Gospel not only to others in the world but also to the powers. Just as in Philippians it is the servanthood of Christ, issuing in the acknowledgement of his cosmic lordship (Phil 2:5–11), that provides the pattern for the Christian's *phronesis* or "mindset" and practice of life, so in Ephesians the manner of Christ's overcoming and reconciling the powers is to be exhibited in and through the style of life—the mutual relations—of the basic unit of the church in the form of its various households or "families" as well as in its gathered assemblies.[28]

It may be noted, in this connection, that the astonishing statement in Eph 3:10 introduces the prayer in 3:15–21. Is this prayer offered on behalf of the readers in respect to their general welfare? Or is it offered in specific reference to their commission to "make known" (*gnoristhe*)[29] the

27. Lincoln, *Ephesians*, 187.

28. The specific terms of the "household code" in Ephesians are experienced as chafing, to say the least, among many in the Christian church today. My own view of such texts emerges in the context of the following considerations: (1) The vision set forth within this code is to be interpreted vis-à-vis the terms and the typical practice of household existence in the first century CE rather than the twenty-first; and (2) in the words of James Russell Lowell's poem, "The Present Crisis," excerpted to appear in some hymn books under the opening line, "Once to every man [*sic*!] and nation," it is the case that "new occasions teach new duties, / Time makes ancient good uncouth." To externalize institutions embedded in the dynamics of time and history is to commit institutional idolatry and, in the conceptuality of the Letter to the Ephesians, to elevate hitherto obedient powers to ultimate status where they become rebellious and oppressive powers. See further below.

29. On the connotations of this verb, compare the comments on its occurrence in Phil 4:6 in chapter 15. I venture to suggest that "making known to the powers the mystery" of the Gospel in Eph 3:10 is, to use a current expression, an act of speaking truth to power. Such an act both empowers the speakers vis-à-vis the Powers and demystifies their claims to power, but, when carried out in the spirit of the Gospel, may also work to draw those powers into *rapprochement*.

PART FIVE: New Testament Afterword

mystery to "the principalities and powers in the heavenly places"—powers, in 6:13, against whom they will have to struggle, for which he says, in 6:12, "be strong [*endynamousthe*] in the Lord and in the strength [*kratei*] of his might [*ischyos*]"? In this latter "empowering" exhortation, one may hear the resumption of prominent elements from the prayer of 3:15-21, elements that in turn resonate with prominent elements in the scene in Daniel 10. I turn to these intertextual elements in the following section.

The Prayer of Empowerment in Ephesians 3:14-21 and the Empowering Vision in Daniel 10

First, then, there is the unusual concluding locution in Eph 3:21, *tou aionos ton aionon*, literally, "for age of ages." As noted already, by others and earlier above, this locution occurs otherwise only in Daniel. That this may signal a specific scriptural background for the prayer is suggested by the following elements in Daniel 10, elements already canvassed in one perspective in chapter 10. When Daniel sees "this great vision" (10:8), "no strength" (*ouk . . . ischys*) is left "within me" (*en emoi*) (10:8; similarly 10:16, 17) and he "retained no strength" (LXX: *ou katischysa*; Thedotion: *ouk ekratesa ischys*) (10:8). Concomitantly, "No spirit [*nešamah*; LXX: *pneuma*; Theodotion: *pnoe*] is left in me [*en emoi*]" (10:17). To Daniel in this state of mind and spirit the figure in his vision says, "stand upright" (RSV), or rather, "stand at your station" (*'amod 'al-'omdeka*; LXX: *stethi epi tou topou* [Thedotion: *te stesei*] *sou*; cf. Eph 6:14, *stete oun*, "stand therefore"). Then the figure touches him "and strengthened me" (LXX: *katischyse me*; Theodotion: *enischysen*) (10:18), addressing him, "fear not, peace be with you; be strong and of good courage" (LXX, Theodotion: *andrizou kai ischye*) (10:19a). And "when he spoke to me," says Daniel, "I strengthened myself [*hithazzaqti*; LXX, Theodotion: *ischysa*]";[30] and he goes on (10:19b), "Let my lord speak, for you have strengthened [LXX: *enischyse*; Theodotion: *enischysas*] me." Then the figure informs Daniel (10:20) that he has come to fight against the (heavenly) "prince" of Persia (cf. on Eph 6:12 above), and in that and such conflict has stood up to "confirm and strengthen [LXX: *enischysai kai andrizesthai*; Theodotion: *eis kratos kai ischyn*] Michael as the heavenly "prince" on behalf of Israel.

In view of Eph 6:12 and the magnitude of the church's commission in 3:10, I take the prayer in 3:14-21 as directly related to that commission. The first petition is "that he would give (*do*) you to be strengthened

30. For "I strengthened myself," see the comment on this verse in Chapter 10.

Redeeming the Expression "Redeeming the Time"

(*krataiothenai*) through *his Spirit* in your inner humanity (*eis ton eso anthropon*), that Christ may dwell in (*en*) your hearts through faith." This petition resounds with the thematics of Daniel 10. The second petition, "that you, being rooted and grounded in love, may have power to comprehend [*exischysete katalabesthai*] with all the saints what is the breadth and length and height and depth," has long puzzled commentators: the breadth and length and height and depth of *what*? The love of Christ mentioned in 3:19? But that verse seems to recur to the theme at the end of 3:17 and develop it further. I venture that we read the verb *katalabesthai* as meaning something like what we find in Daniel 10 where the visionary figure says to Daniel (10:11), "give heed [LXX: *dianoetheti*; Theodotion: *synes* (!)] to the words that I speak to you"; and, acknowledging to Daniel (10:12) that "you set your mind to understand [LXX: *edokas to prosopon sou dianoethenai*; Theodotion: *edokas ten kardian sou tou synienai* (!)]" (10:12), goes on to say (10:14), "I came to make you understand [LXX: *synienai*; Theodotion: *synetisai* (!)[31]] what is to befall your people in the latter days. For the vision is yet [*'od ḥazon*; LXX, Theodotion: *eti* (*gar*) *horasis*] to come."[32] Against the background of this motif in Daniel 10, I propose that the petition in Eph 3:18 refers to the "incomprehensibility" of the eschatological vision that informs not only the substance but (if indeed, in this instance, substance and form can be distinguished) the pleonastic and at times quasi-rhapsodic rhetoric of Ephesians as a whole. That vision concerns the wisdom of God in regard to the creation, its divine governance and, hence, its divinely assured future. If the four-dimensional imagery in Eph 3:18 most closely echoes passages like Job 11:8–9,[33] I take its most proximate scriptural background to be Daniel 10 where the "vision" is presented in reference to the struggles in which Daniel and his friends are then caught up.

The third petition then, in the face of the incomprehensibility of this vision, this wisdom, is that the readers "know the love of Christ, which

---

31. The form *synetisai* is a causative of the verb, *synienai*, frequent in Theodotion's version of Daniel and, as noted by others, echoed in Ephesians in the non-classical expression with the preposition *en*.

32. The phrase, *'od ḥazon*, here and elsewhere in Daniel, derives from Hab 2:2–4—a passage important to Paul. Intertextualist that he clearly was in his scriptural hermeneutics, he will have read Habakkuk and Daniel in close conjunction as backgrounds for his attempts to understand the apocalyptic Gospel of Christ.

33. See, e.g., O'Brien, *Letter to the Ephesians*, 262; more fully, Lincoln, *Ephesians*, 207–13.

surpasses knowledge."[34] It is this "knowledge," which is of the more capacious heart rather than of the more analytical and conceptual mind or *nous* (cf. Phil 4:7), that will enable them to "comprehend" the dimensions of God's working in creation and for its future through the crucified and risen Christ as "head" over all things. It is this latter form of "knowledge," rather than any other systems of speculative knowledge, that will enable the readers, as the last petition puts it, to "be filled [*plerothete*] with all the fulness [*pleroma*] of God." For God's *pleroma*—the quintessential goal of any and all Gnostic quests—is "simply" God's unutterable creation-embracing and creation-reconciling love. It is this vision that informs the ascription in the prayer's concluding doxology, an inscription that indirectly undergirds not only Paul's prayer for his readers but also all of their petitions: "To him who by the power [*dynamin*] at work within [*en*] us is able [*dynameno*] to do far more abundantly than all that we ask or think [*nooumen*]." These words catch up key elements of the preceding petitions and, in celebrating them as an ascription, appropriate them within the bosom of a praise which is itself a, or rather the chief, participation in that power—a participation in that kingdom of Christ which is to endure "to all generations of the age of ages."

(Not to put too fine a point on it, but one wonders whether the expression, "the age of ages," occurring only here, and in Dan 3:90 [LXX] and 7:18 [LXX, Theodotion], does not encapsulate a connotation peculiar to the dreams and visions in Daniel.[35] In them, in one imagery or another, kingdom arises after kingdom—and with each kingdom an age—until the sequence ends in a kingdom, an age, that is, without end. It is, I suggest analogous to expressions such as "the holy of holies," that is, "the quintessentially holy." Here, the eschatological age is like the other ages in being defined by its ruler, but it is unlike them in being temporally undefined in that this rule will never be succeeded by another. It is "the age to end all ages.")

Now the question arises, for those who have come under the commission of Eph 3:10 and the empowering but even more mind-and-heart-expanding prayer of 3:14–21: How in the world, or rather, how in the

---

34. One thinks again of 2 Corinthians 12 where Paul "goes on" to "visions [*optasias*; cf. Theodotions version of Daniel, six times] and revelations of the Lord" in which he "heard 'unutterable utterances' [*arreta rhemata*] which a human may not [or, cannot] articulate [*ouk exon . . . lalesai*]."

35. In 7:18 the Greek expression translates, exactly, Aramaic '*alam 'olmayya*'. The latter construction—the singular of this word followed by its plural—is without parallel in biblical Aramaic or Hebrew.

Redeeming the Expression "Redeeming the Time"

immediate context of one's daily life and one's daily relations, does one, does a given "family" or household, even *begin* to take hold of such a commission?

Here's a story: A seminary student serving a summer church awakens on a Monday morning, feeling entitled by the previous day's ecclesial labors to sleep past sunrise. On awakening, he looks out his kitchen window and sees that the women on both sides of him already have their Monday washing drying in the morning breeze. Taking his scrambled eggs and toast to the table, he opens his Bible to the readings for the day, and eventually arrives at the end of Matthew's Gospel with its Great Commission and its promise, "I am with you always, to the close of the age." He looks again at that commission in its dauntingly comprehensive scope ("into all the world"), and somehow the promise that undergirds it seems as ungraspable in its temporal sweep, its epochal vastness, as the magnitude of the commission "into all the world" is daunting. Here around him are people with such mundane matters as Washday Monday. How is he to appropriate a promise of presence so grand, so diffuse, as "always, to the end of the age"? The scope of that "always" is too big to fit into Washday Monday. He checks the Greek. Aha! Not "always," but *pasas tas hemeras*, like Hebrew, "all the days"! Well! The promise, then, is tailored to fit precisely into each and every day; and Washday Monday is as much such a day as any of those others.[36] Heartened, he returns to his eggs and toast, and then to his Monday schedule.

Here is the sequel: One resident of this little northern Saskatchewan fishing and logging village is Mrs. Hackett (now long since only of blessed memory). A widow, she was a "bluestocking" in her youth, well-educated, a champion of women's rights, a humanist agnostic. She appreciates my visits for the opportunity it gives her to speak of books and ideas. But she is terribly crippled with arthritis, and its pains, coming at regular, short intervals, send such spasms through her that—her eyes seeming almost to shatter like glass with each spasm—any train of thought we have been pursuing is lost to her and she starts off on another tack. At one point she asks me, with the courteous abruptness to which such elegant old age is entitled, "Mr. Janzen, tell me—why do you study Greek? What can you get from the Greek text that the translators have not given us in English?" I narrate my Washday Monday experience; and I am just about to follow

---

36. Note Gershom Scholem, who writes in similar terms of the rabbinic solution to the question of how the Infinite God could be said to dwell in the temple in Zion (*Major Trends*, 260).

the narrative by scoring a theological point with her when another spasm interrupts the conversation, and after it is over she is off on another tack. Wondering whose side God is on here, I resign myself to further conversation on non-theological topics. Some weeks later she tells me that she must go into the city for a routine check-up. She hates to travel alone, so, if she were to pay my bus fare and lunch, would I have time to accompany her. I agree. When the day comes, we set off on the two-hour ride, visiting as usual about this and that during the all-too-brief intervals between spasms. When the bus arrives at its city terminal, she steps down with my help onto the pavement, and toward a cab that by prearrangement is waiting for her there. Just before getting into the back seat she turns back to me and says, "Mr. Janzen, what was it you said the other day about why you study Greek? What was that line that meant so much to you?" I said, "I am with you all the days." "That's it!" she said, and then repeated, "That's it!" and got into the cab.

So here is how Paul funnels the incomprehensible commission of Eph 3:10 into the Washday Mondays and other quotidian activities of the households and "families" among his readers. Here is how, in their mutual relations involving the structural inequalities typical of the age (husband-wife, parent-child, master-slave), they are to model the headship of Christ in his body the Church, and so to make known to the heavenly "families" of principalities and powers how *they* are to practice *their* mutual relations. I tabulate the following principles, though others may need to be added to the list:

1. Eph 5:1: "Be imitators of God, as beloved children."
2. Eph 5:2: "Walk in love, as Christ loved us and gave himself up for us, a fragrant offering and sacrifice to God."
3. Eph 2:8: "Walk as children of light"
4. Eph 4:2: "Bear with one another [*anechomenoi allelon*] in love."
5. Eph 4:32: "Be kind to one another, tenderhearted, forgiving one another, as God in Christ forgave you."
6. Eph 5:21: "Be subject to one another out of reverence for Christ."

Conclusion

In the present study I have been exploring resonances between Ephesians and Daniel. In this connection I want briefly to juxtapose Eph 5:8–14 and Dan 12:1–3, first of all simply by presenting them side by side.

> Dan 12:1–3: "*At that time* shall arise Michael, the great prince who has charge of your people. And there shall be *a time* [LXX: "day"] *of trouble*, such as never has been since there was a nation till *that time* [LXX: "day"]; but *at that time* [LXX: "day"] your people shall be delivered [LXX: *hypsothesetai*, "exalted"], every one whose name shall be found written in the book. And many of those who *sleep* in the dust of the earth shall *awake*, some to everlasting life, and some to shame and everlasting contempt. And those who are *wise* [LXX: *synientes*] shall *shine like the brightness of the firmament*; and those who turn many to righteousness [*maṣdiqe harabbim*], like the stars forever and ever."

> Eph 5:8–14: "Once you were *darkness*, but now you are *light* in the Lord; walk as children of light (for the fruit of light is found in all that is good and right and true [*agathosyne kai dikaiosyne kai aletheia*]), and try to learn what is pleasing to the Lord. Take no part in the unfruitful works of darkness, but instead expose them. For it is a shame even to speak of the things that they do in secret; but when anything is exposed by the light it becomes visible, for anything that becomes visible is light. Therefore it is said, '*Awake*, O *sleeper*, and arise from the dead, and Christ shall give you light.'"

The resonances are generally obvious enough to need no comment. But one item may be lifted up for special notice. In Dan 12:3, I take the two halves of the sentence, as a quasi-poetic parallelism, to be two ways of characterizing the same activity and effect. The common element, the stars in the firmament, may have two aspects of signification. In the ancient world generally—and especially in Babylon and Persia—stars served to orient people in space and to mark times and seasons and special events. The wisdom spoken of here, then, concerns both how to read the signs of the times and how to conduct oneself wisely in a time of troubles or an evil day—precisely the point in Eph 5:16. Each follower, each household, of the ascended and reigning Christ is asked simply to "shine" in his or her or its place in the larger constellation of the universal church, "redeeming the time" one day at a time.

PART FIVE: New Testament Afterword

The household code in Eph 5:22—6:9 would, in its own time, have been understood to choreograph relations between unequals; many aspects of that code, therefore, understandably chafe many in our own time. But the ruling and governing powers in each age tend to consider the structural arrangements and orchestrations of power characteristic of that age to be not only appropriate to that age but embodiments of the very *stoicheia tou kosmou*, the "spiritual building blocks of the universe," and, as such, to be unchangeably enduring. To this degree, they turn epochally limited *stoicheia* into idols. When, under the aegis of the God of Daniel who "changes times and seasons," these idols begin to crack and crumble through their openness to new wisdom and understanding (Dan 2:21), their human servants/lords tend to "tighten the screws" in a vain effort to "strengthen" and "steady" their idols (compare Isa 41:5–7; 45:16; and 46:1–2, 6–7 with Isa 46:3–4, 8–10). Even Mount Zion, with her towers (Ps 48:12, *migdalim*), for all its celebratory focus in praise of its God who is "forever and ever" (Ps 36:14 LXX, *eis ton aiona tou aionos*), does not, as an empirical reality, survive its own eventual hubris, but falls before the armies of Babylon—a city that for its own part also vainly (in both senses of that word) celebrates its own eternity (Isaiah 47). One fundamental lesson of the Old Testament seems to be that, despite human desires and efforts to eternalize the institutional structures and power arrangements of their age, such efforts are an attempt to deny the creaturely, temporal finitude of all such forms.[37] This denial is expressed perhaps most powerfully in the story of Babel in Gen 11:1–9, where humankind, made of dust from the earth (Genesis 2) and held in social cohesion by the language they have in common, attempts to solidify their place on the earth by taking clay and molding it into baked bricks (a patent if implicit attempt to fashion themselves into something autonomously enduringly) to fashion a walled city with a tower (*migdal*) reaching to the heavens. The safety, the enduringness humans seek, says the proverb used as an epigraph to the present volume, lies in no finite, transient form, but only in the everlasting God: "The name of the LORD is a strong *migdal*; the righteous runs into it and is safe" (Prov 18:10; RSV*). Even under the aegis of the risen and ascended Christ, those who follow him will, short of the eschaton, find themselves living, at one and the same time, within that eternal *aion* and "this present evil *aion*."[38]

---

37. For a contemporary analysis, see Becker, *The Denial of Death*.

38. In an intriguing inversion of the theme, Eph 2:22—another one of the epigraphs to the present volume—has it that "Christ is building you into a place where God lives

Redeeming the Expression "Redeeming the Time"

Here, then, is where we begin to get into what it might mean to "redeem the time, because the days are evil." It has to do with learning how to live within the present evil *aion* in the yeasty spirit of the *aionos ton aionon*, the "age to end all ages" (Eph 3:21), whose reigning "head" is the crucified, risen, and exalted Christ. In this already-and-not-yet bifocal existence, one may recall e. e. cummings' aphorism, that "equality is what does not exist among equals."[39] I take cummings to mean that equality in an empirical sense is a myth: no two people are, strictly speaking, equal in all respects; any relationship will display inequalities in power and/or structure. In that sense, equality doesn't exist. Yet, though it does not exist as an empirical *fact*, a *spirit* of equality, such as is variously articulated, for example, in the above six practices or "notes," may inform the way any two unequals "walk" together in time in such a way as to begin to learn, in time, to turn that walk into a "dance."[40]

A pristine example of such an "equal walk between unequals" is the question of the relation of the general rubric to the household code in Eph 5:21 and the relations between husband and wife so fully expressed as the first element in that code. Since debate over the relation between this rubric and this first element—specifically, the call for the wife to be subject to the husband—is fully summarized by O'Brien,[41] I shall comment on only one element in that debate: the precise connotation we should attach to "one another" (*allelois*) in 5:21. Is this a call for symmetrical relations in all instances, and does this general call govern the spirit that should "for this age" inform the observance of the structurally unequal relations to which it is the rubric? Or do the items that follow the rubric limit the sense in which we construe this reciprocal pronoun?[42] The question is perhaps

---

through the Spirit" (so, aptly, the CEB). Can this "place" in any sense function as a refuge in the very midst of all the natural and inhuman shocks that "flesh is heir to"?

39. e. e. cummings, *i: six non-lectures*, 70.

40. Compare the comment of Wall, "Wifely Submission in the Context of Ephesians," 280: "The spiritual hierarchy in effect dismantles the social hierarchy." Analogously, the participation of the household slave, when circumcised, along with the master and family in a Passover that celebrates liberation from Egyptian slavery, may for a long time do nothing more (!) than ameliorate the conditions of slavery in this household at least for that time; but can one deny that the "new leaven" of the Passover meal after so many long centuries has contributed to the dismantling of slavery as a legal institution in the modern world?

41. O'Brien, *Letter to the Ephesians*, 398–405; see also Lincoln, *Ephesians*, 365–94.

42. Against the first construal, O'Brien notes that a "fully reciprocal" meaning in Rev 6:4 ("so that people would slaughter one another [*allelous*]") "does not make sense." He overlooks the scene at the beginning of Heliodorus' *Ethiopean Romance*,

like the question lurking for hermeneuts in Rom 13:8–10: Does the second half of the Jesus-Shema, "you shall love your neighbors as yourselves," "sum up" (*anakephalaio*!) the commandments in the sense that not only the four mentioned, but all other items in the *torah*, are to be carried out not in the letter but in this spirit? Or does the verb *anakephalaio* here have a more robust meaning, such that the love commandment is to *govern* how the *torah* as a whole and in its specific details is to be interpreted and applied to changing conditions in different epochs?[43]

In connection with this issue, I enter the following observations with regard to the verb *proskollaomai* in the quotation in Eph 5:31. In Deuteronomy, the Hebrew verb *dabaq* repeatedly refers to Israel's attaching itself adhesively in covenant loyalty and obedience to God (e.g., Deut 10:20 *kollaomai*; 11:22 *proskollaomai*; cf. Josh 23:8 *proskollaomai*; 2 Kgs 18:6 *kollaomai*; Pss 63:8 *kollaomai*; 119:31 *kollaomai*). When Boaz (inadvertently echoing the verbs "leave" and "cleave" in Gen 2:24, and thereby engaged in a delicious narrative irony over which narrator and reader may wink knowingly) calls on Ruth not to "leave" his field for another but to "cleave" to his maidens, the verb implicitly signals his desire that she come under his patriarchal domain (see the interplay noted in Chapter 15 between the noun in Ruth 2:12 ["wings"] and in Ruth 3:9 ["skirt"]). All the more striking, then, that in Gen 2:24 it is the *man* who "cleaves" to *the woman*! What kind of hermeneutical space is opened up in the semantic tension that becomes visible when Gen 2:24, as quoted in Eph 5:31, is placed alongside

---

where the heroine, sitting in the midst of a field of dead bodies, says to the pirates who have come on the scene but whom she addresses as though ghosts of the slain, "Most of you died at one another's hands [*anakephalaio chersi tais allelon*]." ("An Ethiopian Story," in Reardon, *Collected Ancient Greek Novels*, 355; Greek text in Rattenbury and Lumb, *Heliodore*, 6). O'Brien also overlooks that two or more combatants can indeed inflict mortal wounds on one another—which would befit an apocalyptically dramatic depiction of divine judgment working through self-destructive dynamics.

43. See the discussion in Miller's essay, "The Sufficiency and Insufficiency of the Commandments," where the same issue relates to the relation between the Decalogue as permanent and various bodies of statutes and ordinances as "time and place specific" (23–24). Elsewhere, Miller writes, "[N]one of the specifics (of the law) have the force of the original. They have themselves arisen in different times and under different circumstances, so that the specifications of Deuteronomy are not necessarily like those of Exodus 21–23 and neither of these legal codes are the same as those of Leviticus and Numbers, themselves composed of more than one set of Torah instructions. . . . So in our moral deliberation, we are, like the earlier community of faith at different times in its history, faced with having to determine what God wills of us at our time and how the fundamental guidelines work out in specifics today" ("What the Scriptures Principally Teach," 288–89).

Redeeming the Expression"Redeeming the Time"

the occurrences of this verb in Deuteronomy? Might one read Eph 5:21 as arising in the space opened up by the tensive intertextual juxtaposing of these two covenantal cleavings?[44] And might this suggest a way of understanding how partners may variably dance in time to the music of eternity?[45]

44. One may compare here Wheelwright's comments on the semantic implications of metaphor functioning in the mode of "diaphor." He defines the semantic movement (*phora*) of diaphor as "'through' (*dia*) certain particulars of experience (actual or imagined) in fresh ways, producing new meaning by juxtaposition alone" (*Metaphor & Reality*, 78), and as "the sheer presentation of diverse particulars in a newly designed arrangement" (81). "The essential possibility of diaphor," he writes, "like in the broad ontological fact that new qualities and new meanings can emerge, simply coming into being, out of some hitherto ungrouped combination of elements" (85). I am suggesting that the variable distribution of the verb (*pros*)*kollaomai*, in the Old Testament, when pondered side by side in regard to the relation between the rubric in Eph 5:21 and the following marriage code that follows, opens up a hermeneutical space between the latter two that forestalls open and shut appeals to an "originalist" reading of 5:22–33.

45. If "redeeming the time" consists, as I am suggesting, in learning to "dance in time to the music of eternity," and if the "music of eternity" may be thought of in terms of the Greek Patristic Trinitarian image of *perichoresis* (literally a "dancing around"), Paul's words in Romans 8 about "not knowing what to pray for as is called for" express a concern to become more deeply conformed to the *perichoresis* implicit in that chapter, in the two-fold intercession of the indwelling Spirit and of the exalted Christ to the God the Father. For further reflections on the image of dancing in relation to time and eternity, see Bauckham, "Time and Eternity," 155–226. Note in particular Bauckham's quotation of Moltmann, who writes that "[t]he preferred images for eternal life are . . . dance and music as ways of describing what is yet hardly imaginable in this impaired life" (Moltmann, *Coming of God*, 295; cited in Bauckham, "Time and Eternity," 184). Moltmann's allusion is to the Trinitarian conception of the divine life as a *perichoresis*, literally a "dancing around." Note also Bauckham's section on "Moments of Eternity" (ibid., 187–93), in which he discusses the *experience* of an "eternal now" in which "while it lasts, time ends or stands still," so that "[i]n this present moment, the 'eternal' or the 'eschatological' is experienced immediately, not as a distant prospect along the line of the temporal future" (ibid., 187). See also Bauckham's breathtakingly relevant discussion of the evocation of such moments in Virginia Woolf's *To the Lighthouse*, in his section on "The Eschatological Aesthetics of the Moment," especially the subsection on "The Lighthouse at the End of the World" (ibid., 196–213). The biblical Greek word for such an *eschatological* "moment" is *exaiphnes*, "suddenly," as in Mal 3:1 (LXX); Mark 13:36; Luke 21:34; and, as breaking into the *midst* of time, Luke 2:13; Acts 2:2; 9:3; 22:6. In classical Greek, according to LSJ, the term can refer to the "moment" of death. In Plato, *to exaiphnes*, "the instantaneous," refers to "that which is between motion and rest, and not in the time-series." Precisely—the dance of eternity within time yet not of time. One should not miss Luke's implicit correlation, or *resonance*, between the angelic announcement at Jesus' birth, Paul's Damascus Road experience, and/as the eschaton. Paul's experience of *to exaiphnes* on that road may be taken as paradigmatic of Moltmann's proposal that "the messianic interpretation

PART FIVE: New Testament Afterword

If all this weren't enough, there is one last occurrence of *dabaq/ proskollaomai* whose implications are teasing enough to whet the appetite of a Daniel but daunting enough to move the most intrepid of Nebuchadnezzar's court interpreters to seek to buy time. I refer to the scene in that king's dream which immediately precedes the inbreaking of the final, everlasting kingdom. It is a scene in which the successive kingdoms end in one represented by "legs partly of iron and partly of clay" (Dan 2:34). Daniel's interpretation ends on this note:

> As you saw the iron mixed with miry clay, so they will mix with one another in marriage, but they will not hold together [*la'-lehewon dabeqin denah 'im-denah*; LXX: *ouk esontai de homonoountes oute eunoountes allelois*;[46] Theodotion: *ouk esontai proskollomenoi houtos meta toutou*] just as iron does not mix with clay. (Dan 2:43)

What in the world is this all about? The verb *dabaq*, apart from covenantal and marital contexts, can refer to organic or inorganic substances securely fixed to one another. But the preceding reference to marriage, in a context dealing with kingdoms, suggests marriage covenants between two regimes. Why will the "union" in this instance not hold? Because of the *disparity* between them: one a stronger, one a weaker power. (No political entity likes to be the inferior party to a treaty except in time of great danger; otherwise, at the first opportunity it will seek—as in Psalm 2—to break out of that hegemony.) What is striking for the present discussion is that the verb (which Theodotion translates in accord with the LXX practice attested elsewhere) in this instance identifies the problem as the *lack* of *reciprocity* in their relations, a disparity that the LXX construes in terms of disparate attitudes and disparate dispositions. By this point in the present analysis, in which so many echoes and resonances have emerged between Daniel and Ephesians, I find it difficult to detach the interpretation of this scene (coming as the immediate foil to the everlastiong kingdom) from my reading of the general exhortation in Eph 4:1–16 where the whole passage, framed by "bearing with one another [*anechomenoi allelon*] in love"

---

sees 'the moment' that interrupts time, and lets us pause in the midst of progress, as the power for conversion. At that moment another future becomes perceptible. The laws and forces of the past are no longer 'compulsive.' God's messianic future wins power over the present. New Perspectives open" (*Coming of God*, 45–46, as quoted by Bauckham, "Time and Eternity," 192).

46. Collins, *Daniel*, 152, translates LXX: "They will not be in agreement or well disposed to each other."

and "building itself up in love," emphasizes, on the one hand, a diversity of gifts and functions, and, on the other, a "unity of the Spirit in the bond of peace" (4:3) that gives "the whole body," not the brittleness of a metal, nor the unmeldable disparity of iron and clay, but the organic wholeness and ligamented flexibility of a living body. If the laws and institutions of a given epoch are the ligaments of a society, these ligaments will remain supple only when those laws are not simply perpetuated in originalist fashion, but hermeneutically carried forward in what N. T. Wright, in an essay on authority and the Bible, has called a combination of "both *innovation* and *consistency*."[47]

Such a dance in time to this eternal music will, of course, involve more than just the way in which people orchestrate their own time vis-à-vis the times of others. But a good example of the time-reference of the principles in Ephesians, as they relate to the general rubric of the household code in Eph 5:21 and the mutual submission of husband and wife to each other, is given in Paul's model for conjugal relations in 1 Corinthians 7:[48]

> The husband should give to his wife her conjugal rights, and likewise the wife to her husband. For the wife does not rule over her own body, but the husband does; likewise the husband does not rule over his own body, but the wife does. Do not refuse one another except perhaps by agreement *for a season* [*pros kairon*], that you may devote yourselves [*scholasete*] to prayer; but then come together again, lest Satan tempt you through lack of self-control. (1 Cor 7:3–5)

The urgency of the physiological drive toward conjugal union is one potential basis for one spouse's potential insistence on "now!" over the other. This is especially the case in patriarchal cultures where, all too often, the cultural and even legal code sanctions the husband's peremptory demand for his "conjugal rights." Even where those codes have waned or

---

47. Wright characterizes the Bible as a story of redemption in five acts, in which the New Testament constitutes "the first scene of the fifth act [the act in which we find ourselves], giving hints as well ... of how the play is supposed to end." He likens this to the supposed discovery of a play by Shakespeare missing its fifth and final act. Should the play be staged, he suggests, writing a fifth act "once and for all ... would freeze the play into one form ... Better ... to give the key parts to highly trained, sensitive and experienced Shakespearian actors, who would immerse themselves in the first four acts, and in the language and culture of Shakespeare and his time, *and who would then be told to work out a fifth act for themselves* ... with both *innovation and consistency*" ("How Can the Bible Be Authoritative?" 18–19; his italics).

48. My reading of the relevant passage in 1 Corinthians 7 is informed chiefly by the commentary of Thiselton, *First Epistle to the Corinthians*, 498–510.

putatively been disavowed, because of historically eventuated inequalities in physical strength, psychological and verbal superiority, and so on, such insistence—no doubt predominantly, but not always, on the part of the husband—subjects one spouse's sense of the "timeliness" of such relations in unwelcome fashion to that of the other. In general, for Paul then, "times" for conjugal union are to be orchestrated according to Gospel principles such as I have identified in Ephesians. And it is intriguing how an implicit "redemption of timeliness" (so to speak) comes to explicit articulation, in the expression "for a season," and in the verb *scholasete*. The latter verb—deemed not to merit attention in *TDNT*, and only brief mention in *EDNT*, repays closer study.

According to LSJ, the basic meaning of the verb is "to have leisure or spare time, to be at leisure, have nothing to do."[49] In a telling passage, Plato says of some that "they have no leisure left *hypo polemon*"—their time is taken up in warfare, presumably under the command of those who have declared war. With a dative object, the verb means, "to have leisure, time, or opportunity" for something or someone. Used absolutely, the verb means to devote oneself to learning (to be a "scholar"). And finally, in a spatial reference, the verb means (as in Matt 12:44), "to be vacant, unoccupied," and with a dative object, "to be reserved for." Thus the noun *schole* can refer to a lecture hall (as in Acts 19:9) or "school." All these uses operate within an implicit social structure in which some have varying amounts of leisure or "spare time" by virtue of their status, authority, and power, while others have limited amounts of leisure by virtue of the claims on their time by others over them or their place in the social and economic structure.

In this light, the three occurrences of the verb in the LXX are highly suggestive. In Exodus 5, Moses and Aaron ask Pharaoh for three days of "time off" to go into the wilderness for a feast to their God. Pharaoh takes this request as a sign that Moses and Aaron want to "make them rest [*hišbatem*]" from their burdens (5:5), that is, to stop working. So Pharaoh appoints "taskmasters" to see to it that they keep working, but now also to search out their own straw. "But the number of bricks which they made heretofore you shall lay upon them, you shall by no means lessen it; for they are idle [*scholazousin*]; therefore they cry, 'Let us go and offer sacrifice to our God'" (Exod 5:8; see also 5:17). This passage provides the paradigmatic scriptural portrayal of hegemony understood as enforced control of time over those under that hegemonic power. In such an understanding,

---

49. LSJ, 1747.

the response to those who request "time off" to worship is to take away even such time as the underlings have for their work, by ceasing to provide straw and thereby forcing the underlings to do more work during a given day. The pattern is all too familiar in a time of economic downturn, in a society that has commodified time ("time is money in a sixty-minute hour") as employers "downsize" their work force but demand the same productivity from it. The counterpart to the request of Moses and Aaron comes in Ps 46:10, where the God who is in the midst of Zion causes all wars to cease to the end of the earth, and calls out, indeed to those of whom Plato speaks, "Be still [*scholasate*], and know that I am God. I am exalted among the nations, I am exalted in the earth!"

One more aspect of this word remains for comment. Both as a verb and as a noun its primary connotation is *temporal*, but in both forms it can also connote *space*. This is not an arbitrary shift because the experience of time often has a spatial feel to it. This is nicely illustrated by the first definition of "time" in the OED. Under the most basic meaning of "a space or extent of time," the OED defines time as "the interval between two successive events or acts, or the period through which an action, condition, or state continues." Considered as an *interval*, time is like an empty room (such as a room reserved for lectures) within which an activity can be engaged by those (such as scholars) who have the leisure time to engage in it. Considered as a *period*, time is the temporal dimension of an activity, condition, or state. In a Newtonian universe, time is an empty room in which atoms and larger physical entities hurry and scurry about. In an Einsteinian universe, time is the temporal shape of activity. In the latter conception, activities literally "take time," and insofar as the time they take is extensive throughout a period, activities in this sense "take place." The activities of some parties may take place within a temporal space granted them by those with power or in authority over them, as when Nebuchadnezzar grants Daniel time to tell the dream and make an interpretation of it. But what happens when, as in Exodus 5, the ruling authority refuses to grant time? In that case, whereas the Hebrew slaves simply complain to Moses and Aaron (5:20–21), these two cry out to God in prayer, complaining that Pharaoh "has *done evil* to this people, and thou hast not delivered thy people at all" (5:23). One could thus say that for the Hebrews "the days are evil" under the aegis of this "evil age."

It is obvious by now what I am getting at. If prayer is a form of "taking place" under conditions natural and social that seem to afford one no space to stand, it is also a form—the fundamental form—of "taking time" under conditions natural and social that seem to threaten to deprive one

PART FIVE: New Testament Afterword

of any "time of one's own," such a time being an existential space, so to speak, for enacting oneself, if only by way of freely undergoing what is imposed on one rather than passively, apathetically, and despairingly being crushed by it.[50]

You get the idea. For human experience (only for *human* experience?) time takes various shapes, and it configures human relationships in various ways. In some respects, the "space" of one's own time is imposed on one by factors beyond one's control. But even there, that imposition need not completely disable one's own inner power of existing. Or rather, one need not discover that that inner power is totally defined and delimited by such physical and social constraints. The chief biblical word for such negative constraints is *ṣar* or *ṣarah*, "distress," from the verbal root, *ṣarar*, "bind, tie up, be restricted, narrow, scant, cramped." A classic example is in Ps 18:6. The Greek translation here, in verbal form, is *thlibesthai*, "to be distressed"—literally, to be dis-stressed— just as the noun, *thlipsis*, means "tribulation, distress," and these words are echoed in the Latin *angustia*, "anguish" (think of angina). In David's case, the *ṣar* drove him—like Moses in Exodus 5—to "call upon the LORD." The result, eventually, was that God "brought me forth into a broad place." This language can describe also one's experience of time. And, in the instance of the couple in 1 Corinthians 7, a discovery of what marriage can be like, even within the framework of the conventions of "headship," involves a discovery of how each one's orchestration of the "times" of one's own body can make temporal room for the "times" of the other's embodied existence, and how it is possible to make room—*scholazein*—for a season of prayer. Needless to say, one spouse may find it necessary to "make time" for prayer or other "time off" if the other will not yield it, in which case that may involve crying out to God even as one is constrained by the other. In that case, and to that degree, the powers in the heavenly places (see Eph 3:10; 6:12) think themselves to be confirmed in their understanding of "how things work." The call of Christ, in Ephesians, is quite other. Even where the household or family does not fully exemplify the "codes" of the kingship of Christ, where one or the other or both strive to live by them, and so to "redeem the time," such striving is itself a "making known" to the powers something of the "inexhaustible riches" of the mystery of the kingship of Christ.

---

50. By "enacting oneself," I mean to refer to any action in and through which one *is* oneself. It may involve something as leisurely as writing poetry, making pots, gardening, or woodworking, but it may also involve "working with a will" at a task imposed by another. But, in this latter scenario, even when one is in service to another, insofar as one is "enacting oneself" one does not lose one's soul to the other.

Redeeming the Expression "Redeeming the Time"

What, then, is the difference between the unsuccessful attempt of Nebuchadnezzar's interpreters to "buy time" and Daniel's success in what is, in substance if not in name, the same endeavor? The difference, as Collins notes, is that Daniel *prays*.[51] He prays to the God who "changes times and seasons" (Dan 2:21), that is, to the One who is the true Lord of time. And the God who changes the macro-seasons as indicated in Nebuchadnezzar's dream is also the God who changes the micro-seasons of Daniel's life, in enabling him to find, under Nebuchadnezzar's hegemony, the "time" that the other interpreters could not find. This by no means means that such prayer will always deliver one *from* the temporal constraints of the powers. Even the Jesus of John's Gospel, who has much to say about time and work, light and darkness, and time to finish one's work, dies at the hands of those who finally "overtake" him (to use a third connotation of the verb, *katalambano*, in John 1:5). By rights, that day should be called "Bad Friday." If Jesus, nevertheless, is able to say, "it is finished," this suggests that even under such dire, terminal circumstances, he is still working in unison with the Father, still dancing in time to the music of eternity. So it is not surprising that the day is called by Christians "Good Friday." Not masochistically, nor sadistically! But as the subversive whispering of an open secret—or rather, the whispering of a previously hidden but now opened secret that, as often as it is entered into, remains an unfathomable mystery as of unsearchable riches (see Eph 3:8).

To come, then, finally, to the line that this paper set out to explore. I take it that it does, after all, mean, "redeeming the time, for the days are evil." It is not at all—as current translations and interpretations would have it—that the evil days afford one leisure time—*schole*—for good works, so that one should not stand idly about but take care in good sixty-second-per-minute fashion to exploit whatever opportunities. Part of the evilness of the days is that, as the temporal dimension of this present evil age, they do not grant sufficient time for such good works. But under the aegis of the cross, one can "buy time," through "addressing one another in psalms and hymns and spiritual psalms, singing and making melody to the Lord in your heart, always and for everything giving thanks in the name of our Lord Jesus Christ to God the Father" (Eph 5:19–20). It is in taking the time for such singing, and such praying in the spirit of Eph 3:14–21—singing and praying, if need be, even as one works under the constraints of "the boss"—that one finds oneself empowered to redeem the time. It is a way of coming under Jesus' yoke—or rather, of letting him share one's yoke—and

---

51. Collins, *Daniel*, 157 (on 2:11), 159 (on 2:18).

PART FIVE: New Testament Afterword

learning to pull one's load and bear one's burdens in rhythm with his rhythmic stride, dancing in time to the rhythms of eternity. Come to think of it, perhaps Eph 5:19–20 should be put at the top of the list of fundamental principles—of *stoicheia*—appropriate to the Gospel. If more groups at odds with one another within a church or the Church were simply to sing together more often, perhaps such singing would open up a space within which they might find more opportunity to iron out their differences. It might even recall "the Principalities and Powers in the heavenly places"[52] to that primal scene where, in celebration of the dawning of the first day of creation, "the morning stars sang *together* [*yahad*], / and *all* the children shouted for joy" (Job 38:7).[53]

With such a scenario, as one way of enacting the divine commission given in Eph 3:10, we may compare the call in Ps 29:1–2: "Ascribe to the LORD, O heavenly beings (*bene 'elim*; LXX: *huioi theou*), / ascribe to the LORD glory and strength. / Ascribe to the LORD the glory of his name; / worship the LORD in holy array [or 'splendor']." Here the united congregation in the temple calls upon the heavenly powers to join in their cry, "glory" (29:9). Compare also Ps 138:1, 4: "I give thee thanks, O LORD, with my whole heart; / before the gods (*'elohim*; LXX: *aggelon*) I sing thy praise . . . All the kings of the earth shall praise thee, O LORD, / for they have heard the words of thy [or 'my'] mouth." Whatever the original worshippers may

---

52. Paul's language and conceptuality here would have been familiar to anyone living in his day. Among polytheists, the "heavenly places" were rife with supra-human agents benign and malign; and even for a monotheistic Jew like Paul, such agents functioned either as servants of the one God or as rebellious opponents. A fruitful exploration of how this conceptuality may be explored and appropriated for analysis of current social and political issues of power and justice is provided in Wink, *The Powers That Be*.

53. The Hebrew word *yahad*, originally a noun meaning "unity/unified," almost always in the Bible functions as an adverb, "*together*, of community in action, place, or time" (BDB, 403). In one or two instances it still functions as a noun, and as late as the community of the Dead Sea Scrolls it functions as the designation of that community. The verb *yahad*, means "to be united; to unite," as in Gen 49:6 ("in their assembly [*qahal*] let my glory not be united [*tehad*]"); Isa 14:20 ("You will not be joined [*tehad*] with them in burial"); Ps 86:11 ("Unite [*yahad*] my heart to fear your name"). As "children of God," this primal band is as yet one heavenly "family" (cf. Eph 3:15). Intriguingly, according to *HALOT*, the verb may occur in Job 3:4, 6 where Job, cursing his birth and his conception, cries out, "Let that day be darkness! / May God above not seek it, / nor light shine upon it . . . That night—let thick darkness seize it! / let it not join [*yihad*] with the days of the year, / let it not come into the number of the months." This curse has been taken as a rhetorical rejection of God's very acts of creation as recapitulated in Job's own beginnings. In that case, God's answer to Job is to re-call him to the first dawn of creation and to the song there sung in unison.

have understood themselves to be doing in such heavenward addresses and exhortations, these lines and the commission in Eph 3:10 resonate anew when brought in diaphoric conjunction with each other. If Hannah could pray to God in words that George Herbert called "reversed thunder" (see Chapter 5), we might say, in view of Dan 12:1–3, that here we would have a case of "reversed starlight," such as would constitute one way of "redeeming the time" over which those astral powers were deemed to hold sway in the days of old.

Bibliography

Anderson, Francis I., and David Noel Freedman. *Hosea: A New Translation with Introduction and Commentary.* AB 24. Garden City, NY: Doubleday, 1980.
Appiah, Kwame Anthony. "The Art of Social Exchange." *The New York Times Magazine,* October 22, 2010, p. 22.
Attridge, Harold W. *The Epistle to the Hebrews: A Commentary on the Epistle to the Hebrews.* Hermeneia. Philadelphia: Fortress, 1989.
Austin, J. L. *How to Do Things with Words.* 2nd ed. Edited by J. O. Urmson and Marina Sbisà. Cambridge: Harvard University Press, 1975.
Avnon, Dan. *Martin Buber: The Hidden Dialogue.* Lanham, MD: Rowan & Littlefield, 1998.
Bachelard, Gaston. *The Poetics of Space.* Translated by Maria Jolas. Boston: Beacon, 1994.
Barfield, Owen. *The Rediscovery of Meaning, and Other Essays.* Middletown, CT: Wesleyan University Press, 1977.
Barr, James. *The Semantics of Biblical Language.* 1961. Reprinted, Philadelphia: Trinity Press International, 1991.
Bauckham, Richard. "Time and Eternity." In *God Will Be All in All: The Eschatology of Jürgen Moltmann,* edited by Richard Bauckham, 155–226. Minneapolis: Fortress, 2001.
Becker, Ernest. *The Denial of Death.* New York: Free Press, 1985.
Bollas, Christopher. *The Shadow of the Object: Psychoanalysis of the Unknown Thought.* New York: Columbia University Press, 1987.
Brower, Reuben A. *The Poetry of Robert Frost: Constellations of Intention.* New York: Oxford University Press, 1963.
Brueggemann, Walter. "The Kerygma of the Priestly Writers." *ZAW* 84 (1972) 397–414.
Buber, Martin. *The Prophetic Faith.* New York: Harper, 1960.
Clifford, Richard J. "The Function of Idol Passages in Second Isaiah." *CBQ* 42 (1980) 450–64.
Cohen, Mark E. *The Canonical Lamentations of Ancient Mesopotamia.* 2 vols. Potomac, MD: CDL, 1988.
Coleridge, Samuel Taylor. *Aids to Reflection.* Edited by John B. Beer. The Collected Works of Samuel Taylor Coleridge 9. Bollinger Series 75. Princeton: Princeton University Press, 1993.
———. *Biographia Literaria.* 2 volumes. Edited by James Engell and W. Jackson Bate. The Collected Works of Samuel Taylor Coleridge 7. Princeton: Princeton University Press, 1983.
———. *Collected Letters.* 6 vols. Edited by Earl Leslie Griggs. Oxford: Clarendon, 1956–1971.

Bibliography

———. *Lay Sermons*. Edited by R. J. White, B. Winer, and K. Coburn. The Collected Works of Samuel Taylor Coleridge 6. Princeton: Princeton University Press, 1972.

———. *Lectures 1808–1819: On Literature*. 2 vols. Edited by Reginald A. Foakes. The Collected Works of Samuel Taylor Coleridge 5. Princeton: Princeton University Press, 1987.

———. *The Notebooks of Samuel Taylor Coleridge*. 5 vols. Edited by Kathleen Coburn et al. Princeton: Princeton University Press, 1957–2002.

Collins, Terence. "Decoding the Psalms: A Structural Approach to the Psalter." *JSOT* 37 (1987) 41–60.

Collins, John J. *Daniel: A Commentary on the Book of Daniel*. Hermeneia. Minneapolis: Fortress, 1993.

Cooper, Alan. "Ps 24:7–10: Mythology and Exegesis." *JBL* 102 (1983) 37–60.

Craigie, Peter C. *Psalms 1–50*. WBC 19. Waco, TX: Word, 1983.

Creach, Jerome F. D. *Yahweh as Refuge and the Editing of the Hebrew Psalter*. JSOTSup 217. Sheffield: Sheffield Academic, 1996.

Cross, Frank Moore. *Canaanite Myth and Hebrew Epic: Essays in the History of the Religion of Israel*. Cambridge: Harvard University Press, 1973.

———. *From Epic to Canon: History and Literature in Ancient Israel*. Baltimore: Johns Hopkins University Press, 1998.

———. "Notes on Psalm 93: A Fragment of a Liturgical Poem Affirming Yahweh's Kingship." In *A God So Near: Essays on Old Testament Theology in Honor of Patrick D. Miller*, edited by Brent A. Strawn and Nancy R. Bowen, 73–77. Winona Lake, IN: Eisenbrauns, 2003.

Culler, Jonathan. *The Pursuit of Signs: Semiotics, Literature, Deconstruction*. Ithaca, NY: Cornell University Press, 1981.

cummings, e. e. *i: six non-lectures*. New York: Atheneum, 1971.

Davies, Eryl W. *Numbers*. NCB. Grand Rapids: Eerdmans, 1995.

Davis, Ellen F. "Exploding the Limits: Form and Function in Psalm 22." *JSOT* 53 (1992) 93–105.

Delaney, Kevin. "Using an Electronic Device To Break in a New Violin." *The New York Times*, April 5, 2010, p. B8.

Delitzsch, Franz. *Biblical Commentary on the Psalms*. 3 vols. New York: Funk & Wagnalls, 1883.

Driver, S. R. *Notes on the Hebrew Text and the Topography of the Books of Samuel: With an Introduction on Hebrew Palaeography and the Ancient Versions and Facsimiles of Inscriptions*. 2nd ed. Oxford: Clarendon, 1966.

Dunn, James D. G. *The Epistles to the Colossians and to Philemon: A Commentary on the Greek Text*. NIGTC. Grand Rapids: Eerdmans, 1996.

———. *Romans 1–8*. WBC 38. Dallas: Word, 1988.

Eire, Carlos. *A Very Brief History of Eternity*. Princeton: Princeton University Press, 2010.

Ehrlich, Arnold B. *Randglossen zur Hebräischen Bibel: Textkritisches, sprachliches und sachliches*. 7 vols. Leipzig: Hinrichs, 1908–1914.

Farrer, Austin. *Saving Belief: A Discussion of Essentials*. New York: Morehouse-Barlow, 1965.

Fretheim, Terence E. "Prayer in the Old Testament: Creating Space in the World for God." In *A Primer on Prayer*, edited by Paul R. Sponheim, 52–62. Philadelphia: Fortress, 1988.

Frost, Robert. "Education by Poetry." In *Collected Poems, Prose and Plays*, 723–24. New York: Library of America, 1995.
George, Mark K. *Israel's Tabernacle as Social Space*. AIL 2. Atlanta: Society of Biblical Literature, 2009.
Gottwald, Norman K. *The Tribes of Yahweh: A Sociology of the Religion of Liberated Israel, 1250–1050 B.C.E.* Maryknoll, NY: Orbis, 1979.
Greenblatt, Stephen. "The Circulation of Social Energy." In *Shakespearean Negotiations: The Circulation of Social Energy in Renaissance England*, 1–20. Berkeley: University of California Press, 1988.
———. "Towards a Poetics of Culture." In *The New Historicism*, edited by H. Aram Veeser, 1–14. New York: Routledge, 1989.
Habel, Norman C. *The Book of Job: A Commentary*. OTL. Philadelphia: Westminster, 1985.
Hanson, Paul D. *The Dawn of Apocalyptic: The Historical and Sociological Roots of Jewish Apocalyptic Eschatology*. Philadelphia: Fortress, 1975.
Hartman, Geoffrey H., editor. *New Perspectives on Coleridge and Wordsworth: Selected Papers from the English Institute*. New York: Columbia University Press, 1972.
Hays, Richard B. *Echoes of Scripture in the Letters of Paul*. New Haven: Yale University Press, 1989.
Hays, Richard B., Stefan Alkier, and Leroy A. Huizenga, editors. *Reading the Bible Intertextually*. Waco, TX: Baylor University Press, 2009.
Hayward, Robert. *Divine Name and Presence, the Memra*. Totowa, NJ: Allanheld, Osmun, 1981.
———. "The Holy Name of the God of Moses and the Prologue of St. John's Gospel." *NTS* 25 (1978) 16–32.
Heschel, Abraham J. *The Prophets*. New York: Harper & Row, 1962.
Holladay, William L. *Jeremiah 1: A Commentary on the Book of the Prophet Jeremiah, Chapters 1–25*. Hermeneia. Philadelphia: Fortress, 1986.
Hollander, John. "Wordsworth and the Music of Sound." In *New Perspectives on Coleridge and Wordsworth: Selected Papers from the English Institute*, edited by Geoffrey H. Hartman, 41–84. New York: Columbia University Press, 1972.
Holmes, Richard. *The Age of Wonder: How the Romantic Generation Discovered the Beauty and Terror of Science*. New York: Pantheon, 2008.
Homans, Peter. *The Ability to Mourn: Disillusionment and the Social Origins of Psychoanalysis*. Chicago: University of Chicago Press, 1989.
Hoover, Roy W. "The Harpagmos Enigma: A Philological Solution." *HTR* 64 (1971) 95–119.
Hossfeld, Frank-Lothar, and Erich Zenger. *Psalms 2: A Commentary on Psalms 51–100*. Translated by Linda M. Maloney. Hermeneia. Minneapolis: Fortress, 2005.
———. *Psalms 3: A Commentary on Psalms 101–150*. Translated by Linda M. Maloney. Hermeneia. Minneapolis: Fortress, 2011.
Howard, David M., Jr. *The Structure of Psalms 93–100*. Biblical and Judaic Studies 5. Winona Lake: Eisenbrauns, 1997.
Jacobsen, Thorkild. *The Treasures of Darkness: A History of Mesopotamian Religion*. New Haven: Yale University Press, 1976.
Janzen, J. Gerald. *Abraham and All the Families of the Earth: A Commentary on the Book of Genesis 12–50*. ITC. Grand Rapids: Eerdmans, 1993.
———. *At the Scent of Water: The Ground of Hope in the Book of Job*. Grand Rapids: Eerdmans, 2009.

Bibliography

———. "Creation and New Creation in Philippians 1:6." *HBT* 18 (1996) 27–54.
———. *Exodus*. Westminster Bible Companion. Louisville: Westminster John Knox, 1997.
———. "Habakkuk 2:2–4 in the Light of Recent Philological Advances." *HTR* 73 (1980) 53–78.
———. "Hagar in Paul's Eyes and in the Eyes of Yahweh (Genesis 16): A Study in Horizons." *HBT* 13 (1991) 1–23.
———. "'More Joy in Heaven?': Re-Reading 'God and Satan' alongside Resonant New Testament Texts." In *Reading Job Intertextually*, edited by Katharine Dell and Will Kynes. LHBOTS 574. London: T. & T. Clark International, 2012.
———. "The Lord of the East Wind." *HBT* 26 (2004) 2–47.
———. "Metaphor and Reality in Hosea 11." *Semeia* 24 (1982) 7–44.
———. "Modes of Presence and the Communion of Saints." In *Religious Experience and Process Theology*, edited by Harry James Cargas and Bernard Lee, 147–72. New York: Paulist, 1976.
———. "'(Not) of My Own Accord': Listening for Scriptural Echoes in a Johannine Idiom." *Encounter* 67 (2006) 137–60.
———. "Qohelet on Existence 'Under the Sun.'" *CBQ* 70 (2008) 465–83.
———. "Sin and the Deception of Devout Desire: Paul and the Commandment in Romans 7." *Encounter* 70 (2010) 29–61.
———. "Song of Moses, Song of Miriam: Who is Seconding Whom?" *CBQ* 54 (1992) 211–20.
———. *Studies in the Text of Jeremiah*. HSM 6. Cambridge: Harvard University Press, 1973.
Jaynes, Julian. *The Origin of Consciousness in the Breakdown of the Bicameral Mind*. New York: Mariner, 2000.
Jensen, Joseph. *The Use of tôrâ by Isaiah: His Debate with the Wisdom Tradition*. CBQMS 3. Washington, DC: Catholic Biblical Association of America, 1973.
Jöcken, Peter. *Das Buch Habakuk: Darstellung der Geschichte seiner kritischen Erforschung mit einer eigenen Beurteilung*. BBB 48. Köln: Hanstein, 1977.
Johnson, Luke Timothy. *Hebrews: A Commentary*. NTL. Louisville: Westminster John Knox, 2006.
Johnson, Luke Timothy, and William S. Kurz. *The Future of Catholic Biblical Scholarship: A Constructive Conversation*. Grand Rapids: Eerdmans, 2002.
Kant, Immanuel. *Religion within the Limits of Reason Alone*. Translated by Theodore M. Greene and Hoyt H. Hudson. New York: Harper, 1960.
Kirk, J. R. Daniel. "Reconsidering *dikaioma* in Romans 5:16." *JBL* 126 (2007) 787.
Kraus, Hans-Joachim. *Psalms 1–59*. Translated by Hilton C. Oswald. Continental Commentaries. Minneapolis: Fortress, 1993.
Laing, Dilys. *The Collected Poems of Dilys Laing*. Cleveland: Press of Case Western Reserve University, 1967.
Lefebvre, Henri. *The Production of Space*. Translated by Donald Nicholson-Smith. Cambridge, MA: Blackwell, 1991.
Levenson, Jon D. *Creation and the Persistence of Evil: The Jewish Drama of Divine Omnipotence*. San Francisco: Harper & Row, 1988.
———. *Resurrection and the Restoration of Israel: The Ultimate Victory of the God of Life*. New Haven: Yale University Press, 2006.

Levine, Baruch. *Numbers 1-20: A New Translation with Introduction and Commentary.* AB 4. New York: Doubleday, 1993.
Lincoln, Andrew T. *Ephesians.* WBC 42. Dallas: Word, 1990.
Loewald, Hans W. *The Essential Loewald: Collected Papers and Monographs.* Edited by Norman Quist. Hagerstown, MD: University Publishing Group, 2000.
―――. *Papers on Psychoanalysis.* New Haven: Yale University Press, 1980.
―――. *Psychoanalysis and the History of the Individual.* New Haven: Yale University Press, 1978.
―――. "Psychoanalysis in Search of Nature: Thoughts on Metapsychology, 'Metaphysics,' Projection." *Annual of Psychoanalysis* 16 (1988) 49-54.
―――. *Sublimation: Inquiries into Theoretical Psychoanalysis.* New Haven: Yale University Press, 1988.
Loewenstamm, Samuel E. "Yāpîaḥ, yāpiaḥ, yāpēaḥ." *Leshonenu* 26 (1962-63) 205-8.
Lohfink, Norbert. "Die Universalisierung des 'Bundesformel' in Ps 100,3." *TP* 65 (1990) 172-183.
Lubow, Arthur. "The Sound of Spirit." *The New York Times Magazine*, October 17, 2010, pp. 34-39.
Manchester, Peter. "The Religious Experience of Time and Eternity." In *Classical Mediterranean Spirituality: Egyptian, Greek, Roman*, edited by A. H. Armstrong, 384-407. New York: Crossroad, 1986.
Mays, James Luther. *The Lord Reigns: A Theological Handbook to the Psalms.* Louisville: Westminster John Knox, 1994.
McFarland, Thomas. *Coleridge and the Pantheist Tradition.* Oxford: Clarendon, 1969.
Michel, Walter L. "ṣlmwt, 'Deep Darkness' or 'Shadow of Death'?" *Papers of the Chicago Society of Biblical Research* 29 (1984) 5-20.
Miles, John A., Jr. "Gagging on Job, or The Comedy of Religious Exhaustion." *Semeia* 7 (1977) 71-126.
Milgrom, Jacob. *Numbers: The Traditional Hebrew Text with the New JPS Translation.* JPS Torah Commentary. Philadelphia: Jewish Publication Society, 5750/1990.
Mill, John Stuart. *On Liberty and Other Writings.* Edited by Stefan Collini. Cambridge Texts in the History of Political Thought. Cambridge: Cambridge University Press, 1989.
Miller, Patrick D. "The Blessing of God: An Interpretation of Numbers 6:22-27." *Int* 29 (1975) 240-51.
―――. "Deuteronomy and Psalms: Evoking a Biblical Conversation." *JBL* 118 (1999) 3-18.
―――. "The Hermeneutics of Imprecation." In *The Way of the Lord: Essays in Old Testament Theology*, 193-202. FAT 39. Tübingen: Mohr/Siebeck, 2004.
―――. "Kingship, Torah Obedience, and Prayer: The Theology of Psalms 15-24." In *Neue Wege der Psalmenforschung: Für Walter Beyerlin*, edited by Klaus Seybold and Erich Zenger, 127-42. Herders Biblische Studien 1. Freiburg: Herder, 1994.
―――. "The Ruler in Zion and the Hope of the Poor: Psalms 9-10 in the Context of the Psalter." In *The Way of the Lord: Essays in Old Testament Theology*, 167-77. FAT 39. Tübingen: Mohr/Siebeck, 2004.
―――. "The Sufficiency and Insufficiency of the Commandments." In *The Way of the Lord: Essays in Old Testament Theology*, 17-36. FAT 39. Tübingen: Mohr/Siebeck, 2004.

———. *The Ten Commandments.* Interpretation. Louisville: Westminster John Knox, 2009.

———. *They Cried to the Lord: The Form and Theology of Biblical Prayer.* Minneapolis: Fortress, 1994.

———. "Things Too Wonderful: Prayers of Women in the Old Testament." In *Biblische Theologie und Gesellschaftlicher Wandel: Für Norbert Lohfink SJ,* edited by Georg Braulik, Walter Gross, and Sean McEvenue, 237–51. Freiburg: Herder, 1993.

———. *The Way of the Lord: Essays in Old Testament Theology.* FAT 39. Tübingen: Mohr/Siebeck, 2004.

———. "What the Scriptures Principally Teach." In *Homosexuality and Christian Community,* edited by Choon-Leong Seow, 53–63. Louisville: Westminster John Knox, 1996). Reprinted in Patrick D. Miller, *The Way of the Lord: Essays in Old Testament Theology,* 286–96. FAT 39. Tübingen: Mohr/Siebeck, 2004.

———. "*yāpîaḥ* in Psalm xii 6." *VT* 29 (1979) 495–501.

Moltmann, Jürgen. *The Coming of God: Christian Eschatology.* Translated by Margaret Kohl. London: SCM, 1966.

———. *God in Creation: A New Theology of Creation and the Spirit of God.* Translated by Margaret Kohl. Minneapolis: Fortress, 1993.

———. *The Trinity and the Kingdom: The Doctrine of God.* Minneapolis: Fortress, 1993.

Moran, William L. *The Most Magic Word: Essays on Babylonian and Biblical Literature.* Edited by Ronald S. Hendel. CBQMS 35. Washington, DC: Catholic Biblical Association of America, 2002.

Moule, C. F. D. *The Meaning of Hope: A Biblical Exposition with Concordance.* Philadelphia: Fortress, 1963.

Neville, Robert Cummings. *Eternity and Time's Flow.* Albany: SUNY Press, 1993.

Nussbaum, Martha. *Women and Human Development: The Capabilities Approach.* John Robert Seeley Lectures 3. Cambridge: Cambridge University Press, 2001.

O'Brien, Peter T. *Colossians, Philemon.* WBC 44. Waco, TX: Word, 1982.

———. *The Epistle to the Philippians.* NIGTC. Grand Rapids: Eerdmans, 1991.

———. *The Letter to the Ephesians.* PNTC. Grand Rapids: Eerdmans, 1999.

Perowne, J. J. Sewart. *The Book of Psalms: A New Translation, with Introductions and Notes Explanatory and Critical.* Grand Rapids: Zondervan, 1966.

Price, Lucien. *Dialogues of Alfred North Whitehead.* Jaffey, NH: Godine, 2001.

Pruyser, Paul W. *The Play of the Imagination: Towards a Psychoanalysis of Culture.* New York: International Universities Press, 1983.

Rad, Gerhard von. *Old Testament Theology.* 2 vols. Translated by D. M. G. Stalker. New York: Harper & Row, 1962–65.

Rattenbury, R. M., and T. W. Lumb. *Heliodore: Les Ethiopiques: Tome I.* Paris: Les Belles Lettres, 1935.

Reardon, B. P., editor. *Collected Ancient Greek Novels.* Berkeley: University of California Press, 2008.

Richards, I. A. *Poetries and Sciences.* New York: Norton, 1970.

———. *The Portable Coleridge.* New York: Viking, 1950.

Roberts, J. J. M. "Erra—Scorched Earth." *JCS* 24 (1971) 11–16.

Russell, Bertrand. *Why I Am Not a Christian and Other Essays on Religion and Related Subjects.* New York: Simon & Schuster, 1957.

Sarna, Nahum M. *Exodus: The Traditional Hebrew Text with the New JPS Translation.* JPS Torah Commentary. Philadelphia: Jewish Publication Society, 5751/1991.

Schökel, Luis Alonso. *Diccionario Biblico Hebreo-Español*. Edited by Victor Morla and Vicente Collado. Madrid: Trotta, 1994.
Scholem, Gershom. *Major Trends in Jewish Mysticism*. New York: Schocken, 1995.
Segal, Alan F. *Paul the Convert: The Apostolate and Apostasy of Saul the Pharisee*. New Haven: Yale University Press, 1990.
Sheldrake, Rupert. *A New Science of Life: The Hypothesis of Formative Causation*. Boston: Houghton Mifflin, 1981.
———. *The Presence of the Past: Morphic Resonance and the Habits of Nature*. New York: Vintage, 1989.
Smith, Steven D. *The Disenchantment of Secular Discourse*. Cambridge: Harvard University Press, 2010.
Smith, W. Robertson. *Kinship and Marriage in Early Arabia*. New edition. Edited by Stanley A. Cook. London: Adam & Charles Black, 1903.
Steiner, George. *Real Presences*. Chicago: University of Chicago Press, 1989.
Stendahl, Krister. *Final Account: Paul's Letter to the Romans*. Minneapolis: Fortress, 1995.
Stern, Daniel N. *The Interpersonal World of the Infant: A View from Psychoanalysis and Developmental Psychology*. New York: Basic Books, 1985.
Thiselton, Anthony C. *The First Epistle to the Corinthians*. NIGTC. Grand Rapids: Eerdmans, 2000.
Thompson, Leonard L. "The Jordan Crossing: Ṣidqot Yahweh and World Building." *JBL* 100 (1981) 343–58.
Towner, W. Sibley. "The Poetic Passages in Daniel 1–6." *CBQ* 31 (1969) 317–26.
———. "'Without Our Aid He Did Us Make': Singing the Meaning of the Psalms." In *A God So Near: Essays on Old Testament Theology in Honor of Patrick D. Miller*, edited by Brent A. Strawn and Nancy R. Bowen, 17–34. Winona Lake, IN: Eisenbrauns, 2003.
Vining, Joseph. *The Song Sparrow and the Child: Claims of Science and Humanity*. Notre Dame: University of Notre Dame Pres, 2004.
Wall, Robert W. "Wifely Submission in the Context of Ephesians." *CSR* 17 (1988) 272–85.
Wallace, Catherine M. *For Fidelity: How Intimacy and Commitment Enrich Our Lives*. New York: Vintage, 1999.
———. *Motherhood in the Balance: Children, Career, Me, and God*. Harrisburg, PA: Morehouse, 2001.
Westcott, Brooke Foss. *The Epistle to the Hebrews: The Greek Texts with Notes and Essays*. London: Macmillan, 1928.
Westermann, Claus. *Isaiah 40–66: A Commentary*. Translated by David M. G. Stalker. Old Testament Library. Philadelphia: Westminster, 1969.
Wheeler, Kathleen M. *The Creative Mind in Coleridge's Poetry*. Cambridge: Harvard University Press, 1981.
Wheelwright, Philip. *Metaphor & Reality*. Bloomington: Indiana University Press, 1962.
Whitehead, Alfred North. *Adventures of Ideas*. New York: Free Press, 1967.
———. *Modes of Thought*. New York: Free Press, 1966.
———. *Process and Reality: An Essay in Cosmology*. New York: Macmillian, 1929.
———. *Religion in the Making*. New York: World, 1969.
———. *Science and the Modern World*. New York: Free Press, 1967.
———. *Symbolism, Its Meaning and Effect: Barbour-Page Lectures, University of Virginia, 1927*. New York: Fordham University Press, 1985.

Bibliography

Wilson, Gerald H. *The Editing of the Hebrew Psalter*. SBLDS 76. Chico, CA: Scholars, 1985.
Wink, Walter. *The Powers That Be: Theology for a New Millennium*. New York: Doubleday, 1998.
Winnicott, D. W. *Playing and Reality*. Routledge Classics. London: Routledge, 2005 (orig. pub. 1971).
Winston, David. *The Wisdom of Solomon: A New Translation with Introduction and Commentary*. AB 43. Garden City, NY: Doubleday, 1979.
Wright, N. T. *Paul for Everyone: The Prison Letters: Ephesians, Philippians, Colossians and Philemon*. London: SPCK, 2002.
———. "Poetry and Theology in Colossians 1:15–20." In *The Climax of the Covenant: Christ and the Law in Pauline Theology*, 99–119. Minneapolis: Fortress, 1992.
———. "How Can the Bible Be Authoritative?" *Vox Evangelica* 21 (1991) 7–32.
Zimmerli, Walther. *Ezekiel 2: A Commentary on the Book of the Prophet Ezekiel, Chapters 25–48*. Translated by James D. Martin. Hermeneia. Philadelphia: Fortress, 1983.

Author Index

Alkier, Stefan, 243, 330
Anderson, Francis I., 290
Appiah, Kwame Anthony, 286
Arnold, Matthew, 237
Attridge, Harold W., 328, 344
Auffret, Pierre, 202
Augustine, 103
Austin, J. L., 38, 168, 183
Avnon, Dan, 117, 202

Bachelard, Gaston, 337
Barfield, Owen, 257, 276
Barr, James, 296, 349
Barthes, Roland, 243
Bauckham, Richard, 383–84
Beaumont, George, 269
Becker, Ernest, 380
Bengel, Johann Albrecht, xiii
Bentham, Jeremy, 237
Berkeley, George, 137
Bohm, David, 270
Bollas, Christopher, 57
Brower, Reuben, 8–9, 116, 245, 247, 267
Browning, Robert, 97
Brueggemann, Walter, 38–39, 44–45
Buber, Martin, 116–17, 178, 202

Calvin, John, 85
Carey, Lillian Gilbreth, 349
Carter, Sydney, 292
Clifford, Richard J., 231
Cohen, Mark E., 86

Coleridge, Samuel Taylor, xii, xv–xvi, 5–6, 65, 125, 131, 133, 184–86, 236–37, 241–43, 255–65, 268–69, 272, 274, 278–79, 285–86, 288–89, 295–96, 316, 322, 333, 339
Collins, John J., 356, 365, 372, 384, 389
Collins, Terence, 117
Cooper, Alan, 203, 208
Coverdale, Miles, 166–67, 183, 204
Cowper, William, 149
Craigie, Peter, 110–11, 114
Creach, Jerome F. D., 135, 156
Cross, Frank Moore, xi–xii, 4, 25, 39–40, 45, 80–81, 118
Cullers, Jonathan, 243
cummings, e. e., 381
Cuomo, Mario, 252

Dahood, Mitchell, 188
Davies, Eryl W., 40–41
Davis, Ellen F., 202–3
Debussy, Claude, 121
Delaney, Kevin, 246
Delitzsch, Franz, 166–68, 174, 180
Descartes, Rene, 271, 277
Driver, S. R., 74
Dunn, James D. G., 291, 311, 313
Dylan, Bob, 355

Ehrlich, Arnold B., 218
Eire, Carlos, 304
Epictetus, 313

Author Index

Farrer, Austin, 101, 130, 132–33
Fortes, Meyer, 80, 87
Freedman, David Noel, 189, 208, 290
Fretheim, Terence E., 50, 53
Freud, Sigmund, 86, 138, 160, 251, 255–56, 275–77, 279–81, 342, 354
Frost, Robert, xii, 6, 8–9, 116, 280–81

George, Mark, 245, 333, 335–43
Gerstenberger, Erhard, 298
Gilbreth, Frank, 349
Girard, René, 250
Goethe, Johann Wolfgang von, 7
Gottwald, Norman, 216
Greenblatt, Stephen, 337–39, 341
Griggs, Earl Leslie, 259
Grohman, Deborah, 248

Habel, Norman C., 114
Hanson, Paul D., 237
Hardy, Thomas, 350
Hartman, Geoffrey H., 257
Hays, Richard B., xiii, 242–47, 249, 267, 299, 330, 352, 369
Hayward, Robert, 47, 52–53
Hegel, Georg Wilhelm Friedrich, 295
Heidegger, Martin, 192
Heidt, William G., 100
Heliodorus, 381
Herbert, George, 73, 391
Herodotus, 345
Herschel, John, 117
Heschel, Abraham, 20–21, 23, 146–47, 153, 164, 214, 262, 266
Holladay, William L., 95
Hollander, John, 245, 257, 285
Holmes, Richard, 117, 256
Homans, Peter, 82, 84
Hoover, Roy W., 351

Hossfeld, Frank-Lothar, 116–19, 158, 201–2, 206
Howard, David M., Jr., 103–4, 106, 114, 117, 120, 122–24, 201–2
Huizenga, Leroy A., 243, 330

James, William, 274
Janzen, J. Gerald, xi–xiv, 31, 63, 68, 88, 90, 98, 121, 131, 140, 170–71, 206, 218–19, 221, 229, 245, 249, 259, 273–74, 293, 298, 330, 339, 342
Jacobsen, Thorkild, 90, 105–6, 109, 169–70, 212
Jaynes, Julian, 213
Jensen, Joseph, 221
Jöcken, Peter, 220
Johnson, Luke Timothy, 241, 328, 331, 344
Johnson, Samuel, 245, 251
Jung, Carl, 354

Kant, Immanuel, 333–34
Kethe, William, 99
Kirk, J. R. Daniel, 317
Kraus, Hans-Joachim 200–201
Kristeva, Julia, 243

La Favor, Willie, 248
Laing, Dilys, 23
Leach, Edmund, 293
Lefebvre, Henri, 337–39
Leighton, Robert, 125
Levenson, Jon, 43, 54, 189, 198, 206–8
Levine, Baruch, 41–42
Lincoln, Eric, 360–1, 363–65, 367, 373, 375, 381
Loewald, Hans, xii, 85, 121, 138, 159–60, 192–94, 201, 243, 260, 263–64, 270–82, 342
Loewenstamm, Samuel E., 218
Lohfink, Norbert, 102, 115–16, 120, 126, 128, 250

Author Index

Lowell, James Russell, 373
Lowes, John Livingston, 5
Lubow, Arthur, 287
Lumb, T. W., 382
Luria, Isaac ben, 54, 283
Luther, Martin, 158, 333

Manchester, Peter, 193
Martin, Ralph, 291
Mays, James Luther, 72, 115,
    189–90, 195, 199, 208, 308
McFarland, Thomas, 256–57
Meissner, William, 281
Michel, Walter L., 203
Miles, J. A., Jr., 237
Milgrom, Jacob, 47
Mill, John Stuart, 284
Miller, Patrick D., xii–xiii, xv–xvi,
    48, 63–65, 84, 91–92, 106,
    115, 202, 242, 249–54, 265,
    382
Moltmann, Jürgen, 54–55, 383
Moran, William, 7
Moule, C. F. D., 331

Nevelle, Robert, 193
Nussbaum, Martha, 285

O'Brien, Peter, 291, 323, 360–61,
    363, 375, 381–82

Pärt, Arvo, 287
Perowne, J. J. Stewart, 101, 104
Plato, 383, 386–387
Pope, Alexander, 245, 267
Powell, Anthony, 13, 351
Pruyser, Paul, 82, 281

Rad, Gerhard von, 201, 206,
    212–13
Rattenbury, R. M., 382
Reardon, B., P., 382
Richards, I. A., 237, 255–26, 264
Ricoeur, Paul, 242, 276
Roberts, J. J. M., 90

Robertson Smith, W., 81, 86
Robinson, John, 249
Rolland, Romain, 275–76
Rosenstock, Eliiot, 228
Rosenzweig, Franz, 116, 202
Russell, Bertrand, 161–63, 287–88

Sarna, Nahum, 23
Sartre, Jean-Paul, 100
Schiller, Friedrich, 7–8
Schökel, Alonso, 91
Scholem, Gershom, 54, 377
Scott, R. B. Y., 291
Segal, Alan F., 363
Sheldrake, Rupert, 242, 269–70
Smith, Steven D., 243, 282–88
Sotheby, Thomas, 259–60, 263, 268,
    272
Southwick, Jay, 52
Stein, Jonathan, 228
Steiner, George, 183
Stendahl, Krister, 1
Stern, Daniel N., 87
Strawn, Brent A., xv–xvi

Theodotion, 323–24, 361–64, 366,
    370–72, 374–76, 384
Tillich, Paul, 51
Thiselton, Anthony C., 385
Thompson, Leonard L., 216
Towner, W. Sibley, 188–90, 204, 357
Twain, Mark, 242

Vining, Joseph, 286–87

Wall, Robert W., 381
Wallace, Catherine M., xvi, 242,
    275, 277–78
Weber, Max, 282, 349
Westcott, Brooke Foss, 328, 344
Westermann, Claus, 232
Wheeler, Kathleen M., 261
Wheelwright, Philip, 248–49, 291,
    383

403

Author Index

Whitehead, Alfred North, xii, 50, 82, 84, 86, 97, 121, 201, 214–15, 242, 251, 256, 263, 265–69, 272–73, 283, 304, 338–39
Wilson, Gerald H., 115
Wink, Walter, 390
Winnicott, D. W., 64, 82, 84, 279, 281
Winston, David, 320
Wolff, H. W., 38
Woolf, Virginia, 383
Wordsworth, William, 256–59, 261, 269, 276, 279
Wright, N. T., 291, 349, 385

Zenger, Erich, 116–17, 119, 141, 144, 164, 191, 201–2, 206–7

Scripture Index

Hebrew Bible/Old Testament

Genesis

1–11	40
1	38, 40–41, 43–48, 107, 129, 261, 359
1:2–5	360
1:2	28, 30
1:3	46–47, 51, 53, 55, 290
1:22	39
1:26	39
1:28	27, 39, 46
2–3	106, 113, 231
2	358–59, 380
2:4	39
2:7	153
2:9	113
2:13	82
2:24	382
2:25	325
3:5	106, 113
3:6	113
3:9–10	325
4:1	28
4:7	251
4:25	93
8:17	39
8:21	74
9	45
9:1	39
9:6	39
9:7	39
11	26, 29
11:1–9	56, 380
14:18	335
14:22	33
15:2	29
15:12	152
15:21	171
16:4–5	138
17:20	39
18:22	136
19:27	136
20:16	370
20:18	67
21:22	22
21:33	19
22	131, 293
22:1	129
22:6	293
22:8	293
22:18	154, 293
23:17	191
24:40	22
24:45	74
25:22	139
25:23	169
26:3	22
26:4	154
26:23	29
26:24	22
26:28	22
27:34–38	47
27:41–45	139
27:41	74
27:45	78

Genesis (cont.)

28	90
28:1–4	39
28:3	40
28:13	29
28:15	22
29–35	27
29:31	140
30	27
30:2	29
30:23	140
30:32	139
30:33	139
30:35	139
31:38	78
32:13–21	42
32:20	42
32:22–30	42
33:1	139
33:4–11	42
33:10	42, 139
33:13–14	139
34:3	74
35	45
35:2	22
35:9–11	42
35:11–12	39
36	39
37:2	39
37:9–11	181
39:2	22, 175
39:3	22, 175
41	354
42:16	370
42:36	79, 88
43:14	62, 79, 88
43:29–30	62
44:4	80
45:8	101–2
45:26	222
46:2–4	22
47:27	39
48	29
48:2	154
48:3–4	39
48:4	197
48:15–16	22
49	29
49:6	390
49:22–26	19
49:25	62, 83
50:17	80
50:21	74

Exodus

1	40, 75
1:7	27, 29, 39
1:8	28
2	26, 215
2:23—3:9	266
2:23–25	20–21, 61, 128, 147, 262–63, 289, 342
2:23–24	30
2:23	22
2:24	234
2:25	289
3	10, 19, 22–23, 26–28, 31, 34, 293
3:1–6	20
3:2	22
3:3	22
3:4	22, 129
3:6	29, 31–32
3:7–10	22
3:7–8	21, 31, 147, 263
3:7	22, 147
3:8	147
3:9–10	31
3:10	147, 212
3:12–15	47
3:12	22, 47, 148, 212, 290
3:13	32
3:14–22	40
3:14–15	12, 32, 47, 290
3:14	42, 46–47, 51, 53, 127, 261, 290, 295
3:15	34, 36, 290
3:19–24	40

Exodus (cont.)

4:1–17	34
4:12	290
4:21–23	213
4:22–23	21
5	386–88
5:2	33
5:5	386
5:8	386
5:13	315
5:17	386
5:19	315
5:20–21	387
5:23	387
6:2–8	40
6:3	42
12:12	118
14	143
15	160
15:1–21	61
15:1–18	67, 117, 160, 189
15:1–8	118
15:3	32, 214
15:4–10	143
15:4	143
15:11	109
15:13	190
15:14–16	103
15:17	190
15:18	109, 190
15:19–21	67
15:21	67
16:4	315
18:4	93
19	17
19:3–6	309
19:4	310
19:8	296–97
19:18	293
19:21	18
20:12	176
21–23	382
23:26	78
24:3–8	18
24:9–11	18, 214
25–31	43, 48, 53, 55, 131
25	43
25:2	43, 53, 174, 325
25:3–7	43
25:8	48, 53, 131, 340
25:9	6, 43, 291
25:22	54
25:31	334
28	53, 82
28:8	2
28:12	82
28:29–30	316
28:29	82
28:30	345
31:1–6	290
32–34	41, 124
32:9–14	34
32:9–10	123
32:10	18
32:11–32	123
32:11	45, 149
32:13	149
33–34	42
33:3	17
33:19	36, 40, 42, 45, 127
34:6–7	124–25
34:6	18, 40, 42, 45–46, 127–28
34:9	142
34:10	128
35–40	43, 48, 55, 131
35–39	43
35	43
35:5	43, 342
35:21	43, 342
35:22	342
35:26	342
35:29	43
36:2	342
39:19	310

Leviticus

9:22–23	40, 43
23:37	315

Numbers

6	44, 138
6:22–27	10, 40, 43, 183
6:24–26	38, 40–42, 47, 319
6:24	319
6:25–26	45
6:25	46
6:27	37, 47–48
7:1-2	43
10:35	321
12:7	149, 329
13–14	119, 124
13	9, 151
13:32–33	138
13:33	119
14:11–12	123
14:11	120
14:13–19	123, 149
14:19	142
14:23	120
14:41	120
14:44	120
18:26	96
27:21	297
36:2	96

Deuteronomy

1:13	95
1:31	138
2:7	108
2:30	171, 179–80
3:28	155, 171, 177–78
4:5–8	148
4:10	137
4:20	19
4:26	176, 197
4:28	108
4:31	198
4:40	176
4:49	197
5:5	136
5:16	176
5:20	91
5:33	176
6:2	176
6:4	214
6:10–11	106
6:14	106–7
8	107, 127, 179
8:2	106
8:3	112, 114
8:11–20	113
8:11–19	114
8:12–18	106
8:17	112, 127
8:19	106, 112
10:8	136
10:20	382
11:9	176
11:22	382
12:9	260
14:26	113
14:29	108
15:7	170, 180
16:15	108
17:12	136
17:18–20	138
17:20	176
19:18	92
20:3–4	173
20:8	173
21:29	94
22:17	176
24:19	108
25:15	176
28	235
28:12	108
28:29	175
30:9	108
30:14	315
30:15	197
30:18	176, 197
30:19	197
30:20	176, 197–98
31:6	155, 168, 183
31:7	168
31:20	113, 120

Deuteronomy (cont.)

31:23	155, 168
32:4–6	138
32:8–9	285
32:10–12	310
32:10–11	309
32:10	30
32:11	310
32:13–14	30
32:25	79
32:47	176
33	234
33:12	200
33:26	93
33:28	200

Joshua

1:5	168
1:6	155, 168
1:7	155, 168
1:8	315
1:9	155, 168
1:13	206
1:18	155, 168
2:11	153
5:1	103
9:5	139
10:6	94
10:25	155, 168
13:6	96
13:7	96
21:12	96
23:4	96
23:8	283
24:31	176, 198

Judges

2:7	176
4:12	297
5:2	174
5:10	70
7:2	107
7:29	297
8:11	200
9:15	311
9:46	205
13:17	297
13:25	355
15:2	297
15:5	297
15:9	297
19:3	74
21:15	148

Ruth

1:18	172
1:19–21	78
1:20–21	63, 205
2:12	310–11, 382
2:13	74, 171
3:9	310–11, 382
4:14–17	78
4:16	91

1 Samuel

1–2	63
1	63, 76
1:3–18	61
1:6	73
1:7–8	72
1:8	72
1:10	72, 75
1:11	72
1:12	72
1:13	61, 171
1:15	72
1:16	73
2:1–10	61, 76
2:10	73
3	293
3:4	129
4:4	321
4:11	321
7:9	139
8–12	124

Scripture Index

1 Samuel (cont.)

8:4–8	123
8:11	95
11:10	73
12:23	123
12:24	123
12:25	123–25
15:27	310
15:33	78
17:33–37	89
24:4	310
24:5	354
24:11	310
25:21	79

2 Samuel

6:15	321
6:22	138
7:1–3	320
7:1	206
7:4–17	314, 320
7:12–14	314
7:18–29	314
7:18	320
8	321–22
8:6	321
8:12	321
8:14	321
17:8	89
19:8	74
22:7	181
22:12	94
22:37	181
22:44	368
24:10	354

1 Kings

3:14	176–77
6:24	310
8	309, 312, 314, 316–18, 336
8:12	54
8:27	54
8:28–29	314
8:28	315
8:29	54
8:30	314
8:41–43	54
8:44–59	318
8:44–45	316
8:45	317
8:46–48	316
8:49–50	316
8:49	317
8:51	19
8:54–58	315
8:54	364
8:58	315
8:59	315–18, 345
10:5	137
11:34	94
12:18	172
18:18	101–2
22	213
22:18	169
22:19–22	219
22:21	137
22:38	169

2 Kings

2:19	79
2:21	79
4:33	307
5:11	136
10:18–27	205
14:10	342
18:6	382
18:22	101–2
21:7	335
23:2–3	137
25:4	148
25:30	315

1 Chronicles

16	309, 312
16:11	182

1 Chronicles (cont.)

16:30	111
16:37	315
18:6	321
18:13	321
22:13	155, 168, 172
23:28	137
23:30	137
28:20	155, 168, 172
29	43, 131, 133, 309, 320, 340
29:2	131–32
29:5	44
29:6	44
29:9	44
29:10–19	314
29:10–12	344
29:11	131–32
29:12	132
29:13	132
29:14–15	132
29:14	44, 48, 152, 325
29:15	44
29:16–18	181
29:18–19	344
29:20	364
32:7	155

2 Chronicles

8:14	315
9:4	137
10:18	172–73
11:14–15	172
11:16–17	172
13:1–7	173
24:13	169
25:1	342
26:2	175
29:14	343
30:16	137
30:22	74
31:16	315
32:6	74
32:7	168, 172
34:31–32	137
34:31	341
35:10	137
35:15	137
36:13	171, 180

Ezra

3:4	315

Nehemiah

8:7	137
9:3	137
11:23	315
12:47	315
13:11	137

Esther

4:14	147
13:4	345

Job

1:10	108
2:11	297
3:4	390
3:6	390
3:10	222
3:12	88
3:20	88, 205
3:24	73
4:3–7	77
4:4	172
4:10–11	114
4:13	152
6:4–7	88
6:5	73
6:7	113
6:11	330
7:11	71–72
7:12	114, 121
7:13	71–72
9:8	114, 121

Scripture Index

Job (*cont.*)

9:27	71
10:1	71–72
10:10	290
10:16	114
11:8–9	375
12:12	195
13:3	330
13:6	330
13:15	330
13:16	244
14:9	175
14:13	93
14:15	108
15:4	114
15:6	101–2
16	98
16:5	172
16:8	97
16:18–21	307
16:19–21	97
16:19	96–97
16:21	97, 330
18:15	231
19:26	206
20:12–13	113
21:4	71–72
21:10	78
21:18	136
23:2	71–72
23:4	330
23:7	330
23:13	214
26:2	94
26:12	114
27:3	153
29	72
30	72
30:31	206
31:15	214
31:38	297
34:19	108
34:33	101–2
36:16	181
37:4	73
38:1	206
38:7	206, 263, 295, 390
38:8	82
38:11	114
38:34	293
38:35	129, 293
38:39	113
39:14	83
40:1	63
40:23	82
41:20	82
42:1–6	63
42:3	63
42:5	260
42:8	90

Psalms

1	19, 135–36, 140–42, 160, 191, 308
1:3	175
1:4	136
1:5	142, 306
2	135–36, 308–9, 311, 334, 384
2:6	290
2:7	334
2:8	312, 334
2:11	332
3:1–2	204
3:1	73
3:3–4	205
4:1	41
4:2	180
4:6	41
4:8	205
7:6	159
8:7	108
9–10	110, 112–15, 119
9:15	143
9:20	93
10:2–4	110, 112
10:2	110, 112, 115
10:3–5	110, 112

Psalms (cont.)

10:3	112–13, 120
10:4	112, 119
10:6	111–12
10:11	119
12	94
12:2–3	94
12:2	91–92, 94, 97–98
12:3–4	94
12:3	91
12:4	175
12:5	94
12:6	91–93, 95–98
12:7	98
12:8	98
12:9	98
13	68, 194
13:3	68
13:4	73
13:5–6	68–69
14	114
14:1	119
14:4	114
15–24	200, 202, 204
15	202
15:2–3	140
15:5	111, 141
16	195, 198–203, 205–6, 208
16:5–6	199
16:5	202
16:6	201
16:8–11	203
16:8	111
16:11	202, 205
17:5	111
17:8	310, 319
18	156
18:3	157
18:4–5	142, 156
18:6	156, 388
18:7	171
18:11	1
18:17	169
18:19	156
18:31–33	156
18:33	141–42, 157
18:35–36	156
18:36	157
18:37	181
18:43	332, 368
18:49	157
19:2	108
20:8	142
21:3	196
21:4	196–98
21:6	196–97
21:7	111
21:33	191
22	61, 84–85, 161, 202–4
22:1	83
22:2	73
22:3–5	83
22:3	84, 121
22:9–10	82
22:11	83
22:12–21	204
22:19	83
23	11, 175, 188–90, 195–96, 198–200, 202–4, 206–8
23:2–3	190
23:2	190, 251
23:3–5	190
23:4	203
23:5	190, 200, 202
23:6	9, 190, 198–203, 205
24	202–4, 208
24:3	142
24:7–10	203
25	69
25:1–3	69
25:2–3	69
25:17	166, 180–81
25:19	69
25:20	69, 319
26:12	141–43
27	163, 177, 179, 184, 186, 288
27:2	73

Psalms (cont.)

Reference	Pages
27:4	9, 182, 325, 332
27:5	1, 333
27:8	181–83
27:10	62, 83
27:13–14	185, 332
27:13	179, 329
27:14	11, 165–67, 174, 176–77, 180–83, 185, 187
28:5	108
29	44
29:1–2	390
29:9	390
30	190
30:5	190
30:6	179
30:7	111–12
30:12	190–91
31	177, 179
31:1–2	69
31:1	180
31:7	166
31:8	141–42, 156
31:9–10	180
31:10	332
31:11–18	69
31:15	360
31:16	41
31:20	332
31:21	1
31:23–24	332
31:24	174, 176–77
31:25	166–67, 181
32	324
32:2	73
32:3–4	324
32:5	325
32:6	326
32:7	1
33:4	345
33:6	345
34:5	325
35	88, 90
35:5	136
35:10	86
35:12–14	80
35:12	77, 79–81
35:13–14	77, 80
35:13	81, 85
35:14	77, 81
35:15	77
35:23	159
35:25	86
36:7	310
36:12	142
36:14	380
37:40	94
38:8	222
38:9	73
38:11	79
38:20	79
40	156–57
40:1	157
40:2	142
40:9–10	157
40:11–17	157
40:16	157
41:2	319
41:8	143
41:13	195
42–43	65–66, 74, 158, 163, 182
42	75
42:1	66
42:4	119, 158
42:5	66, 158–59
42:11	66, 119, 158–59
43:4	66
43:5	66, 158–59
44:3	46
44:7	334
44:23	159
46	112
46:1–2	112
46:5	111
46:10	387
48:12	380
51	138

Psalms (cont.)

51:1	49
51:12	44, 49
51:17	326
52	190
52:8	19
52:9–10	190
53	114
53:2	119
54:4	93
55:2	71
55:9	332
55:17	71
57	68
57:1	310
57:3	68
57:7–9	68
59:5	159
60:12	185
61	191
61:2–4	205
61:4	1, 310
62:2	111
62:6	111
63:7	310
63:8	163, 334, 382
64:1	71
65:2	334
66:5–7	19
66:10–12	19
67:1	41
67:2	41
68	363
68:15	363
68:18	363
68:28	363
69	11, 134, 137, 140–42, 144–45, 156–57, 160–61, 164–65, 205, 215, 288, 292, 331–32
69:1–14	141
69:1–2	134
69:2	141–43, 146, 164, 331
69:3	136
69:4	164
69:5–13	141
69:7	141, 144, 331
69:9	141, 164, 299, 331
69:10	141, 331
69:14–15	157
69:14	142–43
69:19	141, 331
69:20	141, 144, 331
69:21	164
69:22–23	164
69:24	164
69:25	164
69:28	164
69:29	157
69:30	158
69:32–36	142
71:1	69
71:6	82
72:12	97
72:17	154
73–83	260
73	205–7
73:3–12	112
73:5	205
73:16	205
73:17	118, 205
73:18–20	205
73:23	205
73:24	205
73:26	205
74:1	102
74:12–17	75, 105
74:13–14	121
76:6	152
76:7	141–42, 148
77	71–74, 158–60
77:1	71
77:2	222
77:3	71, 160
77:4	71, 355
77:5–6	71
77:6	71, 158, 160
77:7–9	71

Scripture Index

Psalms (cont.)

77:11–20	71, 160
77:11–12	71
77:12	71
77:13–20	71, 160
77:14	160
77:21	102
77:24	189
78	189–90, 196
78:19	189
78:42	73
78:52	102, 189
78:54	190
78:60–61	321
78:69	190, 196
78:70–72	189–90
78:71–72	102
78:69	48
79:10	119
79:13	102
80	20, 321
80:3	41, 138
80:6	332
80:7	41, 138
80:8	20
80:14–16	20
80:15	169
80:17	169, 178
80:19	41, 138
84	191
84:1–3	309
84:3	5
84:4	9
84:6	94
84:54	191
86:11	390
88:8	94–95
88:15	222
89:15	46
89:21	171, 178
89:43	143
90:1–2	267, 285
90:2	195
90:8	46
90:17	108
91	2, 5, 205, 308–11
91:1–2	309
91:1	1, 203
91:2	309
91:3	309
91:4	309–10
91:9	309
91:10	309
91:11–12	309
91:12	309
91:16	195
92	191
92:4	108
92:12–14	191
92:12	19
92:13–15	142
92:13	175
93–100	100, 106, 112, 115–18, 121, 124, 126, 129, 135, 201–2
93–99	115–16, 128
93	103, 105, 115, 117–23, 125–28
93:1–2	118, 122
93:1	107, 111–12, 115, 119, 122, 128
93:3	121
93:5	118, 123, 196–98
94	115, 118–20, 123, 127–28
94:2	115, 118–19
94:7	119
95	67, 119–20, 123, 125–28, 206
95:1	67
95:2–3	67
95:6–7	126
95:6	67, 102
95:7	102, 119–20
96	120–23, 126, 128
96:3	330
96:9	123
96:10	107, 111–12, 128
96:11	73, 121

Psalms (*cont.*)

96:13	128
97	121–23, 126, 128
97:1	121, 128
97:3	121
97:4	128
97:7	121, 123
97:10	121, 128
97:11	121–22
97:12	123
98	122–23, 126, 128
98:1	123
98:7	73, 128
98:8	128
99	122–23, 126–27
99:2	128
99:3	123, 127–28
99:4	127
99:5	123
99:6–8	122
99:7	123
99:8	123–26
99:9	123
100	5, 100–101, 105, 115, 126–27
100:3	99–102, 104, 106, 114–19, 126, 128–29, 131, 133
100:5	128
102:25	345
102:26	108
103	67
103:17	195
104	67
104:4	111
104:5	112
104:14	75
104:22	251
104:33–34	67, 70
105:2	70
105:4	182
106:23	142, 148
106:30	142
106:48	195
108	159
108:1–2	159
108:13	185
109:2	94
109:4–5	79–80
109:21	94
109:22	94
109:26–27	94
109:26	94
109:31	94
110	311
110:1	94–95, 311, 334, 336, 368
110:4	335, 345
111:7	108
112	122
112:4	121
113:7	143
115	107–8
115:2	119
115:4	107
116	61
117	103
118:5	180
118:7	93
118:22	368
118:25	175
119	70
119:11	182
119:25	143
119:28	143
119:31	382
119:32	180
119:51	135
119:62	143
119:114	1
121	319
122	321
122:2	142–43
125:1	111
126	122
128:3	19, 79
130	142
130:1	142

Psalms (*cont.*)

130:3	142
130:5	142
130:7–8	142
131	63–66, 74, 82, 84, 319
131:1	63
131:2	64, 84
131:3	65–66
132	196
132:14	196
134:1	142–43, 158
135:2	142–43
137	250–51
137:1	342
137:3	289
137:4	289, 342
138:1	390
138:4	390
138:8	108
139	138, 325
139:13	290
140:5	93
142:2	71
142:7	169
143:5	70, 108
143:6	281
145:5	70
146	67–68
146:5	93
147:14	41
148	67, 129, 263, 295
149	67
150	67

Proverbs

1:33	199
3:2	195
3:5	63
3:16	195
3:34	135
4:20–23	181
6:19	97, 219
6:22	70
9:1–6	89
8	289, 291
8:22–31	289, 291
8:22–25	292
8:23	290
8:28	169
8:30–31	290–91, 295
8:30	291
10:30	111
12:3	111
12:17	97, 219
13:17	97
14:1	88–89
14:5	97, 219
14:25	97, 219
15:5	97
17:12	78, 88
17:13	79
18:10	380
19:5	97, 219
19:9	97, 219
19:28	135
20:1	231
20:6	97
21:3	370
22:17–18	181
23:2	113
23:29	71
24:5	170, 179
26:15	237
28:16	176, 198
30:15–16	20
31:2–9	75
31:17	170, 179

Ecclesiastes (Qohelet)

5:6	108
6:7	113
8:1	46
8:13	176, 198

Song of Songs

4:2	78

Song of Songs (*cont.*)

6:6	78
7:1	108
8:6	90

Isaiah

1:2–3	213
1:2	109
2:1–5	358
2:11–14	109
2:12	109
2:17	109
5:12	108, 223
5:14	113
5:18–19	228
6	293
6:1	109
6:8	129
6:9–10	223
7:9	223
8:2	97
8:5–8	82
8:16–20	219
8:17	228
10	108–9, 129
10:5–19	223
10:12	109
10:13–14	225
10:13	108–9
10:15	107–9
11:1	20
11:6	294
11:7	294
14	109
14:11	110
14:12–14	109–10
16:11	250
17:11	175
17:13	136
19:25	108, 129
21:3	354
22:11	223
22:19	137
24:23	325
25–26	307–8
25:1	307
25:4	307
25:6–9	307
25:8	307
25:29	307
26:1	94–95
26:8–18	307
26:11	307
26:13	307
26:17	307
26:19	188, 307
26:20	307
26:21	307
27:1	105, 121
28:14	231
28:16	223
28:21	233
28:23–29	228
29:5	136
29:8	231
29:9	223
29:20	231
29:23	108
30	227
30:8	219
30:18	228
33:10	109
34	251
35	153
35:3–4	171
35:3	153, 155, 164, 171, 346
35:4	171
35:10	192, 296
37:23	109
40–66	292
40–55	291, 293
40:1–2	74–75, 171
40:2	74, 228
40:5	294
40:8	222
40:10–11	228
40:10	227

Scripture Index

Isaiah (cont.)

40:11	139
40:20	112
40:26	129
40:27	357
40:28	346
40:31	227, 235, 293, 310, 346
41:1	293
41:5–7	107, 380
41:5	136
41:7	112
41:10	178
41:19	294
41:20	294
42:1–4	221
42:4	221
42:9	122
43:1–7	140
43:2	18
43:4	139
43:11	109
43:19	122
44:6	109
44:8	109
44:11	292
44:14	169
44:26	370–71
45:1–13	2
45:1–7	370
45:5	109
45:6	109
45:15	2, 357
45:16	380
45:18–19	89
45:18	43
45:19	357
45:21	101–2, 109
46:1–2	380
46:3–4	346, 380
46:6–7	380
46:8–10	380
46:13	228
47	109, 129, 380
47:8–9	78
47:8	78, 109, 293
47:9	78
47:10	109
47:12–13	293
48:6	294, 357
48:7	294
48:8	294
48:10	18
48:13	129, 293, 295
49:1–6	20
49:4	294
49:5	294
49:8	94
49:14–16	83
49:14–15	62
49:20	78
49:21	78
50:4	294
50:8	292
51:1–3	22
51:4–5	221
51:5	221
51:9–11	105
51:9–10	159
51:9	121
51:10	142
51:11	192, 296
52:1	159
52:8	294
52:9	294
52:13	109
53	22, 357
53:1	233
53:2	20
53:10	176
54	357
54:10	112
55:1–5	89
55:3–5	312
55:10–11	222–23
56:3	19
56:11	113
57:15	109
60:10	94

Isaiah (cont.)

60:15–16	83
60:21	108
61:7	296
63:5	94, 98
63:18	94
64:8	108
65:22	20, 108
65:25	139
66	64
66:10–13	62
66:10–11	83
66:11	65
66:12–13	83
66:12	65
66:22	293

Jeremiah

1	21
1:4–18	212
1:7–8	178
1:7	212
1:8	212
1:10	216
1:13–16	150
1:17–19	70, 150
1:17	150, 155, 178
1:19	212
2:1–6	214
2:8	214
2:13	213–14, 216, 225
3:19	95
3:22	95
4:2	154
4:4	150
4:23–26	30, 46, 49, 216
5:14	150
6:14	71, 125, 214
7:2	145
7:10	145
7:19	145
8:11	125
8:18—9:3	214
9:14	107–8
9:23–24	108
11:3–4	19
11:16	20, 150
11:18—12:6	222
11:18–26	211
11:21–23	80
14:5	83
15:1	124–25, 145
15:7	78
15:10–21	211
15:14	150
15:15	144
15:18	144, 213–14, 222, 229
15:19–21	224, 229
17	70
17:3–8	19
17:4	150
17:12–18	211
17:13	225
17:14–17	69
17:18	69
17:27	150
18:18–23	211
18:20	80, 145
18:21	78
19:14	145
20:1–2	145
20:7–18	211, 213, 215
20:7	213
20:9	150
20:11	32, 214
20:14–18	212
20:18	222
23:1	102
23:5	20
23:6	200
23:18	137, 146, 149
23:22	146, 149
25:6	107
28:5	145
28:11	145
29:7	3
31:31–34	124, 216

Scripture Index

Scripture Index

Jeremiah (cont.)

31:34	142
33:8	142
33:16	200
36:3	142
36:21	145
38:6	144
38:22	144
39	144
42:5	97
44:15	145
44:26–30	216
45:1–5	216
45:4	216
49:19	145
50:2	142
50:9	79
50:44	145
52	144
52:12	145
52:34	315

Lamentations

1:20	79
2:8	297
2:18	222
3:49	222
4:2	311
5:20	196–98

Ezekiel

1:1	362
5:17	78
11:16	2
14:15	78
16:16	139
17:24	20, 175
22:30	148
27:34	142
29:3	104–5, 118
30:25	175
32:2	82
34:31	102
36	78–79
36:5	79
36:12	79
36:13	79
36:14	79
37	208
40:16	96
43:11	331
45:1	96
46:16	96
47:5	109
47:14	96
47:22	96
48:29	96

Daniel

1:4	361–62
1:5	315
1:17	361–62
2	323–24, 352–53, 360, 362, 364, 369–70, 373
2:1	354
2:5	370
2:8	349, 351–52, 356, 360, 370
2:9	370
2:11	358–89
2:15	356
2:16	356
2:18	361, 389
2:19	357, 361–62 389
2:20–23	367, 368
2:21	380, 389
2:22	357, 360, 362
2:23–25	369, 371
2:24	358
2:27–30	361, 389
2:28–30	362
2:28	357, 362
2:34–35	360
2:34	384
2:35	369
2:43	369, 384
2:44–45	360

Daniel (cont.)

2:44	369
2:45	370
2:47	361, 389
3	22
3:14	371
3:23	22
3:29	357, 362
3:30	357, 362
3:47	357, 362
3:90	364, 376
4:5	354
4:19	358
4:29–30	367
4:34	365, 367
5:6	354, 358
5:10	354, 358
6	373
6:6–10	365
6:11	364
6:12	371
7:15	354
7:17–18	364
7:18	364, 376
7:28	151, 354, 358
8–10	151
8:1	151
8:17	151
8:18	151–52, 363
8:19	151
8:26	151, 371
9:1–2	151
9:3–19	151
9:12–13	371
9:13	361–62
9:20–27	151
9:23	361–62
9:27	175
10	156, 362, 374, 375
10:1–6	152
10:1	357, 361–62
10:7	361–62
10:8–9	152
10:8	361–62, 374, 375
10:9	363
10:11	152–53, 361–62
10:12	152, 375
10:13	366
10:14	151–52, 375
10:16–17	152, 361–62, 374
10:17	153, 361–62, 374
10:18	153, 374
10:19	153–55, 374
10:20	366, 374
10:21	366, 372
11:2	371
11:35	357, 362
12:1–3	366, 379, 391
12:2	207–8
12:3	188, 379

Hosea

1:9	290
2:16	74
3:5	235
4	281
6	93
6:5	221
9:11–14	79
9:12	79
9:14	78
11:1	213
11:10	235
12:2–14	90
13:3	136
13:7–8	90
13:8	78
13:9	93
13:13	90

Amos

2:14	170, 179
5:7	92
5:10	92
7:1–3	21
7:4–6	21
7:7–9	21

Scripture Index

Amos (cont.)

8:1–3	21
8:10–17	21
9:1	21

Jonah

3:6	281

Micah

4:10	82

Nahum

2:1	155, 170, 179
2:7	331

Habakkuk

1:1–4	98, 220
1:1	220, 231
1:2–4	220–24, 226, 228
1:2–3	222, 225, 235
1:2	222, 230
1:3	222, 225, 231
1:4	220, 222–23, 225, 227, 229, 233, 235
1:5–11	223–24, 226–27, 230, 232–33, 235
1:5	224, 226, 228–29, 231, 233
1:6–11	223–24
1:7	225, 231
1:8	231
1:11	225
1:12–17	224, 226, 232
1:12	224–25, 233, 235
1:13–17	226–27, 235
1:13–16	230
1:13	225–26, 229–31
1:14–17	98
1:14–16	230
1:15–16	230
1:16	225, 231
1:17	226, 230
2	155, 232
2:1–4	226, 329–30
2:1	155, 226, 231, 233, 330–31
2:2–20	232
2:2–4	98, 218–22, 226–27, 230–32, 235, 237, 331, 375
2:2	231, 330–31
2:3	98, 151, 218–19, 223, 234–35, 330
2:4	218–19, 226, 228–29, 231, 233, 235, 329–30
2:5–20	230–32, 235
2:5	230–31
2:6	230
2:9–11	231
2:9	231
2:13–17	230
2:13	231
2:15	231
2:16	231
3	98, 156, 220, 232
3:1–19	232
3:2	232–33, 235
3:3–15	234–35
3:3	227, 234
3:6	231
3:7	231
3:16	156, 235
3:17–19	12, 156, 235
3:18–19	331

Zephaniah

2:2	136

Zechariah

14:9	214

Malachi

1:9	45

Malachi (cont.)

3:1	383
3:2	362–63
3:11	78–79

~

NEW TESTAMENT

Matthew

4:1–11	308
4:1	309
4:5	310
4:11	310
5:37	308
6	318
6:2	306
6:5	306–7
6:6	306–7
6:16	306
6:32	323
7:24–27	306
11:5	307
11:28–30	280
12:44	386
13:9	308
13:38	308
13:52	303
14:23	308
16	311
16:13–20	37
16:18	26
16:24	313
23:35	307
23:37–38	310
26:30	308
26:39	308
26:41	308
26:42	308
26:44	308
27:52	307

Mark

1:35	308
2:6–8	75
4:28	169
6:1–6	37
7:34	57
8:34	313
12:34	163
13:36	383
14:6	318
14:36	56–57, 133
15:23	164
15:34	57
15:36	164

Luke

1:38	129, 293
2:13	383
2:19	182
2:51	182
7:22	307
21:34	383
21:35	307
24:32	204
24:45	204

John

1:1–2	52, 55
1:5	389
1:14	55
3:4	90
4:38	270
5:28	307
14:2	285
14:27	319
16:21	307
17	55
17:5–6	55
17:11	55
17:20–23	55
18:17–18	130
20:11–16	84
22:1–23	37

Scripture Index

Acts

1:14	297
1:20	164
2:2	383
2:14	164
2:24	203
2:25–28	203
2:46	297
4:24	297
5:12	297
7:57	297
8:6	297
9:3	383
12:20	297
15	288
15:8	288
15:22	288
15:25	164, 288, 296–97
15:28	288
18:12	297
19:9	386
19:29	164, 297
22:6	383
22:17	363

Romans

5	318
5:12–21	317–18
5:14	317
5:16	317–18
5:17	317
5:18	317–18
5:21	317
6	304
6:4	304
8	56–57, 76, 133, 312–13, 317–18, 322, 329, 336, 383
8:2	57
8:6	322
8:9–11	312
8:9	57
8:14	21
8:15–39	313
8:15–28	313
8:15–16	56–57, 312–13, 317–18
8:15	324
8:16	288, 313
8:17	58, 313
8:20–22	56
8:22	313
8:23	56, 313
8:25–26	318
8:26–27	57, 312, 316–17, 322, 324
8:26	56–57, 65, 288, 312–14, 317, 326
8:27	56, 313, 327
8:28	58, 313
8:29	52, 57, 292, 313
8:30	52, 313
8:32	313
8:33	292
8:34	56, 311–13, 316–18, 336
8:35–39	58
8:35	313
8:37	319
8:38–39	317, 351
11:9–10	164
11:36	245, 314, 318
12:1–2	367
12:2	360
13:8–10	382
14:17	322
15:3	299
15:4	246, 299
15:5–6	299
15:6	297–98
15:8–13	129
15:13	164, 322

1 Corinthians

1:18–25	367
1:30	367
2:6	360

1 Corinthians (*cont.*)

2:7–8	367
2:7	360
7	385, 388
7:3–5	385
10:4	246
14:28	76
15:34	307

2 Corinthians

1:3–4	217
3:18	192
4:6	45
4:17–18	192
12	363, 376
12:1	363
12:4	363

Galatians

1:3	359
1:4	351, 360
1:12	363
1:16	363
3:13	351, 359
4:3	351
4:4–5	351, 367
4:4	312
4:5	359
4:9–10	359
4:9	351
4:10	351
4:26	304
5:22–23	322

Ephesians

1–3	368
1:3–14	364, 368, 372
1:3	363
1:7	14, 365, 367
1:9–10	368
1:9	361, 372
1:10	368
1:14	365
1:15—3:21	368
1:17	365
1:20–23	366, 368
2:1–3	360, 366
2:2	360, 365
2:3–14	367
2:6	321, 326
2:8	363, 377
2:15–22	367
2:22	380
3:1	350, 364
3:2	364
3:3–5	367
3:3	372, 362, 361
3:4	361–62, 364
3:5	372
3:8	389
3:9–10	285, 372–73
3:9	361
3:10	305, 372–74, 376–77, 388, 390–91
3:12	364
3:13	364
3:14–32	376
3:14–21	364, 374, 389
3:14	364
3:15–21	373–74
3:15	368, 373, 390
3:17	375
3:18	375
3:19	375
3:21	364, 374, 381
4–6	368
4:1–16	368, 384
4:2	377
4:3	322, 385
4:8	363
4:14	370
4:15	368–69, 372
4:16	369
4:18	366
4:30	365

Scripture Index

Ephesians (cont.)

4:32	377
5:1	377
5:2	377
5:8–14	367, 378
5:8	360
5:15–20	75
5:16	274, 349, 351–53, 359–60, 366, 379
5:19–20	389–90
5:19	76
5:21—6:9	373
5:21	377, 381, 383, 385
5:22—6:9	380
5:22–23	383
5:23	368
5:31	382
5:32	361
6:12	305, 374, 388
6:13	360, 374
6:14	374
6:19	361, 372

Philippians

1:6	245, 259
1:19	244
1:22–24	326
1:24–25	326
2	304
2:1	322
2:5–11	313, 373
2:6–11	298–99
2:12–13	313
2:16	330
3	304
3:10–11	313
3:10	322
3:12	330
3:13–14	304
3:14	330
3:19	305
3:20	305, 321–22
4:3	164
4:6–8	323, 326
4:6–7	319
4:6	314, 326, 373
4:7	321–22, 376
4:8	322, 326
4:9	322

Colossians

1	289
1:1–4	318
1:9	295
1:15–20	291, 296
1:15	292
1:17	243, 291–92, 294–96
1:18	292
1:19	295
1:28	295
2:1–4	304
2:3	295
2:8	20
2:9	295
2:12	304–5
2:23	295
2:29	351
3:1–4	303–5, 311, 318, 329
3:1–3	346
3:1	305
3:2	305
3:3	305
3:16	295

1 Thessalonians

1:7–12	85
1:7	85

2 Thessalonians

3:2	307

2 Timothy

2:19	307

Hebrews

1	248, 334
1:2–3	344
1:2	345
1:3	334–335
1:5	334
1:8	334
1:13	334–335
2:9	344
2:17	346
3:1	344
3:2	346
3:5–6	329, 346
3:6	329
4:12–13	13
4:14	334
4:16	335–336
5:5	334
5:6	335
5:10	335
6:11	329
6:19	335
6:20	335–36, 344–45
7:1	335
7:3	345
7:10	335
7:15	335
7:17	335, 345
7:19	335
7:21	345
7:22	344
7:24	345
7:25	345, 336
7:28	334, 345
8:1	335
8:2	335
8:5	335
9:11	335
10:1	345
10:10	335
10:12	335, 345
10:13	331
10:14	345
10:19	344
10:20	335
10:22	329, 345
10:23	164, 329, 346
10:27	307
10:32	329
10:33	331
10:36	329
10:37–38	329
11	329–31
11:1	164, 329–31, 333, 347
11:11	346
11:26	164, 331
12:1–12	331
12:1	329–30
12:2	12, 328–33, 335, 344
12:3	329, 331
12:7	329
12:12–13	346
12:12	164
12:29	17
12:24	344
13:8	346
13:12	344
13:13	164, 331

James

5:16	334

1 Peter

5:7	323

1 John

4:1	254

Revelation

3:5	164
6:4	381
7:9–12	285
7:17	307
8:1	12, 348
8:13	307

Scripture Index

Revelation (*cont.*)

12:2	307
13:8	164
14:1–3	285
15:21–27	164
16:1	164
17:8	164
20:12	164
20:14	307
21:14	307

DEUTEROCANONICAL AND PSEUDEPIGRAPHICAL BOOKS

Judith

3:6	320

Wisdom of Solomon

13:2	360
17:15–16	320

Sirach

34:4	370

Prayer of Azariah

11–13	22

1 Esdras

4:56	320
5:47	297
5:58	297
9:38	297

2 Maccabees

6–7	22

3 Maccabees

3:11	345
3:22	345
4:4	297
4:6	297
4:16	345
5:50–51	297
6:28	345
6:39	297
7:6	363

4 Maccabees   22

4 Ezra

5:22	361
14:40	361
14:47	361

Testament of Levi

2:3	361
18:1	361
18:7	361

Jubilees

22:6–9	367

www.ingramcontent.com/pod-product-compliance
Lightning Source LLC
Chambersburg PA
CBHW021927290426
44108CB00012B/747